HOW PROFESSIONALS
MAKE DECISIONS

Expertise: Research and Applications
Robert R. Hoffman, Nancy J. Cooke, K. Anders Ericsson, Gary Klein,
Eduardo Salas, Dean K. Simonton, Robert J. Sternberg,
and Christopher D. Wickens, **Series Editors**

Hoc/Caccibue/Hollnagel (1995) • *Expertise and Technology: Issues in Cognition and Human-Computer Interaction*

Mieg (2001) • *The Social Psychology of Expertise: Case Studies in Research, Professional Domains, and Expert Roles*

Montgomery/Lipshitz/Brehmer (2004) • *How Professionals Make Decisions*

Noice/Noice (1997) • *The Nature of Expertise in Professional Acting: A Cognitive View*

Salas/Klein (2001) • *Linking Expertise and Naturalistic Decision Making*

Schraagen/Chipman/Shalin (2000) • *Cognitive Task Analysis*

Zsambok/Klein (1997) • *Naturalistic Decision Making*

HOW PROFESSIONALS MAKE DECISIONS

Edited by

Henry Montgomery
Stockholm University

Raanan Lipshitz
University of Haifa

Berndt Brehmer
National Defence College, Sweden

 LAWRENCE ERLBAUM ASSOCIATES, PUBLISHERS
2005   Mahwah, New Jersey   London

Copyright © 2005 by Lawrence Erlbaum Associates, Inc.
 All rights reserved. No part of this book may be reproduced in
 any form, by photostat, microform, retrieval system, or any other
 means, without the prior written permission of the publisher.

Lawrence Erlbaum Associates, Inc., Publishers
10 Industrial Avenue
Mahwah, New Jersey 07430

Cover design by Kathryn Houghtaling Lacey

Library of Congress Cataloging-in-Publication Data

How professionals make decisions / edited by Henry Montgomery, Raanan Lipshitz,
  Berndt Brehmer.
    p. cm.
  Includes bibliographical references and indexes.
  ISBN 0-8058-4470-8 (alk. paper) — ISBN 0-8058-4471-6 (pbk. : alk. paper)
  1. Decision making. 2. Problem solving. 3. Professional employees. I. Montgomery,
Henry. II. Lipshitz, Raanan. III. Brehmer, Berndt.

HD30.23.H68 2004
153.8′3—dc22   2003066708
                                                                            CIP

Books published by Lawrence Erlbaum Associates are printed on acid-free paper,
and their bindings are chosen for strength and durability.

Printed in the United States of America
10 9 8 7 6 5 4 3 2 1

Contents

Preface

This book has been realized as a result of the 2000 Naturalistic Decision Making Conference in Stockholm. This event provided the basis for the chapters of this book. We are indebted to the sponsors of this conference: Swedish Research Council, The Bank of Sweden Tercentenary Foundation, and the Swedish National Defence College. We also wish to thank Peter Thunholm from the Swedish Defence College who organized the conference together with editors of this book and, in addition, made the conference work practically.

Many thanks also go to the contributors of this volume for their efforts to bring this project to a successful end.

Finally, we wish to acknowledge Robert Hoffman, series editor for this book, for his support and helpful advice in planning this book project, and Kristin Duch and Sondra Guideman of Lawrence Erlbaum Associates for their help and professionalism with editorial and production matters.

Henry Montgomery
Raanan Lipshitz
Berndt Brehmer

1

Introduction:
From the First to the Fifth Volume of
Naturalistic Decision-Making Research

Henry Montgomery
Stockholm University

Raanan Lipshitz
University of Haifa

Berndt Brehmer
National Defence College, Sweden

A decade has passed since the volume of the first conference of naturalistic decision making (NDM) was published (Klein, Orasanu, Calderwood, & Zsambok, 1993). In this introductory chapter, we discuss the contributions made by the 27 subsequent chapters of this volume in light of the research presented in the Klein et al. (1993) volume. We present some key findings and attempt to address the following three questions. To what extent has progress been made with respect to issues presented in the Klein et al. (1993) volume? Have new issues been raised? What are the challenges for future NDM research?

The book is organized into three sections: Individual Decision Making, Social Decision Making, and Advances in Naturalistic Decision-Making Methodology. In the following, we discuss each section separately.

INDIVIDUAL DECISION MAKING

In their chapter in Klein et al.'s (1993) volume, Orasanu and Connolly (1993) listed the following eight factors that characterize decision making in naturalistic settings:

1. Ill-structured problems.
2. Uncertain dynamic environments.
3. Shifting, ill-defined, or competing goals.
4. Action/feedback loops.
5. Time stress.
6. High stakes.
7. Multiple players.
8. Organizational goals and norms.

This list of factors has now become a standard in descriptions of naturalistic decision making (Lipshitz, Klein, Orasanu, & Salas, 2001). At the same time, it serves as a basis for critiques of what is lacking in the "traditional" research in judgment and decision making (Cohen, 1993). To what extent are these factors visible in the chapters on individual decision making in this volume? Do these chapters give new knowledge with respect to each of these factors? Have new themes emerged in these chapters concerning views of naturalistic decision making since the Orasanu and Connolly (1993) list was presented?

Taking the nine chapters on individual decision making as a whole, the answers to all three questions is "yes," although the chapters vary in their focus across and within the three questions. A common theme across all chapters is that they concern fairly ill-structured decision problems in which it is an open question what the options are and/or how they and their environment should be described. However, the examined problems differ with respect to how close they were in other respects to the prototypical NDM studies or if they are theoretically oriented. We begin by discussing two empirical contributions that concern decision making in typical NDM tasks.

In his case study of decision processes involved in a collision between two ships (chap. 2, this volume) Burns highlights how the construction of an appropriate mental model of the task environment is crucial for the possibility of making an adequate decision. Moreover, Burns' chapter illustrates how decision makers need to update their mental models as a function of the changes in the decision situation (Factor 2 in Orasanu & Connolly's, 1993, list), partly resulting from action/feedback loops (Factor 4) in a situation involving high time stress (Factor 5) and high stakes (Factor 6) and multiple players, that is, the commanders of the colliding ships (Factor 7). In chapter 2 (this volume), Burns outlines a "mental model" theory of decisions made by experienced people under complex dynamic and uncertain situations.

Although the decision task examined by Omodei, Wearing, McLennan, Elliott, and Clancy (chap. 3, this volume)—command control of fire fighting—involved a laboratory simulation, it showed many of the typical NDM qualities (action/feedback loops, time stress, etc.). Four experiments showed that—in contrast to conventional wisdom—more information may deteriorate the performance. It was concluded that if a resource is made available, commanders feel compelled to use it even when their cognitive system is overloaded.

Thunholm (chap. 4, this volume) outlines a prescriptive model of military tactical decision making under time pressure. The model specifies a number of decision-making steps that can be deliberately practiced in military training. The model is similar to Klein's (1993) recognition-primed decision (RPD) theory in that it assumes that decisions are made by recognition of typical situations and courses of action rather than by comparing given options. Thunholm develops the RPD model by giving a more detailed account of decision-making steps involved in an NDM situation. In line with Burns' model (chap. 2, this volume), Thunholm stresses the role of an adequate understanding of the situation and the need of updating this understanding in relation to the information that floods a contemporary military staff.

Three chapters report empirical studies that examine particular aspects of NDM by using tasks in which participants were presented with hypothetical scenarios rather than being involved in interaction with a dynamic task environment. Sterman and Sweeney (chap. 5, this volume) illustrate the importance of an adequate understanding of dynamic situations for taking relevant actions. They describe results from a "system thinking inventory" examining the ability of people to understand the most basic elements of complex dynamic situations. It was found that participants from an elite business school have a poor understanding of stock and flow relations and time delays, even in highly simplified settings. More generally, Sterman and Sweeney's research illustrates people's difficulties in understanding action/ feedback loops (Orasanu and Connolly's, 1993, factor 4), especially when there are time delays.

Lipshitz and Pras (chap. 6, this volume) asked participants to think aloud while solving more and less well-defined problems concerning interpersonal issues. Their results show how the understanding of the situation—or to use the term commonly used by NDM researchers, *situation awareness*, is the critical element in decision making in a situation that differs from the type of situations typically studied by NDM researchers, that is, without high stakes, pressure of time, and expert decision makers. Even so, participants' decision processes, in line with RPD theory, exhibited a two-cluster phase structure (processes preceding choice vs. elaboration of choice) in contrast to classical decision theory that is focused on the choice between given options. Lipshitz and Pras conclude that the principal factor that determines the use of RPDs is the mode of problem presentation, specifically, absence of choice alternatives and a visual display.

Cesna and Mosier (chap. 7, this volume) also used a task that differed from the prototypical NDM decision task. Critical care nurses were studied, comparing skill levels and decision making. They used a prediction paradigm asking participants to predict what an expert would do in given decision points within a complex clinical scenario. Again, high stakes and time pressure were not present, although expert decision making was highlighted. In line with RPD theory, it was found that expert nurses, as compared to less experienced nurses, were more accurate and confident in their decision making and generated fewer options for each decision point.

Similar to all the empirical chapters in this section, Montgomery's (chap. 8, this volume) contribution deals with how professionals make decisions. However, in Montgomery's study, the time scale is years instead of minutes and seconds, which typically is the case in NDM studies. Montgomery used data on actual and forecasted gross national products for analyzing accuracy for and psychological mechanisms underlying the forecasts. A number of judgmental biases (optimist, anchoring and adjustment, and availability bias) were identified in the forecasts, suggesting a possibility to correct economic forecasts to improve their accuracy.

Both the two remaining chapters in this section are mainly theoretical and shed light on decision making of professionals. Goitein and Bond (chap. 9, this volume) highlight individual differences in professional decision making. They describe a taxonomy of seven decision-making styles among professionals, which they validate by self-rating data from managers, engineers, and accountants. In contrast to the other chapters, Goitein and Bond's contribution focuses on the role of organizational goals and norms in professional decision making (Factor 8 in Orasanu and Connolly's, 1993, list of NDM criteria). Ericsson (chap. 10, this volume) puts the acquisition of high-quality expert performance into a broad framework anchored in research on biological and cognitive systems. Ericsson reviews evidence on the effects of practice and adaptation and shows that under certain external conditions, biological systems—including our own—are capable of dramatic change down to the cellular model. Ericsson has an optimistic view of people's ability to develop expertise through deliberate practice. However, it is an optimism that is coupled with a tough message to the would-be expert: Expert performance is typically acquired over years and after thousands hours of deliberate training.

To summarize, the nine chapters on individual decision making by professionals are related to the list of NMD factors outlined in the first NDM volume. The chapters illustrate that knowledge about professional decision making has advanced since the inception of the NDM movement about 10 years ago. More is known about how professionals make their decisions (e.g., with respect to the role of mental models and concerning the applicability of RPD theory). More is also known about biases in professional decision making. Finally, two of the chapters in this volume show that NDM research may benefit by studying variations in professional decision making across individuals and across time as a result of deliberate practice. However, although knowledge has accumulated with respect to and beyond the Orasanu and Connolly (1993) list of NDM criteria, we think that present state of knowledge of professional decision making gives interesting bits and pieces but is quite far from a complete picture of the decision process. Presently, there exists only a fragmented picture of the interplay among the key factors in professional decision making, such as situation characteristics, the decision maker's skill and knowledge, how the skill and knowledge is used in a given situation and driven by motivational factors, and how the decision maker acquires and develops expertise. To get better understanding of the "full picture," it is recommendable to collect concurrent data of decision processes in naturalistic situations or in close simulations of such situations.

SOCIAL DECISION MAKING

All decisions, including those made by professionals, are made in a social context. People are never completely alone. The social perspective is especially clear when the focus in on how teams (as opposed to single individuals) make decisions. The study of team decision making has been a major theme in NDM research since its beginning around 1990. In Klein et al.'s (1993) volume, two chapters were devoted to team decision making (Orasanu & Salas, 1993; Duffy, 1993). In the following, we use the chapter by Orasanu and Salas as a point of departure for the discussion of the nine chapters in the section on social decision making. However, we also show that several of these chapters bring up the role of social factors in a more general sense by attending to the social context surrounding individuals' and groups' decision making.

The key idea in Orasanu and Salas' (1993) chapter was the notion of shared mental models. More specifically, Orasanu and Salas found that effective teams appear to share a mental model of their members' knowledge, skill, anticipated behaviors and needs. Orasanu and Salas also reported research showing that failure in team decision making was due to ill-functioning or lacking shared mental models.

Several of the chapters in this volume allude to shared mental models explicitly or implicitly. Two chapters aspire to clarify the concept of shared mental models by listing different features and dimensions of shared mental models. Kline (chap. 11, this volume) discusses the characteristics of intuitive decision making in teams on the basis of the results from a study of five experienced teams. These characteristics include routines, team norms, team goals, and team identity. Kline outlines a model of intuitive decision making. A key assumption in the model is that when faced with a decision event, team members intuitively compare the event to the knowledge in the shared mental models. In a study of performance by teams of naval cadets, Brun, Eid, Johnsen, Laberg, and Ekornås (chap. 12, this volume) used a questionnaire measuring "hard" (task-oriented) categories and "soft" (person-oriented) categories of shared mental models: equipment model and task models on one hand and team interaction model and team model on the other hand. It was expected that the agreement across team members should be higher for hard than for soft models. However, the team members appeared to share all models to an equal and fairly high extent. Another surprising finding was that participating in a course that tried to enhance the shared mental models of the team had no effects on the actual use of mental models. Brun et al. speculate that this pattern of results might be due to the fact the teams had been working together for 2 years before participating in the study.

In a pilot study of a joint mobile command system (ROLF Joint Mobile Command and Control Concept), Johansson, Granlund, and Waern (chap. 13, this volume) examined two factors that may affect the use of mental models, namely, verbal communication (via e-mail) versus visual communication (given on a map). They found that verbal communication was more effective than visual communication in creating a well-functioning, shared mental model. This result may be surprising because visual communication in this case probably gave more direct access to the relevant information (about attempts to extinguish a forest fire) than was true for verbal communication. Perhaps a shared model that is given (positions on a map) may be less effective than a mental model that is actively construed by the team members through verbal communication.

McLennan, Pavlou, and Omodei (chap. 14, this volume) present data that may shed light on differences between good and poor teams in their use of (shared) mental models. Twelve fire officers' decision-making performance was assessed by a panel of highly experienced observers. Superior performance was associated with more effective self-control of task-focused cognitive activity. Perhaps a parallel may be drawn to Johansson et al.'s data that suggest that active construal of a shared mental model has positive effects on team performance.

Allwood and Hedelin (chap. 15, this volume) demonstrate how a shared mental model may be construed in professional decision making in the business world. They report that "selling-in," that is, anchoring and creating commitment for a decision, is a major aspect of the decision making in Swedish companies.

Berggren (chap. 16, this volume) presents a case of deficient shared mental models. He presents data showing that as pilot mental work load increases, assessments of discrepancies between self-assessment of pilot performance and those of an observer likewise increase.

The six chapters just reviewed sharpen one's understanding of possible features and roles of shared mental models as well as about the possibilities and limitations of learning and using shared mental models efficiently. However, much more research needs to be done, using more extensive sets of data, to get solid knowledge about these issues.

The three remaining chapters on social decision making illustrate how professional decisions are dependent on cultural factors characterizing the context in which the decisions are made. The role of culture in professional decision making has attracted little attention in previous NDM volumes. Perhaps we witness the emergence of a new subarea in NDM research.

Klein (chap. 17, this volume) describes a number of cultural dimensions for decision making: dialectical reasoning, hypothetical reasoning, independence, power distance, and uncertainty avoidance and discusses how they may lead to problems in interpersonal understanding. Note that Allwood and Hedelin's chapter 15 (this volume) may illustrate how the organizational culture in Sweden and Scandinavia (low power distance) affects professional decision making in this part of the world.

Vaughan (chap. 18, this volume) analyzes the Challenger disaster from a cultural perspective. Vaughan describes how an organizational culture developed at NASA in which decision makers became habituated to normalize potential signals of potential danger. This process was helped by a political culture that stressed commercial aspects of the space shuttle program and by a bureaucratic culture that became pathologically fixated on desk work and filling out forms. Another contributing factor was structural secrecy resulting from division of labor, hierarchy, complexity, and geographical dispersion of parts. Structural secrecy undermined the ability of people situated in one part of the organization to fully understand what happens in other parts. It is interesting to note that structural secrecy can be seen as being equivalent to the absence of a shared mental model.

Ideas about what is ethically correct and incorrect and about the solving of ethical problems is an important component of an organizational culture. Such ideas are not always explicitly formulated, implying that the ethical competence may be improved by making them explicit. Kavathatzopoulos (chap. 19, this volume) describes the development of methods for assessing the ethical competence of an organization and the implementation of training methods for helping decision makers to cope effectively with difficult ethical problems.

ADVANCES IN NDM METHODOLOGY

The common concern of the three chapters on methodology in Klein et al.'s (1993) volume was how to conduct NDM studies in a way that will contribute to the construction of generalizable findings and models. Hammond

(1993) proposed a method for predicting general characteristics of decision makers' cognitive processes (e.g., intuitive vs. analytical styles) from general task characteristics (e.g., number of cues and their mode of display); Christensen-Szalanski (1993) promoted the importance of effect size information for the application of experimental results to naturalistic environments; and Woods (1993) reviewed a variety of data collection and analysis techniques that are particularly applicable to naturalistic settings, under the general heading of process tracing. Only Woods' chapter can be related directly to the nine chapters on methodology in this volume, indicating that the more basic orientation of Hammond and Christensen-Szalanski is incompatible with the concerns of NDM researchers.

Three chapters in this volume deal with process-tracing methodologies. Andersson (chap. 20, this volume) describes a computer-driven simulation designed to study how experienced loan officers make decisions concerning loans to small firms and to train novice loan officers. The chapter illustrates how the ecological validity of simulations can be improved by input from available research and content area experts and the application of process tracing to diagnose the results of particular individuals and test general decision models. Ranyard and Williamson (chap. 21, this volume) propose a method for generating information-search data through a conversation-like interaction between experimenter and participant. The chapter compares the proposed method with alternative question-asking and think-aloud methods and demonstrates the application of process-tracing analyses to identify search patterns and risk defusing operators. Finally, Hoc and Amalberti (chap. 22, this volume) outline a verbal protocols coding scheme for analyzing the cognitive processes of human operators working in complex dynamic situations (e.g., air traffic controllers). The scheme's unique feature is that it is driven by an underlying theoretical framework that integrates four basic cognitive activities: information gathering, interpretation, decision making, and metacognition. Using standard basic categories ensures high intercoder reliability and allows comparisons across different work environments. Anchoring the scheme on a general theoretical framework enables researchers with good domain knowledge to make inferences that go beyond purely data-driven analysis.

The next three chapters in the methodology section are concerned with cognitive task analysis (CTA). CTA is a generic label for a variety of methods that attempt to identify how experts perform complex cognitive tasks. Three criticisms of traditional CTA methodologies are that they require extensive experience for effective application, are too time consuming for application in many organizational settings, and tend to produce voluminous data that are difficult to analyze systematically. In response to these criticisms, Militello, Hutton, Pliske, Knight, and Klein (1997) developed Applied Cognitive Task Analysis (ACTA), a set of structured modules that simplifies the conduct of CTA and the analysis of its results, thereby making it accessible to inexperienced practitioners. Klein and Militello (chap. 23, this volume) describe the development process and current version of one of ACTA's modules, the knowledge audit interview. The purpose of the audit is to elicit from subject matter experts the different aspects of expertise required to perform a task skillfully. The simplification of this task was achieved by surveying the literature on expertise and identifying eight probes that are likely to elicit key types of cognitive information (e.g., perceptual skills and self-monitoring) across a broad range of domains. In chapter 24 (this volume) Gore and Riley describe their experience in the application of the complete ACTA procedure to analyze the task of recruitment and selection by hotel human resource professionals. The chapter is valuable for providing detailed account of the actual experience of the application of ACTA by members of its target population—practitioners with no previous experience or training in CTA. Finally, Peterson, Stine, and Darken (chap. 25, this volume) describe a failed attempt to apply CTA based on Klein critical decision-making method (Hoffman, Crandall, & Shadbolt, 1998; Klein, Calderwood, & MacGregor, 1989) as a basis for developing training programs for tactical land navigators. Peterson et al. (chap. 25, this volume) provide a vivid description of the difficulties of applying CTA in a field setting and remind researchers that even the most flexible methodology will yield poor quality data unless some basic requirements specific to the method are met.

The two next two chapters in the methodology section are concerned with observation in a naturalistic setting. Lipshitz (chap. 26, this volume) is concerned with the problem of how observation can be performed rigorously without the controls that maximize the rigor of experimental methodologies. Lipshitz proposes rigorous observation should be construed in terms of the plausibility of the assumptions underlying the transition from data to conclusions and the tightness of the argumentation employed in the process. In addition, he suggests a two-step procedure that operationalizes this notion.

Roth and Patterson (chap. 27, this volume) assert that observation is invaluable for the discovery (rather than hypothesis testing) phase of research, implying that such studies should be evaluated in terms of the insightfulness rather than the validity of their results. They illustrate their assertion in two case studies that describe the introduction of new support systems in which observations provided critical information during the design and implementation of new systems.

Finally, Norros and Klemola (chap. 28, this volume) present and demonstrate a participant-centered methodology in which operators' task and task performance are analyzed from the vantage of the operators' construction of the situation. Their solution to the problem of moving from data that is gathered with minimal intrusion to conclusion is particularly interesting because it is anchored in conceptual frameworks that are probably unfamiliar to most NDM researchers.

Collectively, the chapters on methodology in this volume suggest several conclusions about the current state of NDM methodology as reflected by them.

1. The methodologies the NDM researchers employ or develop are focused primarily on solving concrete problems and only secondarily on testing or developing general theories or models.
2. Considerable effort is put in basing new methods on relevant theories and experimental findings.
3. NDM methodologies must be relatively simple, possess visible face validity, and forgo traditional methods of safeguarding rigor to gain the cooperation of practitioners to be studied.
4. NDM still lacks generally accepted criteria of rigor that are suitable for its particular nature. Such criteria are essential for guiding the design of good research and the evaluation of research outcomes.
5. The challenge of demonstrating that NDM methodologies can produce general and testable models in addition to valuable insights on how decisions are actually made looms as large today as it was a decade ago.

CONCLUDING REMARKS

The first NDM volume (Klein et al., 1993) aspired to present a paradigm shift in the study of decision making. This volume shows that a new paradigm indeed has been established and become normal science. Research has proceeded along the lines staked out in the first volume. In addition, new areas have emerged such as the role of cultural factors for NDM. However, it may be seen as surprising that not many "full-blown" NDM studies have been conducted attempting to grasp the complex web of factors shaping decision making in field settings. To conduct research along these lines and to develop a methodology for field studies are challenges for future research on NDM.

REFERENCES

Christensen-Szalanski, J. J. J. (1993). A comment on applying experimental findings of cognitive biases to naturalistic environments. In G. A. Klein, J. Orasanu, R. Calderwood, & C. E. Zsambok (Eds.), *Decision making in action: Models and methods* (pp. 252–261). Norwood, NJ: Ablex.

Cohen, M. S. (1993). The naturalistic basis of decision biases. In G. A. Klein, J. Orasanu, R. Calderwood, & C. E. Zsambok (Eds.), *Decision making in action: Models and methods* (pp. 51–99). Norwood, NJ: Ablex.

Duffy, L. (1993). Team decision making biases: An information processing perspective. In G. A. Klein, J. Orasanu, R. Calderwood, & C. E. Zsambok (Eds.), *Decision making in action: Models and methods* (pp. 346–359). Norwood, NJ: Ablex.

Hammond, K. R. (1993). Naturalistic decision making from a naturalistic viewpoint: Its past, present, future. In G. A. Klein, J. Orasanu, R. Calderwood, & C. E. Zsambok (Eds.), *Decision making in action: Models and methods* (pp. 205–227). Norwood, NJ: Ablex.

Hoffman, R. R., Crandall, B., & Shadbolt, N. (1998). Use of the critical decision method to elicit expert knowledge: A case study in the methodology of cognitive task analysis. *Human Factors, 40,* 254–276.

Klein, G. A. (1993). A recognition-primed decision model of rapid decision making. In G. A. Klein, J. Orasanu, R. Calderwood, & C. E. Zsambok (Eds.), *Decision making in action: Models and methods* (pp. 138–147). Norwood, NJ: Ablex.

Klein, G., Calderwood, R., & MacGregor, D. (1989). Critical decision method for eliciting knowledge. *IEEE Transactions on Systems, Man, and Cybernetics, 19,* 462–472.

Klein, G. A., Orasanu, J., Calderwood, R., & Zsambok, C. E. (Eds.). (1993). *Decision making in action: Models and methods.* Norwood, NJ: Ablex.

Lipshitz, R., Klein, G., Orasanu, J., & Salas, E. (2001). Taking stock of naturalistic decision making. *Journal of Behavioral Decision Making, 14,* 331–352.

Militello, L. G., Hutton, R. J. B., Pliske, R. M., Knight, B. J., & Klein, G. (1997). *Applied cognitive task analysis (ACTA) methodology* (Final Rep. contract No. N66001–94–C–7034 prepared for Navy Personnel Research and Development Center). Fairborn, OH: Kledin Associates Inc.

Orasanu, J., & Connolly, T. (1993). The reinvention of decision making. In G. A. Klein, J. Orasanu, R. Calderwood, & C. E. Zsambok (Eds.), *Decision making in action: Models and methods* (pp. 3–20). Norwood, NJ: Ablex.

Orasanu, J., & Salas, E. (1993). Team decision making in complex environments. In G. A. Klein, J. Orasanu, R. Calderwood, & C. E. Zsambok (Eds.), *Decision making in action: Models and methods* (pp. 327–345). Norwood, NJ: Ablex.

Woods, D. D. (1993). Process-tracing methods for the study of cognition outside of the experimental psychological laboratory. In G. A. Klein, J. Orasanu, R. Calderwood, & C. E. Zsambok (Eds.), *Decision making in action: Models and methods* (pp. 228–251). Norwood, NJ: Ablex.

INDIVIDUAL DECISION MAKING

2

Mental Models and Normal Errors

Kevin Burns
The MITRE Corporation

Perrow (1984) provided the following account of what he called a "normal" accident (see Fig. 2.1):

On a beautiful night in October 1978, in the Chesapeake Bay, two vessels sighted one another visually and on radar. On one of them, the Coast Guard cutter training vessel Cuyahoga, the captain (a chief warrant officer) saw the other ship up ahead as a small object on the radar, and visually he saw two lights, indicating that it was proceeding in the same direction as his own ship [Fig. 2.1, (t1), top panel]. He *thought* [italics added] it possibly was a fishing vessel. The first mate saw the lights, but saw three, and estimated (correctly) that it was a ship proceeding toward them [Fig. 2.1, (t1), bottom panel]. He had no responsibility to inform the captain, nor did he think he needed to. Since the two ships drew together so rapidly, the captain *decided* [italics added] that it must be a very slow fishing boat that he was about to overtake [Fig. 2.1, (t2), top panel]. This reinforced his incorrect interpretation. The lookout knew the captain was aware of the ship, so did not comment further as it got quite close and seemed to be nearly on a collision course. Since both ships were traveling full speed, the closing came fast. The other ship, a large cargo ship, did not establish any bridge-to-bridge communication, because the passing was routine. But at the last moment the captain of the Cuyahoga *realized* [italics added] that in overtaking the supposed fishing boat, which he assumed was on a near-parallel course, he would cut off that boat's ability to turn as both of them approached the Potomac River [Fig. 2.1, (t3), top panel]. So he ordered a turn to the port [Fig. 2.1, (t4), top panel]. This brought him di-

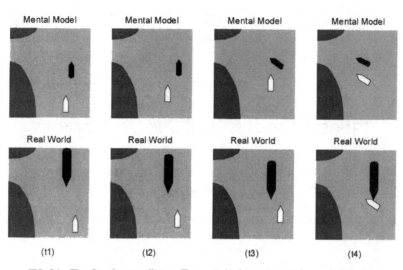

FIG. 2.1. The Cuyahoga collision. Top panels depict captain's mental model and bottom panels depict real world. Small black boat (top panels) denotes fishing vessel and large black boat (bottom panels) denotes freighter. White boat denotes Cuyahoga in top and bottom panels. Time epochs are labeled (t1), (t2), (t3), and (t4).

rectly in the path of the oncoming freighter, which hit the cutter [Fig. 2.1, (t4), bottom panel]. Eleven coastguardsmen perished. (p. 215)

In retrospect, the question is (Perrow, 1984), "Why would a ship in a safe passing situation suddenly turn and be impaled by a cargo ship four times its length?" (p. 217). Based on his analysis of this and other accidents in complex systems, Perrow concluded that the answer lies in the construction of a "faulty reality" (i.e., situation awareness; see Sarter & Woods, 1991). In Perrow's (1984) words, ". . . they [captains, operators, etc.] built perfectly reasonable *mental models* [italics added] of the world, which work almost all the time, but occasionally turn out to be almost an inversion of what really exist" (p. 230). This is a common theme among researchers of both human error (Reason, 1990) and human power (Klein, 1998) and has influenced numerous theories of human decision making (see Lipshitz, 1993). My objective is to extend these theories with a more formal (computational) analysis of how (exactly) decision makers construct mental models to achieve situation awareness.

In this chapter, I outline a formal (Bayesian) framework and perform a detailed (computational) analysis of the Cuyahoga collision. The new contribution provided by my framework/analysis is to show that this prototypical "error" in situation awareness can be characterized as normal (normative) in a mathematical sense. I discuss the theoretical advantage of my framework in unifying the heuristics and biases proposed by previous researchers and the practical advantage of my framework in guiding the design of "decision support systems."

MENTAL MODELS AND SITUATION AWARENESS

The idea that human reasoning employs internal models is attributed to Craik (1943; see Johnson-Laird, 1983) whose original theory highlighted three facets of these models. First, mental models are internal belief structures that represent an external reality. Second, mental models are constructed by computational processes to accomplish cognitive tasks. Third, mental models are adapted to the natural context of practical situations. Craik also highlighted three functions of mental models (also noted later by Rouse & Morris, 1986), that is, to describe, explain, and predict the world. Thus, influenced by Craik's original theory and informed by subsequent theories, I offer the following as a working definition: *Mental models* are adaptive belief constructs used to describe, explain, and predict situations.

According to this definition, mental models are representational structures, not computational processes. However, it is vacuous to talk about representations without also talking about how these representations are used to accomplish specific tasks. Thus, my formal framework includes representational structures (which I call *models*) as well as the computational processes (which I call *modules*) that operate on these structures (Fig. 2.2).

Situation awareness can be characterized as the dynamic construction of mental models (via mental modules) to describe, explain, and predict an evolving situation. This is in general agreement with the following definition proposed by Endsley (1988; also see Sarter & Woods, 1991): "[Situation awareness is] the perception of the elements in the environment within a volume of time and space, the comprehension of their meaning and projection of their status in the near future." Note that Endsley's three facets of situation awareness correspond to the three functions of mental models identified by Rouse and Morris (1986) and others, namely, describing (elements in the environment), explaining (their meaning), and predicting (their future status).

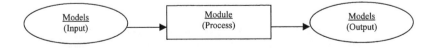

FIG. 2.2. A mental module (process) operates on input models and generates output models.

Most researchers of mental models have suggested that the mind represents information at different levels of abstraction. For example, Johnson-Laird (1983) distinguished between different levels of linguistic information, Rasmussen (1983) distinguished between different levels of system representation, and Moray (1990) distinguished between different levels of Aristotelian causation. However, the levels that have been proposed by these researchers are typically limited to deterministic representations and do not address the probabilistic representations that are required to reason under uncertainty (see Burns, 2001; Gigerenzer, Hoffrage, & Kleinbölting, 1991; Johnson-Laird, 1994).

Here I argue that cause and effect are fundamental concepts in human reasoning (also see Moray, 1990; Pearl, 2000; Rasmussen, 1983) and that reasoning about causality requires both deterministic and probabilistic models. As such, I propose three fundamental levels of mental representation (Fig. 2.3), which I characterize as Confidence (C), Hypothesis (H), and Evidence (E). I also propose (similar to other researchers; see previous) that more complex (domain-specific) representations are constructed by stacking this fundamental trio into strata (see Fig. 2.3). In the remainder of this chapter, I focus on the three levels C, H, and E within a single stratum.

To give an example for the case of the Cuyahoga, the captain's initial model of E was "two lights." This is a deterministic representation because there was no doubt in the captain's mind (by Perrow's, 1984, account)

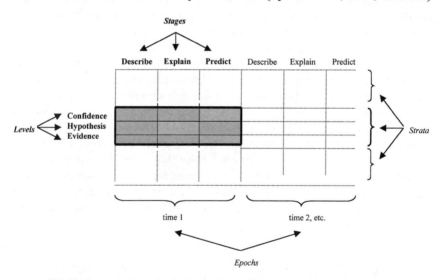

FIG. 2.3. In my framework, the *levels* refer to different types of mental *models* (i.e., representations) and the *stages* refer to different types of mental *modules* (i.e., computations). The levels are further grouped into *strata* (of domain abstraction) and the stages are further grouped into *epochs* (of scenario evolution).

about seeing two lights. One H (H_1) was that of a "fishing vessel moving in the same direction." The captain thought it was possibly a fishing vessel and "possibly" represents a measure of C in the H_1, $C(H_1)$. Although it is difficult to assign precise quantitative values to terms such as *possible* (see Zadeh, 1996), I propose that people mentally represent at least rough probability estimates (see Burns, 2001; Hertwig, Hoffrage, & Martignon, 1999; Juslin & Olsson, 1999). For example, in this analysis of the Cuyahoga collision, I assume that C can take on one of the following order-of-magnitude values: Impossible (0), unlikely (ε), possible (η), likely (ρ), or certain (1).

MENTAL MODULES AND BAYESIAN INFERENCE

Besides different levels of representation, situation awareness also involves different "stages" of computation (Fig. 2.3). Here I use a Bayesian formalism to characterize the various stages (modules) that are needed to construct various levels (models). In the following, I outline the computational details of each stage (see Table 2.1), and later I use this formal framework to analyze the Cuyahoga collision.

Describing

The process of describing a situation can be characterized as representing information about the observed E (effect) and possible $\{H_1, H_2, H_3\}$ abbreviated to $\{H_i\}$ (causes) of this E. I propose that only those H_i with high $C(H_i)$ are active at a given time (due to attention/memory limitations), and that C in each H, $C(H_i)$ is represented in order-of-magnitude terms by probabilistic mental models (i.e., C = 0, ε, η, ρ, or 1). I also propose that E is represented by a deterministic mental model.

TABLE 2.1
Mental Models and Mental Modules in Situation Awareness

Stage	*Module*	*Input Models*	*Output Models*
Describing	Represent evidence	$\{E\}$	E
	Activate hypotheses	$\{H\}$	$\{H_i\}$
	Estimate confidence	$\{C(H)\}$	$\{C(H_i)\}$
Explaining	Estimate likelihood	E, H_i	$C(E \mid H_i)$
	Update confidence	$C(H_i)$, $C(E \mid H_i)$	$C(H_i \mid E)$
	Generate hypotheses	H_i, $C(H_i)$	H'_i, $C(H'_i)$
Predicting	Project hypotheses	$H_{i(t)}$, $C(H_{i(t)})$	$H_{j(t+1)}$, $C(H_{j(t+1)})$, $E_{k(t+1)}$
	Estimate likelihood	$E_{k(t+1)}$, $H_{j(t+1)}$	$C(E_{k(t+1)} \mid H_{j(t+1)})$
	Anticipate evidence	$C(H_{j(t+1)})$, $C(E_{k(t+1)} \mid H_{j(t+1)})$	$C(E_{k(t+1)})$

Note. E = evidence; H = hypothesis; C = confidence; i is an index for hypotheses at the current state t, j is an index for hypotheses at a future state t + 1, k is an index for evidence at a future state t + 1.

Explaining

The process of explaining a situation can be characterized as identifying the most likely (represented by H) of the observed E. I propose that this process is governed by Bayes Rule (see Burns, 2001; Knill & Richards, 1996).

$$C(H_i | E) \sim C(H_i) * C(E | H_i),$$

where each H_i is an active H, $C(H_i)$ is the *prior* (C in H_i before E is observed), $C(E | H_i)$ is the *likelihood* (of observing E, given H_i) and $C(H_i | E)$ is the *posterior* (C in H_i given E). Note that the term $C(E | H_i)$ on the right hand side of this equation is "natural" in the sense that it models causality in the direction that nature works, from causes (represented by H) to effects (represented by E). Bayes Rule then specifies how this causal knowledge can be used to make the reverse inference of $C(H_i | E)$, that is, to explain the possible causes of E.

When E and H_i are familiar to a decision maker, an estimate of $C(E | H_i)$ may be stored in memory and therefore be available for recall (see Tversky & Kahneman, 1974). In other cases, a decision maker must estimate $C(E | H_i)$ by a process commonly referred to as *mental simulation* (see Biel & Montgomery, 1986; Dougherty, Gettys, & Thomas, 1997; Hendrickx, Vlek, & Oppewal, 1989; Johnson-Laird, 1994; Jungermann & Thüring, 1987; Kahneman & Tversky, 1982; Kahneman & Varey, 1982; Klein, 1998; Klein & Crandall, 1995; Klein & Hoffman, 1993; Thüring & Jungermann, 1986; Tversky & Kahneman, 1982). Similarly, a decision maker may need to change the "frame of discernment" as a situation unfolds by generating new hypotheses (i.e., "stories"; see Klein, 1998; Pennington & Hastie, 1993). For example, if the posteriors do not adequately distinguish between C in different Hs, then the decision maker may generate a new (more discriminating) set of Hs {H'_i} and corresponding Cs {$C(H'_i)$}.

Predicting

The process of predicting a situation can be characterized as identifying what E is most likely to be observed in the future. Mental simulation (see previously) also plays a key role in this stage of situation awareness. That is, to predict future E, a decision maker must project (i.e., via mental simulation or perhaps via recall from memory) how a hypothesized state i at time t, $H_{i(t)}$, will evolve to a new state j at time t + 1, $H_{j(t + 1)}$. Next, the decision maker must simulate (or recall) what E_k at time t + 1 is likely to be caused by this new state, that is, to estimate $C(E_{k(t + 1)} | H_{j(t + 1)})$. Finally, the decision maker can estimate C in the expected E, $E_{k(t + 1)}$ by summing over all Hs that can possibly give rise to that E:

$$C(E_{k(t+1)}) = \Sigma_j\ [\ C(H_{j(t+1)})\ *\ C(E_{k(t+1)}|H_{j(t+1)})\ ]$$

Repeat

Three previously mentioned stages (i.e., describing, explaining, and predicting) are repeated at each epoch (time step) as new evidence is acquired from an evolving situation. Thus, in this formal framework (Table 2.1), situation awareness is characterized as the dynamic construction of mental models via iterative application of mental modules.

ANALYZING THE CUYAHOGA COLLISION

Here I use this formal framework (Table 2.1) to analyze the Cuyahoga collision. I focus on the captain's mental models and step through the scenario illustrated in Fig. 2.1. Each step of the analysis corresponds to a single stage of situation awareness (i.e., describing, explaining, or predicting) at a given time epoch, that is, (t1), (t2), and so forth.

Describing (t1): The Captain Sees Two Lights. At (t1), the captain's mental model of the E can be characterized as $E_{(t1)}$ = two lights. Based on Perrow's (1984) account, I assume that the captain entertained the following four Hs:

$H_{1(t1)}$ = "fishing vessel moving in same direction."
$H_{2(t1)}$ = "fishing vessel moving in opposite direction."
$H_{3(t1)}$ = "other vessel moving in same direction."
$H_{4(t1)}$ = "other vessel moving in opposite direction."

Assuming that the captain considered all $H_{i(t1)}$ roughly equally likely (prior to observing any evidence), then $C(H_{1(t1)}) = C(H_{2(t1)}) = C(H_{3(t1)}) = C(H_{4(t1)}) \sim \eta$.

Explaining (t1): The Captain Infers That the Other Ship Is Proceeding in the Same Direction. Based on domain causality (i.e., same direction causes two lights, different direction causes three lights), I assume the following likelihood for each H:

$$C(E_{(t1)}|H_{1(t1)}) \sim \rho;\quad C(E_{(t1)}|H_{2(t1)}) \sim \varepsilon;\quad C(E_{(t1)}|H_{3(t1)}) \sim \rho;\quad C(E_{(t1)}|H_{4(t1)}) \sim \varepsilon,$$

where ρ means "likely" and ε means "unlikely." Applying Bayes Rule, I use these likelihoods to update the priors $C(H_{i(t1)})$ from Describing (t1) previously and compute the posteriors $C(H_{i(t1)}|E_{(t1)})$:

$$C(H_{1(t1)} | E_{(t1)}) \sim \eta * \rho; \quad C(H_{2(t1)} | E_{(t1)}) \sim \eta * \varepsilon;$$
$$C(H_{3(t1)} | E_{(t1)}) \sim \eta * \rho; \quad C(H_{4(t1)} | E_{(t1)}) \sim \eta * \varepsilon$$

Thus, the two most likely (posterior) explanations at time (t1) are $H_{1(t1)}$ and $H_{3(t1)}$. In words, the most plausible explanation is that the other ship is moving in the same direction. Because the posteriors at (t1) will act as priors at (t2), I now set the following:

$$C(H_{1(t2)}) \sim \eta * \rho; \quad C(H_{2(t2)}) \sim \eta * \varepsilon; \quad C(H_{3(t2)}) \sim \eta * \rho; \quad C(H_{4(t2)}) \sim \eta * \varepsilon$$

Predicting (t1): The Captain Expects That the Ships Will Draw Together Rapidly. At this point, the captain mentally simulates (or recalls from memory) what will happen in the next epoch (t2) and constructs mental models of the resulting (expected) E. I summarize the possible E at (t2) as the following:

$E_{rapidly(t2)}$ = "ships appear to draw together rapidly."
$E_{slowly(t2)}$ = "ships appear to draw together slowly (or not at all)."

I then characterize the likelihood of observing each $E_{k(t2)}$ given each $H_{j(t2)}$:

$$C(E_{rapidly(t2)} | H_{1(t2)}) \sim \rho; \quad C(E_{rapidly(t2)} | H_{2(t2)}) \sim \rho; \quad C(E_{rapidly(t2)} | H_{3(t2)}) \sim \varepsilon;$$
$$C(E_{rapidly(t2)} | H_{4(t2)}) \sim \rho; \quad C(E_{slowly(t2)} | H_{1(t2)}) \sim \varepsilon; \quad C(E_{slowly(t2)} | H_{2(t2)}) \sim \varepsilon;$$
$$C(E_{slowly(t2)} | H_{3(t2)}) \sim \rho; \quad C(E_{slowly(t2)} | H_{4(t2)}) \sim \varepsilon$$

In words, the expectation for H_1, H_2, and H_4 is that the ships will draw together rapidly, and the expectation for H_3 is that the ships will draw together slowly (or not at all). Combining these likelihoods with the priors $C(H_{j(t2)})$ from Explaining (t1) previously and summing over all active Hs, yields the following:

$$C(E_{rapidly(t2)}) \sim \Sigma_j [C(H_{j(t2)}) * C(E_{rapidly(t2)} | H_{j(t2)})]$$
$$= \eta * [(\rho * \rho) + (\varepsilon * \rho) + (\rho * \varepsilon) + (\varepsilon * \rho)]$$

$$C(E_{slowly(t2)}) \sim \Sigma_j [C(H_{j(t2)}) * C(E_{slowly(t2)} | H_{j(t2)})]$$
$$= \eta * [(\rho * \varepsilon) + (\varepsilon * \varepsilon) + (\rho * \rho) + (\varepsilon * \varepsilon)]$$

Because $C(E_{rapidly(t2)}) > C(E_{slowly(t2)})$, the most likely expectation for (t2) is that the ships will draw together rapidly.

Describing (t2): The Ships Appear to Draw Together Rapidly. At (t2), the new E (which confirms the captain's expectation; see previously) is that the ships indeed appear to draw together rapidly. That is, $E_{(t2)} = E_{rapidly(t2)}$.

Explaining (t2): The Captain Infers That It Is a Fishing Vessel Moving in the Same Direction. Given the new E, $E_{(t2)} = $ "ships appear to draw together rapidly," the likelihoods $C(E_{rapidly(t2)} | H_{j(t2)})$ from Predicting (t1) previously can be used in Bayes Rule to update the $C(H_{j(t2)})$ from Explaining (t1) previously:

$$C(H_{1(t2)} | E_{(t2)}) \sim (\eta\rho) * \rho; \; C(H_{2(t2)} | E_{(t2)}) \sim (\eta\epsilon) * \rho; \; C(H_{3(t2)} | E_{(t2)}) \sim (\eta\rho) * \epsilon;$$
$$C(H_{4(t2)} | E_{(t2)}) \sim (\eta\epsilon) * \rho$$

In words, the only likely H at this time is that of a "fishing vessel moving in same direction," $H_{1(t2)}$. Note that although Perrow (1984) called this "reinforcing the incorrect interpretation," and Hutchins (1995) called it "confirmation bias," my analysis shows it is actually a normative (Bayesian) result of updating the modeled Hs with the modeled E.

Predicting (t2): The Captain Expects the Other Ship to Turn and Cause a Collision. Focusing on the only likely H, $H_{1(t2)}$, the captain uses his domain knowledge of fishing boats to mentally project (i.e., simulate or recall) a new model at time (t3), $H_{1(t3)}$. This new model represents a scenario in which the Cuyahoga will cut off the fishing vessel's ability to turn as it approaches the river (Fig. 2.1, (t3), top panel).

Acting: The Captain Turns His Ship to Avoid a Collision. Based on his situation awareness (see previously), the captain decides to take action to prevent a collision. Note that this decision is reasonable in light of the captain's mental models (as analyzed here), including his mental simulation of how fishing boats behave in this part of the Chesapeake Bay (i.e., his projected model $H_{1(t3)}$). The fundamental problem was that the most likely H at time (t2), $H_{1(t2)}$, did not match reality in a significant way. The unfortunate consequence was a collision, which was exactly what the captain's action was intended to prevent!

DISCUSSION

The case of the Cuyahoga is a classic example of what is often called a *decision error*. However, contrary to this label, my analysis shows that the captain's decision was rational in the context of his perception of the situation (i.e., his mental models). This is similar to the conclusion reached by

Perrow (1984) and others. However, my analysis goes further in demonstrating that the computational processes (i.e., mental modules) underlying the construction of these mental models were also rational. That is, the captain's probabilistic perception of the situation was also consistent with the norms of Bayesian inference, given his deterministic models of H and E. This is an important insight because confirmation bias is one of the most commonly cited causes of human error in the command and control of complex systems (see Perrow for more examples in transportation systems, nuclear power, and other domains; see Klein's, 1998, account of the Vincennes shootdown of an Iranian airbus for an example in the domain of air defense).

Normal Errors

Perrow (1984) called such accidents normal because he believed that they are inevitable (although not necessarily anticipated; also see Wagenaar & Groeneweg, 1987) in complex human–machine systems. My analysis suggests that the associated errors are also normal in a mathematical sense because the underling computations can be characterized as normative (Bayesian). The catch, of course, is that these Bayesian computations operate within the bounded context of the decision makers' mental models.

For example, a perfect (unbounded) Bayesian would have considered at least two things that a human (bounded) Bayesian like the captain of the Cuyahoga apparently did not. First, unbounded Bayesians would have considered the possibility that their original model of the E (i.e., two lights) was incomplete. That is, the E at one stratum of abstraction (see Fig. 2.3) is actually one of several (perhaps many) Hs (with associated Cs) at a lower stratum of abstraction. Second, unbounded Bayesians would have retained all Hs about the other ship and used this information in their subsequent decision to act. That is, they would have retained the unlikely H that the other ship might be moving in the opposite direction and they would have considered the high cost of their contemplated action (to turn) in light of this possibility.

Because of these limitations, some might characterize the captain as non-normative (biased) relative to an unbounded Bayesian. However, given the finite resources available to naturalistic decision makers, I suggest that it is more accurate (and more useful) to characterize the captain as a Bayesian bounded by mental models.

Heuristics and Biases

At first glance, my claims about the normative nature of human reasoning (previously) may appear to be at odds with the vast literature on non-normative heuristics and biases (see Kahneman, Slovic, & Tversky, 1982;

Sage, 1981). It is not. For example, the widely cited heuristics known as anchoring/adjustment, availability, representativeness, and simulation (see Tversky & Kahneman, 1974) are all implicitly captured by my computational modules (Table 2.1). Anchoring/adjustment is inherent in the module "update confidence" in which a prior anchor is updated with a likelihood adjustment. Availability is inherent in the module "activate hypotheses" in which $C(H_i)$ is a formal measure of availability used to determine which H_i are active at a given time. The simulation heuristic is inherent in the modules "generate hypotheses" and "project hypotheses," which consider the deterministic construction of new models (H'_i and $H_{j(t+1)}$) via mental simulation. Finally, representativeness is inherent in the module "estimate likelihood" in which $C(E|H)$ is a formal measure of how representative E is of H. Thus, rather than being in disagreement with previous findings, I believe that the formal framework (Table 2.1) can clarify and unify many of the isolated heuristics proposed by previous research.

I also believe that this approach can shed new light on the biases that result from these heuristics. In particular, the main finding of my analysis is that a prototypical case of confirmation bias (i.e., the Cuyahoga collision) can be explained in terms of normative computations operating on cognitive representations. Gigerenzer et al. (1991) made a similar claim with regard to two other widely cited biases, that is, the overconfidence effect and the hard–easy effect. Here I stress that my results stem from an integrated analysis of computational modules and representational models in a naturalistic context (see Burns, 2001; Richards, 1988; Richards, Jepson, & Feldman, 1996). This is an important shift in emphasis because research on heuristics and biases has typically focused on computations (i.e., heuristics) while making nonnatural assumptions about representations (i.e., not adequately characterizing the participants' tacit assumptions; see Nickerson, 1996). My approach focuses on mental models in a natural context simply because the performance of any computation (module) is bounded by the representations (models) available to it.

Decision Support Systems

As discussed previously, models in the mind (i.e., mental models) are used by decision makers to describe, explain, and predict situations. Similarly, models of the mind are needed by decision scientists to describe, explain, and predict cognitive competence. Thus far, most theories of naturalistic decision making (see Lipshitz, 1993) have been limited to qualitative models that only describe human performance. My goal is to develop more quantitative models (also see Miao, Zacharias, & Kao, 1997) that can go further to explain and predict human performance in naturalistic decision making.

My work is motivated by a desire to improve the design of computer systems that support human decision making in practical applications. My claim is that the design of external systems provided to decision makers must consider the internal models used by these decision makers. That is, computational models of how decisions should be made (optimally, in support systems) must be developed in concert with computational models of how decisions are made (actually, in human minds). I believe that the formal framework presented in this chapter can help bridge this gap and thereby improve both the scientific understanding of human decision making and the engineering development of computer support systems.

CONCLUSION

My objective was to specify, in computational terms, how decision makers construct mental models to achieve situation awareness. I began by outlining a formal framework comprising three levels of mental representation and three stages of mental computation. I then used this framework to analyze a case study in naturalistic decision making. My main insight is that prototypical decision errors can be attributed to normal (Bayesian) mental modules operating on natural (bounded) mental models.

REFERENCES

Biel, A., & Montgomery, H. (1986). Scenarios in energy planning. In B. Brehmer, H. Jungermann, P. Lourens, & G. Sevón (Eds.), *New directions in research on decision making* (pp. 205–218). Amsterdam: North-Holland.

Burns, K. (2001). Mental models of line drawings. *Perception, 30*, 1249–1261.

Craik, K. (1943). *The nature of explanation.* Cambridge, England: Cambridge University Press.

Dougherty, M., Gettys, C., & Thomas, R. (1997). The role of mental simulation in judgments of likelihood. *Organizational Behavior and Human Decision Processes, 70*, 135–148.

Endsley, M. (1988). Design and evaluation for situation awareness enhancement. *Proceedings of the 32nd Annual Meeting of the Human Factors Society*, 97–101.

Gigerenzer, G., Hoffrage, U., & Kleinbölting, H. (1991). Probabilistic mental models: A Brunswikian theory of confidence. *Psychological Review, 98*, 506–528.

Hendrickx, L., Vlek, C., & Oppewal, H. (1989). Relative importance of scenario information and frequency information in the judgment of risk. *Acta Psychologica, 72*, 41–63.

Hertwig, R., Hoffrage, U., & Martignon, L. (1999). Quick estimation: Letting the environment do the work. In G. Gigerenzer & P. M. Todd (Eds.), *Simple heuristics that make us smart* (pp. 191–208). New York: Oxford University Press.

Hutchins, E. (1995). *Cognition in the wild.* Cambridge, MA: MIT Press.

Johnson-Laird, P. (1983). *Mental models: Towards a cognitive science of language, inference and consciousness.* Cambridge, MA: Harvard University Press.

Johnson-Laird, P. (1994). Mental models and probabilistic thinking. *Cognition, 50*, 189–209.

Jungermann, H., & Thüring, M. (1987). The use of mental models for generating scenarios. In G. Wright & P. Ayton (Eds.), *Judgmental forecasting* (pp. 245–266). Chichester, England: Wiley.

Juslin, P., & Olsson, H. (1999). Computational models of subjective probability calibration. In P. Juslin & H. Montgomery (Eds.), *Judgment and decision making: Neo-Brunswikian and process-tracing approaches* (pp. 67–95). Mahwah, NJ: Lawrence Erlbaum Associates.

Kahneman, D., Slovic, P., & Tversky, A. (Eds.). (1982). *Judgment under uncertainty: Heuristics and biases.* Cambridge, England: Cambridge University Press.

Kahneman, D., & Tversky, A. (1982). The simulation heuristic. In D. Kahneman, P. Slovic, & A. Tversky (Eds.), *Judgment under uncertainty: Heuristics and biases* (pp. 201–208). Cambridge, England: Cambridge University Press.

Kahneman, D., & Varey, C. (1982). Propensities and counterfactuals: The loser that almost won. In W. M. Goldstein & R. M. Hogarth (Eds.), *Research on judgment and decision making: Currents, connections and controversies* (pp. 322–341). Cambridge, England: Cambridge University Press.

Klein, G. (1998). *Sources of power: How people make decisions.* Cambridge, MA: MIT Press.

Klein, G., & Crandall, B. (1995). The role of mental simulation in naturalistic decision making. In P. Hancock, J. Flach, J. Caird, & K. Vicente (Eds.), *Local applications of the ecological approach to human-machine systems* (pp. 324–358). Mahwah, NJ: Lawrence Erlbaum Associates.

Klein, G., & Hoffman, R. (1993). Seeing the invisible: Perceptual-cognitive aspects of expertise. In M. Rabinowitz (Ed.), *Cognitive science foundations of instruction* (pp. 203–226). Mahwah, NJ: Lawrence Erlbaum Associates.

Knill, D., & Richards, W. (Eds.). (1996). *Perception as Bayesian inference.* Cambridge, England: Cambridge University Press.

Lipshitz, R. (1993). Converging themes in the study of decision making in realistic settings. In G. A. Klein, J. Orasanu, R. Calderwood, & C. E. Zsambok (Eds.), *Decision making in action: Models and methods* (pp. 103–137). Norwood, NJ: Ablex.

Miao, A., Zacharias, G., & Kao, S. (1997). A computational situation assessment model for nuclear power plant operations. *IEEE Transactions on Systems, Man and Cybernetics—Part A: Systems and Humans, 27,* 728–742.

Moray, N. (1990). A lattice theory approach to the structure of mental models. *Philosophical Transactions of the Royal Society of London, B 327,* 577–583.

Nickerson, R. (1996). Ambiguities and unstated assumptions in probabilistic reasoning. *Psychological Bulletin, 120,* 410–433.

Pearl, J. (2000). *Causality: Models, reasoning and reference.* Cambridge, England: Cambridge University Press.

Pennington, N., & Hastie, R. (1993). A theory of explanation-based decision making. In G. A. Klein, A. J. Orasanu, R. Calderwood, & C. E. Zsambok (Eds.), *Decision making in action: Models and methods* (pp. 188–201). Norwood, NJ: Ablex.

Perrow, C. (1984). *Normal accidents: Living with high-risk technologies.* New York: Basic Books.

Rasmussen, J. (1983). Skills, rules and knowledge; Signals, signs and symbols, and other distinctions in human performance models. *IEEE Transactions on Systems, Man and Cybernetics, 13,* 257–266.

Reason, J. (1990). *Human error.* Cambridge, England: Cambridge University Press.

Richards, W. (Ed.). (1988). *Natural computation.* Cambridge, MA: MIT Press.

Richards, W., Jepson, A., & Feldman, J. (1996). Priors, preferences and categorical percepts. In D. Knill & W. Richards (Eds.), *Perception as Bayesian inference* (pp. 93–122). Cambridge, England: Cambridge University Press.

Rouse, W., & Morris, N. (1986). On looking into the black box: Prospects and limits in the search for mental models. *Psychological Bulletin, 100,* 349–363.

Sage, A. (1981). Behavioral and organizational considerations in the design of information systems and processes for planning and decision support. *IEEE Transactions on Systems, Man and Cybernetics, 11,* 640–678.

Sarter, N., & Woods, D. (1991). Situation awareness: A critical but ill-defined phenomenon. *The International Journal of Aviation Psychology, 1,* 45–57.

Thüring, M., & Jungermann, H. (1986). Constructing and running mental models for inferences about the future. In B. Brehmer, H. Jungermann, P. Lourens, & G. Sevón (Eds.), *New directions in research on decision making* (pp. 163–174). Amsterdam: North-Holland.

Tversky, A., & Kahneman, D. (1974, Sept. 27). Judgement under uncertainty: Heuristics and biases. *Science, 185*, 1124–1131.

Tversky, A., & Kahneman, D. (1982). Causal schemas in judgments under uncertainty. In D. Kahneman, P. Slovic, & A. Tversky (Eds.), *Judgment under uncertainty: Heuristics and biases* (pp. 117–128). Cambridge, England: Cambridge University Press.

Wagenaar, W., & Groeneweg, J. (1987). Accidents at sea: Multiple causes and impossible consequences. *International Journal of Man–Machine Studies, 27*, 587–598.

Zadeh, L. (1996). Fuzzy logic = computing with words. *IEEE Transactions on Fuzzy Systems, 4*, 103–111.

3

"More Is Better?": A Bias Toward Overuse of Resources in Naturalistic Decision-Making Settings

Mary M. Omodei
Jim McLennan
Glenn C. Elliott
La Trobe University, Melbourne

Alexander J. Wearing
Julia M. Clancy
University of Melbourne

Most military and emergency services organizations appear to believe that "more is better" with regard to resources available to a commander, such as (a) access to detailed information, (b) access to reliable information, (c) direct control over assets and personnel, and (d) amount of communication input from subordinates (e.g., Cardinal, 1998; Seal, 1998). There is general acknowledgment that there are limits to how much of any such resources a commander can effectively process under the time pressure that characterizes a typical military engagement or emergency incident. However, it appears to be assumed that it is not necessary to impose external limits on a commander's freedom, allowing the commander to select from the available resources those that he or she believes can be effectively managed. That is, military and emergency services organizations operate as if their trained leaders, as rational decision makers, can optimally allocate their cognitive capacities adaptively to deal with any situation. It seems to be assumed that commanders are able to effectively regulate their task-related cognitive workload with respect to using each of the aforementioned types of resources. Recent research into decision making in complex environments has suggested that commanders may in fact be disadvantaged by the provision of multiple information sources (Seagull, Wickens, & Loeb, 2001) and multiple opportunities for action (Elg, 2001). In this chapter, we report the results of four experiments that tested the conventional wisdom account that the more information and control resources one can provide a commander, the better the result will be. In the experiments reported here, we systematically manipulated each of the four resources listed previously to find out their effect on decision-making performance and commanders' associated experiences.

There has been a skepticism in the naturalistic decision-making community of the value of attempts to capture in the laboratory the essential characteristics of decision making as this occurs in naturalistic situations. Team-based (networked) computer simulations are now increasingly recognized by practitioners as having adequate fidelity for training and for decision support in operations (cf. Elliott, Dalrymple, Regian, & Schiflett, 2001). One of the outcomes of this recognition is that such simulations are also seen by researchers as a suitable platform for investigating the key psychological processes involved in multiperson dynamic decision tasks (Weaver, Bowers, Salas, & Cannon-Bowers, 1995).

In the experiments reported here, we have used the Networked Fire Chief (Omodei, Taranto, & Wearing, 1999) microworld generating program to create a complex forest fire-fighting task environment. Networked Fire Chief allows for detailed recreation of landscapes of varying composition, flammability characteristics, and asset values. Participants are asked to control the spread of these simulated fires by dispatching fire trucks and helicopters to drop water on the fires and to refill with water at specified lake locations. To maintain a focus on strategic, as distinct from tactical, decision making for research participants, the simulated appliances are programmed to automatically locate and extinguish any fires within a three landscape-segment radius of their current position. Furthermore, if dispatched to fill with water, appliances automatically return to resume fire fighting in their original location.

The Networked Fire Chief microworld generating program allows for the creation of multiparticipant decision-making scenarios and the flexible implementation of a range of command, control, communication, and intelligence structures. Although a wide range of scenarios can, in principle, be created, forest fire fighting has been selected as the specific microworld scenario in all our experiments to date because we have been able to recruit as participants persons who (although technically not experts) bring to the task a well-developed understanding of forest fire behavior and methods for its control.

In our Networked Fire Chief experimental scenarios, we have been concerned to implement decision-making scenarios that possess the deep structure of command and control of emergency incidents rather than being concerned with superficial surface structure (thus, we aimed for psychological, as distinct from physical, fidelity). We argue, therefore, that our findings are not limited to fire fighting but transfer to a range of time-pressured, high-stakes, multiparticipant dynamic decision-making settings: military operations, transport control, plant operations, and so forth.

In the experiments reported following, participants were in teams of three, one individual being assigned the role of the fireground commander in control of a simulated forest area divided into two geographical sectors, each under the control of one of the other participants. As detailed following, in all four experiments, sector controllers were research confederates trained to act consistently across participants and experimental manipulations. Communications between commanders and subordinates were via aviation-type headsets.

In all experiments (with the exception of Experiment 3), the research participant, assuming the role of fireground commander, was unable to directly deploy fire-fighting appliances to the simulated fires, having to rely on verbal commands to the subordinates to control fire-fighting activities. Note that in Experiment 3, removal of this constraint constituted the experimental manipulation. In all experiments, subordinates were instructed to act only as ordered by their commander and were not allowed to initiate strategic-level decision making.

In all experiments, participants were given approximately 3 hr of training prior to their experimental trials. Experimental trial durations were approximately 15 min. With respect to the resource manipulation in each study, the resource itself was normatively of considerable value. That is, the additional resource would confer a considerable advantage under conditions of unlimited cognitive capacity and/or lack of time pressure.

The primary dependent variable was the commander's decision-making skill as measured by the overall value of landscape assets (as a percentage) saved from destruction by fire at the end of each experimental trial. In the scenarios created for this set of four experiments, fire development was such that a good performance resulted in over 90% of assets saved, whereas a poor performance resulted in just over 80% of assets saved. If scenarios were allowed to run with no attempt to halt the spread of fires, baseline performance scores were on the order of 60% of assets saved. Adequate between-trial reliability has consistently been demonstrated with respect to this performance measure ($r > .7$). With respect to the assessment of participant expectations concerning the outcome of the experimental manipulations, these were obtained only on completion of all experimental trials (thereby avoiding the introduction of demand characteristics).

EXPERIMENT I

In this experiment, there were two groups of participants. In one group, the commanders had to rely on subordinates to provide, on request, much of their information. In the other group, complete information was continuously provided on-screen to the commanders.

Method

The experimental conditions were as follows:

Incomplete information condition: Incomplete landscape features (only major features, no wind details, and no fire warnings visible). Verbal reports from subordinates could be requested by the commander about these details.

Detailed information condition: All landscape features, wind details, and immediate fire warnings always visible on-screen to the commander.

There were 18 teams of 3 persons in each of the two experimental conditions. In all teams, the participant was assigned the role of commander and two trained research assistants occupied the two (confederate) subordinate, sector controller roles.

Results and Discussion

Average team performance was significantly better in the incomplete information condition, $t(35) = 1.88$, $p < .05$ (see Fig. 3.1). Note that this cannot be explained simply in terms of expectancy effects, as the findings were in spite of commanders' expectations that detailed on-screen information (more) would result in better performance.

Two possibilities arose for the observed performance degradation associated with the passive presentation of full and accurate on-screen information to the commander. First, perhaps participants were less able to prioritize the information available when that information was automatically provided rather than actively, and therefore also selectively, requested. Second, perhaps information inspection took precedence (a) over information integration and forecasting and (b) over intention formation and action generation (i.e., participants were trapped in preaction phases of cognitive activities relating to situation assessment).

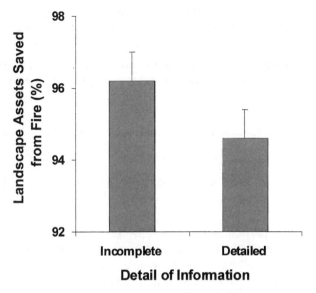

FIG. 3.1. Continuous provision of detailed information.

EXPERIMENT 2

In this repeated measures experiment, all participants experienced trials under two conditions: one in which all information was reliable and one in which critical information was known to be unreliable (approximately 50% reliability).

Method

The experimental conditions were as follows:

> Unreliable information condition: Of the five wind forecasts that were programmed to occur during the 15-min. trials, only two occurred as predicted. For the remainder, the wind conditions either did not change at all or changed to a strength or direction that differed from that forecast.

Verbatim instructions to participants before trials in this condition were as follows:

- As you know, weather conditions are sometimes such that it is not possible to predict what the weather will do. For the next trial, I'd like you to assume that the weather bureau is not able to reliably predict the wind conditions, in fact they'll get it right about half of the time.

Completely reliable information condition: Of the five wind forecasts that were programmed to occur during the 15-min. trials, all occurred as forecasted.

Verbatim instructions to participants before trials in this condition were as follows:

• In this trial the wind forecast is 100% reliable, that is, any forecast changes will occur exactly as forecast.

There were 31 participants in the repeated measures design (order of presentation of unreliable vs. reliable trials was counterbalanced). The participant was assigned the role of commander, and two trained research assistants occupied the two (confederate) subordinate roles.

Note that participants (commanders) were given equivalent training experience of both experimental conditions.

Results and Discussion

Average team performance was significantly better in the unreliable information condition, $t(30) = 8.16$, $p < .01$ (see Fig. 3.2).

Note that this cannot be simply explained as an expectancy effect because performance degradation was in spite of commanders' expectations that performance with reliable information (more) would be superior.

Participants' reports of their experiences of the two experimental conditions suggests (a) a felt pressure to work harder to take into account the more reliable information together with (b) a sense of relief when the information was known to be unreliable, providing an "excuse" not to have to work so hard. This suggests a tendency to totally discount information of known dubious reliability in conjunction with a tendency to try to take into careful account all information believed to be reliable. As we note in the General Discussion later, there was no evidence that participants expended differing levels of overall mental effort across the two conditions. This suggests that the apparent overutilization of information in the reliable information condition was probably at the expense of either (a) selective quality use of this information or (b) allocating attention to other important task-related information such as keeping track of fire locations.

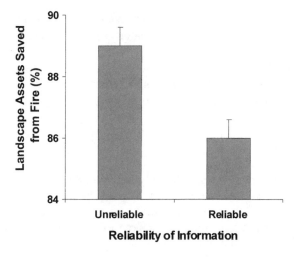

FIG. 3.2. Provision of reliable information.

EXPERIMENT 3

An experiment was conducted to investigate performance effects of the amount of control a commander has over the operational resources available to his or her team. In this experiment, there were two groups of participants. In one group, the standard experimental protocol used in Experiments 1 and 2 reported previously was employed in which the commander (participant) was unable to physically direct fire-fighting appliances, having to rely on his or her subordinate sector controllers to do so following explicit instructions. In the other group, the commander (as well as the sector controllers) was able to directly deploy fire-fighting appliances via his or her computer terminal. That is, if the fireground commander desired, he or she did not have to rely solely on subordinate sector controllers for operational activity. Note the availability of sector controller subordinates to implement any (or all) of the commander's orders was fully retained. That is, the commander was provided with the opportunity (not necessity) for micromanagement in that he or she could choose to bypass a level of command to control directly the (simulated) crews of the fire trucks and fire helicopters.

Method

The experimental conditions were as follows:

Indirect control condition: All commanders' actions were implemented via communication of orders to subordinates (delegation).

Direct control condition: Some (or all) commanders' actions could be implemented via direct control of simulated appliances (micromanagement).

There were 18 teams of 3 persons in each of the two experimental conditions. In all teams, the participant was assigned the role of commander and two trained research assistants occupied the two (confederate) subordinate roles.

Results and Discussion

Average performance was significantly better in the indirect control condition, $t(35) = 1.88$, $p < .05$ (see Fig. 3.3). As participants were not constrained in the direct control condition to actually issue direct commands; an examination of the performance implications of those who chose to do so is informative. Large individual differences were noted across the 18 participants in the direct control condition, with more commander micromanagement being associated with worse performance: $r = -.41$.

Note that this finding cannot be simply explained as an expectancy effect because commanders' expectations were that direct control (more) would result in better team performance.

EXPERIMENT 4

This experiment was conducted to investigate effects on performance of the amount of communicative input from subordinates to commanders. In

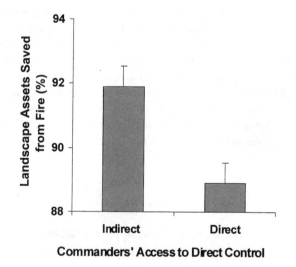

FIG. 3.3. Access to directional control of operational resources.

this repeated measures experiment, all participants experienced trials under two conditions. One condition was the standard experimental protocol used in all the experiments outlined previously in which subordinates were able to communicate with their commanders insofar as it was necessary to clarify instructions or to provide information as requested by the commander. The comparison condition employed a one-way communication condition in which subordinates were prevented from talking to their commander.

Method

The experimental conditions were as follows:

> One-way condition: Subordinates were unable to respond to their commander, that is, communication channels were only one-way—commander to subordinates.

Verbatim instructions to participants were as follows:

- The Incident Controller can initiate communication with his/her Sector Controllers at any time. SC's, your job is to follow the IC's orders at all times. As the IC has an overview of the whole area, they must assume total decision making responsibility.
- With the way that the communications equipment is set up, the IC will not be able to hear the SC's at all. This communication structure can be advantageous in time pressured situations as the IC's time is not wasted with unnecessary communications.

Two-way condition: Subordinates were able to respond to but not challenge/question commanders.

Verbatim instructions to participants were as follows:

- The Incident Controller can initiate communication with his/her Sector Controllers at any time.
- SC's your job is to follow the IC's orders at all times. As the IC has an overview of the whole area, they must assume total decision making responsibility. It has been found that any questioning and challenging of the IC's orders can be time consuming and can interfere with the IC's ability to give appropriate orders to both Sector Controllers. You have to be careful as SC's not to assume the same level of decision making as the IC. Therefore you must not question the orders that the IC gives you at any time.

There were 31 participants in the repeated measures design (order of presentation of one-way vs. two-way trials was counterbalanced). The participant was assigned the role of commander, and two trained research assistants occupied the two (confederate) subordinate roles.

Note that participants (commanders) were given equivalent training experience of both experimental conditions.

Results and Discussion

There was no difference in performance between one-way and two-way conditions, $t(30) = 0.46$, $p > .05$ (see Fig. 3.4). Commanders' expectations were that the two-way (more) would be better than the one-way condition.

A plausible interpretation of the no differences finding is that any advantages that might accrue to commanders by being able to get verbal input from subordinates is at some cost. Part of this cost might be the necessary attentional and cognitive resources that must be allocated to processing any subordinate communicative utterance. Such processing of subordinate communications might be at the expense of (a) clarity in the commanders' own communications to these subordinates, (b) overall time spent on strategy formation, and (c) periods of uninterrupted time for such strategy formation. Also, commanders might institute more vigilant monitoring of subordinate behavior in the absence of verbal feedback.

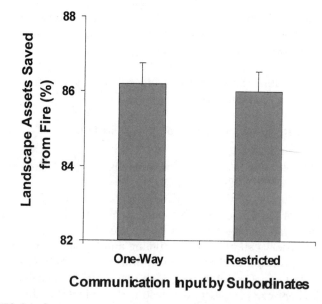

FIG. 3.4. One-way versus two-way communication input by subordinates.

GENERAL DISCUSSION

The findings across all of our experimental manipulations cast serious doubt on any uncritical assumption that more is better with respect to the provision of access to resources by a commander in a time-pressured, distributed dynamic command environment characterized by uncertainty. It appears that if a resource is made available, commanders feel compelled to use it. Such a bias to overutilize resources appears to be nonspecific, applying to all types of resources, whether this be information gathering, opportunities for action, or communication input. In other words, our commanders seemed to be unable to optimally regulate their levels of both information acquisition and control activity. That is, in a resource-rich environment, commanders will utilize resources even when their cognitive system is so overloaded as to result in a degradation of performance. Other deleterious effects of such cognitive overload might include (a) decrement in maintenance of adequate global situation awareness, (b) impairment of high-level strategic thinking, and (c) diminished appreciation of the time scales involved in setting action in train.

Our participants' self-reports suggest that any such overutilization of resources is not accompanied by an increase in overall cognitive effort across all resources provided in the environment. That is, there is likely to be a tradeoff between quantity of resources utilized and the quality of the decision-making process underlying such use. One possibility suggested by our findings for such a difference in the quality of cognitive activity is that the amount of strategic thought might be greater in a relatively underresourced environment. Another somewhat related possibility suggested by our findings is that there is an increase in action implementation over information acquisition in a relatively underresourced environment.

Possible Underlying Psychological Processes

Patterns of behavior observed in our participants, together with six participants' own self-reports during posttask interviews (cued by replaying their previous simulation trial) suggest that the bias to overutilize resources operates outside of conscious awareness and as such may represent the activation of one or more higher order generic cognitive schema (cf. Reason, 1990). Precisely what might characterize such schematic processing is unclear. Our data suggest the following possibilities:

1. A preference for errors of commission rather than omission. In most emergency situations, time is short, so there could very well be a general task bias for action over delay. There is some degree of self-protection in being able to identify clearly specified attempts to use resources, however un-

productive such attempts might be. The findings in the literature on judgment and choice with respect to the effects of accountability (Tetlock, 1988) in general and errors of commission in particular (Kerr, MacCoun, & Kramer, 1996) warrant careful attention for their applicability to complex dynamic systems.

2. An illusory sense of greater control via activity. Schmitt and Klein (1996) argued that decision makers, to avoid subjective uncertainty, act in such a manner as to achieve an illusory sense of cognitive control over the environment. That is, activity regardless of its adequacy provides a sense that one is having some desirable effect in the problematic situation.

3. An illusory sense of greater self-competence via activity. Doerner (1990) argued that decision makers act so as to guard against a sense of personal incompetence. Doerner's account proposes a primarily affective orientation to protect one's sense of self-worth (i.e., to maintain ego integrity) as distinct from Schmitt and Klein's (1996) cognitive orientation to experience direct control.

4. An overestimation of personal ability. A somewhat related self-serving bias that could account for the pattern of findings is an overestimation of personal ability with respect to both speed of information processing and amount of information that can be concurrently managed in working memory. There is a substantial body of literature demonstrating a general overconfidence in the quality of one's own decision making, an overconfidence bias that is believed to be cognitively as well as motivationally based (Camerer & Johnson, 1991).

Conclusion

If it is the case that any of the preceding four processes were underlying the behavior of participants in the experiments outlined previously (we suggest all four may operate), the basic problem is one of inadequate metacognitive control (self-regulation with respect to attentional and cognitive task control activity). As such, these findings have substantial implications for how one conceives of command and control decision making in human factors terms and for how one designs systems and provides aids to optimize such decision making.

ACKNOWLEDGMENTS

The research was supported by research grants from the Australian Defence Science and Technology Organisation and the Australian Research Council.

REFERENCES

Camerer, C. F., & Johnson, E. J. (1991). The process-performance paradox in expert judgment: How can experts know so much and predict so badly? In K. A. Ericsson & J. Smith (Eds.), *Toward a general theory of expertise: Prospects and limits* (pp. 195–217). Cambridge, England: Cambridge University Press.

Cardinal, R. (1998, April). The firefight. *Wildfire, 7,* 36–37.

Doerner, D. (1990). The logic of failure. *Philosophical Transactions Of The Royal Society Of London, 327,* 463–473.

Elg, F. (2001, November). *Frequency matching in a microworld setting.* Paper presented at the first international workshop on cognitive research with microworlds, Granada, Spain.

Elliott, L. R., Dalrymple, M., Regian, J. W., & Schiflett, S. G. (2001, October). *Scaling scenarios for synthetic task environments: Issues related to fidelity and validity.* Paper presented at the Human Factors and Ergonomics Society 45th annual meeting, Minneapolis.

Kerr, N. L., MacCoun, R. J., & Kramer, G. P. (1996). Bias in judgment: Comparing individuals and groups. *Psychological Review, 103,* 687–719.

Omodei, M. M., Taranto, P., & Wearing, A. J. (1999). Networked Fire Chief (Version 1.0) [Computer program]. La Trobe University, Melbourne.

Reason, J. (1990). *Human error.* New York: Cambridge University Press.

Schmitt, J. F., & Klein, G. A. (1996). Fighting the fog: Dealing with battlefield uncertainty. *Marine Corps Gazette, 80,* 2–12.

Seagull, J. F., Wickens, C. D., & Loeb, R. G. (2001, October). *When is less more? Attention and workload in auditory, visual, and redundant patient-monitoring conditions.* Paper presented at the Human Factors and Ergonomics Society 45th annual meeting, Minneapolis.

Seal, T. E. (1998, December). Building for the future: A 'combat development' approach. *Wildfire, 7,* 12–16.

Tetlock, P. E. (1988). Losing our religion: On the precariousness of precise normative standards in complex accountability systems. In R. M. Kramer (Ed.), *Power and influence in organizations* (pp. 121–144). Thousand Oaks, CA: Sage.

Weaver, J. L., Bowers, C. A., Salas, E., & Cannon-Bowers, J. A. (1995). Networked simulations: New paradigms for team performance research. *Behavior Research Methods: Instruments & Computers, 27,* 12–24.

4

Planning Under Time Pressure: An Attempt Toward a Prescriptive Model of Military Tactical Decision Making

Peter Thunholm
National Defence College, Sweden

Military science literature, as well as army regulations of different armies, contain both explicit and implicit statements and assumptions about how the battlefield environment affects the military decision situation. The battlefield environment contains (a) uncertainty (Clausewitz, 1976) due to a lack of information and due to unreliable or false information; (b) time pressure (Smith, 1989) due to the need for a high tempo to be able to seize, and keep, the initiative in a battle (Arméreglemente del 2 Taktik [AR2], 1995; Dupuy, 1984); and (c) high complexity (Clausewitz, 1976; Van Creveld, 1985) because the battlefield activities are affected by a lot of factors, not least humans, at different levels and because the number of units and activities that should be coordinated is large, at least at the higher levels of command (i.e., brigade, division, and above). The battlefield decision problem is also a dynamic one, which means that (a) it requires a series of decisions, (b) these decisions are not independent, (c) the state of the problem changes by its own power and by the actions of the decision maker, and (d) decisions must be made in real time (Brehmer, 1992). An important consequence of the dynamics is that the quality of any decision can only be judged in retrospect. What is right and wrong is a result of when it is done and what the opponent does.

To be able to manage the military decision-making process (MDMP), modern Western armies have tried to structure it. The result is prescriptive, highly formalized step-by-step models in which decisions are supposed to develop linearly (AR2, 1995; Scott, 1992; U.S. Army Field Manual

(FM) 101–5, 1997). The models are consistent with conventional-phase models of human problem solving, dividing the process into sequences of problem definition, alternative generation, refinement, and selection.

The current Swedish army tactical decision-making process (AR2, 1995, pp. 185–190) consists of 22 steps divided into 6 separate main steps. These main steps are shown in Table 4.1. The model is not described in great detail in the Army regulation, but there is a very detailed practice about how to use the model. This practice is passed on to new officers in military schools by oral tradition and unofficial manuals.

In Step 1, the mission is analyzed. In Step 2, the current and possible future situations are analyzed in several aspects such as enemy forces, own forces, terrain, and so forth. This is done in a certain order in which conclusions about requirements for the plan are drawn. The main emphasis in the training of officers is placed on Step 3 in which the planning staff officers are supposed to develop at least two possible enemy courses of action (COAs) and three of their own forces' different and rather detailed COAs that describe the development of events during the whole battle until the objectives of the higher commander have been met. Next, one should compare one's own COAs with a set of selection criteria and make a suggestion to the commander about which COA to choose. The last event in Step 3 is deciding on a COA. The commander normally makes this decision. In Step 4, an overall description of the whole operation is formulated and that description or "concept of operations" forms the base for a more detailed description in Step 5.

Although military officers are subjected to substantial amounts of training in the application of the model, it is often not used, especially when there is time pressure. When referring to traditional models like the U.S. Army model, Schmitt and Klein (1999) stated that "these models are inconsistent with the actual strategies of skilled planners, and they slow down the decision cycle. As a result, the formal models are usually ignored in practice, in order to generate faster tempo" (p. 510).

TABLE 4.1
The Swedish Army Model for Tactical Planning and Decision Making

Planning Model

Planning		
Step 1.	The mission	
Step 2.	The combat estimate	
Step 3.	Considerations	
Step 4.	Concept of operations	Combat Plan
Step 5.	Development of concept of operations	
Execution		
(Step 6.)	Basis for the execution of operations	

The purpose of this chapter is to present an alternative prescriptive military decision-making model rooted in the naturalistic decision-making tradition. The new planning under time pressure (PUT) model views the mental activity in the MDMP as a problem-solving activity. In the PUT model, only one course of action is fully developed, and that course of action is developed from having the eyes open to many possibilities at the beginning of the decision process. The stress in the PUT model is on planning, modifying, and refining this course of action and not, as in traditional military decision-making models, to choose between alternative courses of action. Compared to the current Swedish army tactical decision-making process, the main advantage of the PUT model is that it is more in line with contemporary research results about problem-solving and decision-making processes.

The focus of the model that is presented in this chapter is to optimize the decision process to increase the decision quality compared with today's (Swedish) planning model (AR2, 1995, pp. 185–190). An important input to be able to do this, however, is to determine what constitutes decision quality in a military tactical decision. There does not appear to be consensus in this question. (Hitchins, 1998, p. 654; Roberts, 1991, p. 22). In the following section, I discuss a number of quality criteria in military decision making that are all addressed in the PUT model.

QUALITY CRITERIA IN MILITARY DECISION PROBLEMS

To be able to judge if one decision model is better than another, it is necessary to make certain assumptions as to which factors contribute to quality and how the need for quality should direct the formation of a prescriptive model for military decision making. What assumptions regarding criteria for decision quality can then be made? I describe six quality criteria briefly following. They have been deemed subjectively to be the most important to consider when different prescriptive military decision methods are compared.

The importance for a military unit to be able to be quicker than the opponent has long been known. This was noted already by Clausewitz (1976, p. 383). Since World War II, there has been a focus on speed as a "force multiplier" (Scott, 1992, pp. 10–16). Concepts such as "decision loop" (e.g., Smith, 1989, pp. 42–54) point out how the need for speed has come to include the decision process itself. If speed is seen as an important factor in military planning, it becomes highly relevant to place a cost–benefit perspective on those events that are a part of the military planning process. Only those events that are deemed to be cost efficient should be a part of the process. The ability to leapfrog the opponent's decision cycle is seen as being of crucial importance to be able to seize and keep the initiative in battle (AR2, 1995, p. 42). The time needed for making decisions is, therefore, an important quality criterion when comparing different prescriptive military decision models.

Both civilian and military management philosophy stress the importance of the commander or manager clearly communicating his or her will to subordinates (e.g., AR2, 1995, pp. 76–78; Bruzelius & Skärvad, 1995, pp. 102–104). This could probably be explained by the fact that a common goal of the members of an organization is a necessary precondition for coordinating the efforts. A common vision also empowers subordinate commanders to take initiatives and act in the spirit of the commander even in unexpected circumstances and communication breakdowns. Clarity in the formulation of the vision thus becomes a second important quality criterion when comparing different decision models.

Certain guidance in the discussion of quality in military decisions was given by Hitchins (1998, pp. 654–655) who stated that both an outcome and a process perspective can be applied. The *process perspective* stresses the importance of taking all available information into consideration and being logical and rational to exploit force multipliers and refuse the opponent any advantages. The *outcome perspective* stresses that bold, unpredictable decisions that are made quickly and decisively contribute to slowing down the opponent and force them to react rather than act. In this way, the decisive commander can turn a chaotic situation into a linear one that can be controlled. Hitchins stated, however, that the most important criterion for quality in this type of decision is emotional. The most important success indicator would then be that the decision inspires ones' own forces and deters the opponent. "An inspiring decision both encourages and binds a force together: if at the same time it dismays the enemy, it can shift the balance of belief from one force to the other" (Hitchins, 1998, p. 655). According to Hitchins, bold, unpredictable, and quick decisions contribute to inspiration and self-confidence. Analogous to Hitchins, emotional aspects must also be factored in. This is not primarily because Hitchins relied on research results but because the reasoning has high face validity. The perception that the plan/decision has high credibility, can be executed, and is necessary for those who will carry out the decision is likely to facilitate the execution, especially when it requires sacrifices. This is particularly true for military decisions on the battlefield. This criterion can be concluded by saying that the decision must "feel right" both for the decision maker and those who are carrying out the decision (staff and subordinate commanders). When the decision feels right, that feeling is likely to inspire. The perception of credibility, execution, and necessity thus becomes the third quality criterion when comparing different decision models.

In most military doctrines, surprise is considered an efficient means for success (e.g., AR2, 1995, p. 45). This is because the actions that an opponent has been able to predict ought to have less success than the unpredictable actions. Especially for the one who has fewer resources and thus lower endurance than the opponent, it becomes crucial to act unpredictably or the chances of becoming superior on the battlefield are reduced. New and unexpected actions are also likely to generate uncertainty for the opponent. Thus, degree of originality in the decision becomes the fourth quality criterion.

Efficient impact on the opponent is of course more likely if several systems can be deployed simultaneously against them. Such concentration of forces requires coordination in time and space. Thus, the degree of systems coordination becomes the fifth decision criterion.

Capturing real experience of war is probably necessary to make the training realistic and to prepare a unit for battle. This capturing of experience is done regularly in the Swedish officers' training. The experiences must, of course, be adapted to the current state of technology to be useful. To integrate earlier war experiences when making a military decision and to adapt those experiences to the current situation is also seen as a contributing factor to increased decision quality in military decisions. The degree of leveraging earlier experience thus becomes the sixth and final quality criterion to consider when comparing different military prescriptive decision models.

NEW PERSPECTIVES ON MILITARY TACTICAL PLANNING

During the 1980s and 1990s, some of the research on decision making moved away from the laboratories and into the field and began to describe professional decision making in real-life decision situations. A great discrepancy has been shown between the traditional prescriptive models for military decision making and the way in which experienced officers actually make most of their decisions during military exercises in the field (e.g., Kaempf, Klein, Thordsen, & Wolf, 1996; Klein, 1989). A conclusion that can be made is that because the prescriptive models are not much used, military commanders and staff lack a useful model for decision making in real situations. It also turns out that different officers use different strategies for their decision making. The most common ones are probably the so-called naturalistic strategies (Pascual & Henderson, 1997). Klein (1989) reported that military decision making in field settings could be modeled by the recognition-primed decision (RPD) model, and Schmitt and Klein (1999) took the step from description to prescription in their recognitional planning model (RPM). Schmitt and Klein's model is aimed at military operations planning and is built on previous research on how military battle planning and decision making actually is executed. The model also aims to be a "cognitively correct" model for such decision making. The main purpose is to speed up the military planning process. For a detailed description of the model, see Schmitt and Klein (1999, pp. 514–520). The RPM model consists of five separate phases that should occur in order but will normally also take place with some kind of parallelism in that the planners will frequently loop back to earlier phases as well as forward to later phases. The first phase is identify mission and conceptualize a COA. In this phase, the decision maker tries to understand the mission in terms of a rough solution. The result of this phase is a decision formulated as a tentative concept of operations. This phase seems to be similar to the recognition phase of the RPD model that ends with the generation of an action. The following phase in the RPM is analyze/operationalize COA. This is the phase in which the staff experts detail and refine the rough concept COA into a rather detailed plan. The next phase in the RPM is to war-game the COA. In this step, the plan is validated with respect to probable or possible enemy COAs, and when the COA is validated, it becomes "the plan." These two phases of the RPM correspond to the mental simulation and will-it-work and modify phases of the RPD model. The last phase of the RPM model is develop operational orders. This phase has no direct correspondence in the RPD model but could be viewed as the implementation phase.

THE PUT MODEL

The PUT model is primarily intended for higher tactical army units in the field (Brigade Division) where there is a substantial planning capacity. However, the model is probably also useful at lower levels even if the degree of complexity there often is lower. The PUT model contains three separate main steps. Each step aims at answering one main question: What must be achieved? (Step 1); How can this be achieved? (Step 2); and How should this be achieved? (Step 3). Each main event contains subprocesses. The events should be followed in the designated order, as this is considered to increase the likelihood of original solutions, but iterations between the events (and subprocesses) will have to occur. As can be seen in Fig. 4.1, one of the subprocesses in Step 3 is open ended. The evolving battle calls for dynamic decision making, which means that adjustments and further progressing of the plan must go on until the whole mission is accomplished.

The PUT model is not a static model. It could be changed according to the needs of the situation and the planners' individual preferences. If the situation is very well known, situation assessment could be reduced to a minimum. If there is also severe time pressure, the whole of Step 2 could be excluded from the process. Severe time pressure could also reduce the simulation in Step 3 to a mental simulation conducted in the head of a planner when developing a credible plan.

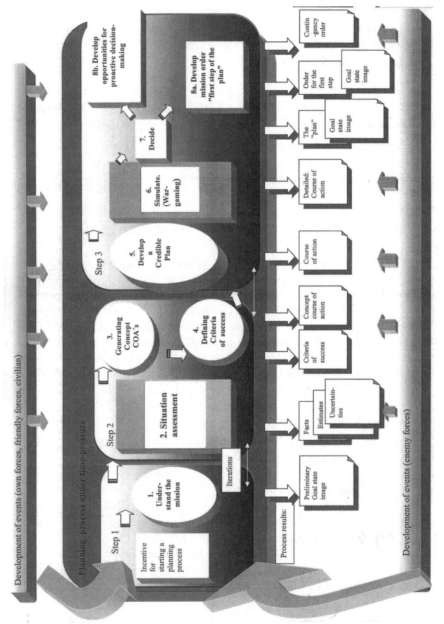

FIG. 4.1. Planning under time pressure (PUT) model. COA = courses of action.

49

Phases of Military Decision Making According to the PUT Model

Incentive for Starting the Planning Process. Analysis of the ongoing activity should be made continuously by the staff and encompass the following:

1. Consideration: Do our actions give the intended results? Are the things that happen a consequence of our actions or of other reasons?
2. Are the assumptions on which the current combat plan rests still valid?
3. Trend analysis, for example, are there trends in the information flow that indicate something unexpected?

This continuous analysis, though it is not really a part of the decision process itself, gives incentives to initiate a new decision process.

1. Understand the mission. This process aims at clarifying what must be achieved in the light of a given order or situation that has arisen. It is desirable to think in terms of capabilities rather than military units, although this may be difficult. Freeman and Cohen (1996) showed that better tactical solutions could be reached if the decision maker avoids adopting a potential interpretation of the situation early on in the decision process. Under time pressure, there is also a tendency to "focus on the most critical" and to optimize the solution against this instead of trying to think in a more integral way (Thunholm, 1997; Zakay, 1993, p. 60). The outcome of the process is depicted on a map overlay or on a sketch with an explanatory text. As such it constitutes a preliminary goal state vision[1] overlay.

2. Situation assessment. The process aims at integrating new information into a holistic picture of the situation that has arisen. Klein (1989, p. 59) concluded that experienced decision makers (experts), more than novices, place a strong emphasis on a thorough understanding of the situation. A part of the difference between experts and novices is that the experts have a substantially higher ability to discern type situations within their area of expertise. It is thus important to make an effort to understand the current situation and its inherent possibilities and threats. The outcome of the process is a holistic mental representation of the current situation and what could evolve. Uncertainties are to be identified.

---

[1]The concept "end state" is not used in the Swedish army field manuals and there is no other comparable term either. The term *goal state vision*, which is used here, resembles the term *end state*, but the establishment of a goal state vision is more to be seen as a military goal state that directs the planning process and is intimately connected to a certain mission. When the mission changes so will the goal state vision.

3. Generating possible concept COAs. This process aims at generating possible concept COAs. All the concepts that "come to mind" are visualized only with pen and paper (the concepts could be sketched with a few pen strokes). It is important that the decision maker tries to keep an open mind and allows all ideas to emerge without making reality assessments.

4. Define criteria of success. This process aims at capturing previous war experience in the planning process. This is done by listing general and specific criteria of success. The foundation for these factors is partly earlier war experience that has been condensed in concepts such as "surprise," "concentration of forces," "freedom of action," and "local superiority" and partly conclusions that have been made earlier or that can be drawn from the analyses that have been made up to this point.

5. Develop a credible plan. The process aims at developing a tentative combat plan. At this point in the process there is a good base for the decision maker to intuitively sense what the solution should look like. The term *credible* refers to the feeling of the decision maker. When the plan feels right to the commander, it should be easier to inspire subordinates. There is usually no need to compare the plan with another plan before a final decision is made. Drillings and Serfaty (1997, p. 74) referred to a series of studies concerned with generating alternatives in connection with decision making in the field. Drillings and Serfaty stated that there is no support for the notion that it is good to develop several different courses of actions prior to making the decision. In the vast majority of cases, the decision maker chooses the alternative that was first generated anyway (see also Klein & Crandall, 1996.) A good reason, however, to develop more than one plan is if some fundamental uncertainty exists. In that case, different plans could be adapted to different outcomes of the uncertainty. The credible plan should be described in such a detailed way that, together with the goal state vision, it can form the basis of a graphically represented order to the directly subordinate commanders. The criteria of success are used in the detailing work.

6. Simulate. This process aims to confront the solution with at least one and ideally several possible scenarios (enemy alternatives) and thereby gain an understanding for required modifications and identify the need for (a) coordination, (b) further information, and (c) future decisions that will have to be made. Simulation could be done formally with the aid of computerized war games or on maps. It could also be done informally as a mental simulation that will mainly take part when the credible plan is developed.

7. Decide. This process aims at deciding how the task should be carried out and to start the execution. In the case prescribed by the model, namely, that there is only one alternative, the commander should formally decide on establishing the goal state vision and the credible plan that has passed the simulation and also decide to start the execution of the first step of the plan. In the case of two or more existing plans, the decision is made by comparing

the alternatives based on the most important criteria of success (which then become criteria for selection).

8a. Develop mission order: "Step one of the plan." This process aims at transforming the decision into an order for fulfilling the primary task to directly subordinate units.

8b. Develop opportunities for proactive decision making. This process is initiated parallel to the order work and aims at preparing, as far as is possible, quick command of the unit's action once it has been initiated by the given order. Once the first hand tasks are executed by the directly subordinate commanders, the work of developing a plan for future follow-up of the action continues in the staff. The purpose is to define in advance times, places, or actions in which decisions concerning the continued sequence of events can be predicted. The purpose is also to prepare these decisions as far as the time allows so that delays in the combat caused by commanders' need for planning are avoided (to try to leapfrog the decision cycle of the opponent).

Unique Features of the PUT Model

In this section, I describe the differences and similarities between the PUT model and the RPM model. I also describe how the implications of the military battlefield environment and the military decision problem are realized in the PUT model and how the model is a better base for high decision quality than the AR2 model is.

The PUT model is different from the RPM model (Schmitt & Klein, 1999) in some important aspects. It is difficult to exactly appreciate these differences because the RPM model is not described in a very detailed way. The main difference seems to be that the RPM model is based primarily on what planners actually do when they engage in military planning. Schmitt and Klein obviously saw no need to specify activities that will enhance the possibilities of innovative thinking. Therefore, the RPM model lacks a step similar to Step 2 in the PUT model. The RPM model prescribes that the decision or the conceptualization of a COA should already emerge in the first step of the process when the commander identifies the mission. Such emergence could of course take place in some situations, but real combat, fortunately, is relatively rare, which means that even experienced officers seldom have expert knowledge when it comes to leading the combat in the field. Their ability to intuitively recognize a real combat situation and instantly be able to generate a good COA should not be exaggerated. The PUT model prescribes that the goal state is thoroughly elaborated and established in terms of capabilities, not in terms of military units, before a tentative solution is put forward. Another difference is that Schmitt and Klein argued (p. 513) that situation assessment is a necessary base for planning, but that it is not a part of the planning process. I argue that the situation must always be understood in light of the given inputs, for example, a new task. In real situations, the decision maker is likely to have a good understanding of the situation once the planning process begins. However, the information contained by a new or modified task must be integrated into the current understanding. This is likely to mean that the previous situation is looked upon with "new eyes," as goals and expectations affect what is being considered in the amount of information that constantly floods a contemporary military staff (see, for instance, Endsley, 1997, p. 276). As mentioned previously, there are possibilities to make shortcuts in the PUT model also. If Step 2 is left out, there is no great difference between the RPM and the PUT models. The decision-making models of an army will always be subjected to training by officers during their military education. A good prescriptive model could then shape the thought process in a desirable way and probably be a base for sound thinking and decision making when it comes to real battlefields.

The aforementioned assumptions regarding the decision problem and the environment have generated certain major requirements for the PUT model that I describe following.

A general requirement, in line with one quality criterion, is that the model should quickly lead to an acceptable solution. As the PUT model primarily is built for use on the battlefield (uncertainty, time pressure, high complexity, and dynamics), a cost–benefit perspective has been applied to the planning activities on which time is spent. Only events that add something crucial to the quality of the decision up front should be a part of the decision process; this will always be a matter of offsetting time loss against the need for analysis. In the PUT model, compared to the AR2 model, time is gained by (a) explicitly involving the commander in several steps of the process; (b) routinely developing only one COA; (c) developing the vision and the COA early in the process, which enables the directly subordinate commanders to initiate their own planning process concurrently; (d) the directly subordinate commanders participating in a vital part of the planning process; and (e) detailed planning, prior to giving the order, being limited to what is necessary for the initiation of acting the first step of the plan.

Another requirement for the PUT model is that the uncertainty of the battlefield must be dealt with. In accordance with Janis (1989), the PUT model prescribes that uncertainties are identified. However, no technique to handle that is prescribed. The technique to quantify uncertainty by assigning subjective probabilities (e.g., as in multiattribute utility theory) is not considered a viable path because sufficient experience and knowledge to make such valuations is generally lacking in officers. Obviously, the uncertainty must be dealt with by the decision maker, for example, by detailing reserves, avoiding irreversible actions, or collecting additional information (Lipshitz & Strauss, 1997), but there is not considered to be any one generic technique for handling uncertainty that can be built into a military tactical decision model for field use.

A third requirement of the PUT model that is linked to the high complexity of the decision problem is that all steps in the decision process cannot be purely analytical. Military planning fulfills the criteria that according to Hammond (1993, pp. 211–225) generate a large portion of intuition within the decision process. If a time dimension is introduced, a complex decision process will always sway between intuition and analysis with relatively regular intervals, according to Hammond. Even if a train of thought, for example, an inclination to act in a certain way with a unit in combat, cannot be derived consciously, this does not disqualify it per se. The PUT model allows for the natural iterations that exist in a complex problem-solving process and that have been shown in studies of military decision making (e.g., Schmitt & Klein, 1999, p. 513).

The dynamics of the decision problem also imply quick and step-by-step decisions that are sequentially released to subordinate units according to the development of the battle. In the PUT model, orders to subordinates are released step by step, although the planning process results in a goal state vision that expects the accomplishment of the whole mission. The wargaming phase is also seen as a preparation phase for foreseeable later decisions.

When it comes to the remaining quality criteria, besides time needed for planning, clarity in the formulation of the vision is ensured in the PUT model by the initial definition, and the continuous refinement, of the goal state vision. The goal state vision is graphically visualized and explicitly guides the whole planning process in a clearer way than in the AR2 model. The perception of credibility, execution, and necessity of the plan is ensured in the PUT model in two different ways: first, by prompting the decision maker to develop a plan that feels right; second, by thoroughly testing (war-gaming) the plan before it is decided. None of these measures are prescribed by the AR2 model. On the contrary, the AR2 model leaves no room for intuition in the planning process. The plan should be based entirely on analytic thinking. Originality in the plan is encouraged in the PUT model by thoroughly and critically analyzing the situation and avoiding premature and routine-based processing. By trying to see the gestalt of the situation before making conclusions about the solution, creative solutions stand a better chance to come to the mind of the decision maker. Claxton (1999, p. 61) concluded that there are reasons to also believe that a short period of incubation (5 to 15 min.) can enhance the possibilities of creative problem solving considerably. In the AR2 model, the decision maker is encouraged instead to draw far-reaching conclusions about the plan in a piecemeal manner before a holistic picture of the situation is established. In the PUT model, a high degree of systems coordination is accomplished in the plan by prompting subordinate commanders to take part in the war-gaming. In the AR2 model, no war-gaming is prescribed. Leveraging earlier experience is accomplished in the PUT model by defining and prioritizing criteria of success. In the AR2 model, criteria of selection are identified and used to choose a COA. These criteria of selection are, however, not defined in the same way as the criteria of success in the PUT model.

CONCLUSIONS

The PUT model is intended to support the military planning and decision-making process in real settings but also to be an aid in the training of officers to be good decision makers. Preliminary results from testing the model suggest that it is considered as a much better instrument for planning under time pressure than the AR2 model, at least by the officers tested, but a lot more testing needs to be done. One important task is to study military planning in different situations when time pressure, uncertainty, and complexity could be varied systematically. Another task is to study military decision making at different hierarchical levels. The planning and decision-making process will probably look different at different levels. Today the PUT model is developed for use at the army brigade and division levels. The model will probably need modification to be a useful tool at all levels of the military organization. Finally, one also needs to study how individual differences among military decision makers should be considered in the prescriptive military decision-making models to make them more flexible and more used than today's traditional military models.

REFERENCES

Arméreglemente del 2 Taktik [Army Regulations Part 2 Tactics]. (1995). Stockholm: Försvarets bok- och blankettförråd.

Brehmer, B. (1992). Dynamic decision making: Human control of complex systems. *Acta Psychologica, 81*, 211–241. New York: North-Holland.

Bruzelius, L. H., & Skärvad, P. H. (1995). *Integrerad organisationslära* [Integrated organizational Theory]. Lund, Sweden: Studentlitteratur.

Clausewitz, C. Von. (1976). *On war* [M. Howard & P. Paret, Eds. and Trans.]. Princeton, NJ: Princeton University Press.

Claxton, G. (1999). *Hare brain, tortoise mind: Why intelligence increases when you think less.* Hopewell, NJ: Ecco Press.

Drillings, M., & Serfaty, D. (1997). Naturalistic decision making in command and control. In C. E. Zsambok & G. Klein (Eds.), *Naturalistic decision making* (pp. 71–80). Mahwah, NJ: Lawrence Erlbaum Associates.

Dupuy, T. N. (1984). *The evolution of weapons and warfare.* New York: Plenum.

Endsley, M. (1997). The role of situation awareness in naturalistic decision making. In C. E. Zsambok & G. Klein (Eds.), *Naturalistic decision making* (pp. 269–283). Mahwah, NJ: Lawrence Erlbaum Associates.

Freeman, J. T., & Cohen, M. S. (1996, June). *Training for complex decision-making: A test of instruction based on the recognition/metacognition model.* In Proceedings of the 1996 Command and Control Research and Technology symposium. Monterey, CA: Naval Postgraduate School, pp. 260–271.

Hammond, K. R. (1993). Naturalistic decision making from a Brunswikian viewpoint: Its past, present, future. In G. A. Klein, J. Orasanu, R. Calderwood, & C. E. Zsambok (Eds.), *Decision making in action: Models and methods* (pp. 205–227). Norwood, NJ: Ablex.

Hitchins, D. H. (1998, September). *Effective decisions in emerging conflicts.* In Proceedings of fourth international symposium on Command and Control Research and Technology. Stockholm: National Defence College, Sweden, pp. 287–295.

Janis, I. L. (1989). *Crucial decisions.* New York: Free Press.

Kaempf, G. L., Klein, G., Thordsen, M. L., & Wolf, S. (1996). Decision making in complex naval command-and-control environments. *Human Factors, 38,* 220–231.

Klein, G. (1989). Strategies of decision making. *Military Review,* May, 56–64.

Klein, G., & Crandall, B. (1996). *Recognition-primed decision strategies.* Alexandria, VA: U.S. Army Research Institute for the Behavioral and Social Sciences.

Lipshitz, R., & Strauss, O. (1997). Coping with uncertainty: A naturalistic decision-making analysis. *Organizational Behavior and Human Decision Processes, 69,* 149–163.

Pascual, R., & Henderson, S. (1997). Evidence of naturalistic decision making in military command and control. In C. E. Zsambok & G. Klein (Eds.), *Naturalistic decision making* (pp. 217–227). Mahwah, NJ: Lawrence Erlbaum Associates.

Roberts, N. C. (1991). *New directions for military decision making research in combat and operational settings.* Monterey, CA: Naval Postgraduate School.

Schmitt, J., & Klein, G. (1999). How we plan. *Marine Corps Gazette, 83*(10), 18–26.

Scott, H. D., Jr. (1992). *Time management and the military decision making process.* Fort Leavenworth, KS: School of Advanced Military Studies, United States Army Command and General Staff College.

Smith, K. B. (1989). Combat information flow. *Military Review,* April, 42–54.

Thunholm, R. P. (1997). *Erfarenhet tidspress och beslutsfattande—en experimentell studie av arméofficerare i en taktisk beslutssituation* [Experience, time pressure and decision making—An experimental study of army officers in a tactical decision situation]. Unpublished master's thesis, Department of Psychology, Stockholm University, Stockholm.

U.S. Army field manual (FM) 101–5. (1997). *Staff organization and operations.* Department of the Army, Washington, DC.

Van Creveld, M. (1985). *Command in war.* Cambridge, MA: Harvard University Press.

Zakay, D. (1993). The impact of time perception processes on decision making under time stress. In O. Svenson & J. Maule (Eds.), *Time pressure and stress in human judgement and decision making* (pp. 170–178). New York: Plenum.

5

Managing Complex Dynamic Systems: Challenge and Opportunity for Naturalistic Decision-Making Theory

John D. Sterman
MIT Sloan School of Management

Linda Booth Sweeney
Harvard Graduate School of Education

Naturalistic decision-making (NDM) theory emphasizes the effectiveness and evolution of decision making in real-life settings (e.g., Gigerenzer & Goldstein, 1996; Gigerenzer et al., 1999; Klein, Orasanu, Calderwood, & Zsambok, 1993; Lipshitz, Klein, Orasanu, & Salas, 2000; Zsambok & Klein, 1997). Many NDM researchers argue that human performance in the naturalistic setting is often quite good even when performance on similar tasks in the laboratory is poor. The difference is often explained as the result of unfamiliar and unrealistic laboratory settings rather than cognitive "errors and biases." The naturalistic argument is at least implicitly dynamic and evolutionary: High-performing decision-making heuristics may arise and persist in a population through both learning and selection. At the same time, the evolutionary argument helps explain why the heuristics people use may not work well outside of the specific context in which they evolved. A heuristic may depend, for its effectiveness, on unique features of the particular setting; if that setting remains sufficiently stable, that heuristic may propagate and evolve to high efficacy for that environment even if it is ineffective in other settings, such as a laboratory task that has the same underlying logical structure but is otherwise unfamiliar.

Although the naturalistic perspective is highly useful in many situations, research in the naturalistic tradition, as in much of the laboratory work it critiques, has focused primarily on simple and static tasks. These tasks are often one-shot affairs such as the classic Wason card task (Wason, 1960), or choice under uncertainty (e.g., Lipshitz & Strauss, 1997), or repeated tasks with outcome feedback. Whether laboratory-based or naturalistic, these tasks are often unrealistic because they do not capture the interactions between decision makers and the environment. Many important decision-making settings are intrinsically dynamic. Decisions made today alter the state of the system, thus changing the situation we face tomorrow—that is, there is feedback between our decisions and the systems in which we are embedded. These feedbacks go far beyond the familiar outcome feedback through which we can (potentially) correct errors or update expectations. In complex dynamic systems, our actions alter the state of the system: The situation itself changes endogenously. Such systems are said to have a high degree of dynamic complexity (Sterman, 1994).

Systems with high degrees of dynamic complexity are growing in importance. More important, most of the changes we now struggle to comprehend arise as consequences, intended and unintended, of our own past actions. Many times our best efforts to solve a problem actually make it worse. Our policies may create unanticipated side effects, and our attempts to stabilize the system may destabilize it. Forrester (1971) called such phenomena the "counterintuitive behavior of social systems." The common result is *policy resistance*, the tendency for interventions to be delayed, diluted, or defeated by the response of the system to the intervention itself (Sterman, 2000).

For example, a firm's decision to cut prices may boost its market share in the short run, but the extra business may overburden its order fulfillment and customer service capability, slashing growth later on. Lower prices might also trigger a price war with competitors. Slower sales growth and lower margins will feed back to the firm's market valuation and hence its ability to raise capital, retain talented employees, launch new products, and provide customer service. The decision environment the management team faces next month, next quarter, and next year will be very different than the one they face (and think they learned from) today—and to a large extent, the turbulence the managers experience is caused, directly and indirectly, by their own past actions.

These recursive causal interactions involve both negative (self-correcting) and positive (self-reinforcing) feedbacks, long time delays, significant accumulations (stock and flow structures), and nonlinearities. Dynamic complexity arises in diverse situations and across many orders of magnitude in temporal, spatial, and organization scale—from the dynamics of viral replication within an individual to the global HIV pandemic, from real-time process control of an oil refinery to the boom-and-bust cycles afflicting world energy markets, and from the climate controls of an automobile to the dynamics of the global climate. Such systems should be a prime subject for research in NDM. Yet, despite some notable exceptions (e.g., Brehmer, 1992; Dörner, 1996;

Frensch & Funke, 1995; Omodei & Wearing, 1995a, 1995b), research on decision making in complex dynamic systems, both in the laboratory and in the field, remains rare. Most dynamic decision-making research has centered on tasks unfolding in minutes to days, such as industrial process control, combat, or fire fighting. Such tasks are of undoubted importance. However, the conditions for learning and evolution in these tasks, with their comparatively short delays, frequent repetition, rapid accumulation of experience, and opportunities for experimentation, are much better than in business, economic, and public policy settings in which the delays are much longer, decisions are often irreversible, and experiments are difficult or impossible.

Research suggests people have great difficulty understanding and managing dynamically complex systems. These studies generally show that performance deteriorates rapidly (relative to optimal) when even modest levels of dynamic complexity are introduced and that learning is weak and slow even with repeated trials, unlimited time, and performance incentives (e.g., Diehl & Sterman, 1995; Paich & Sterman, 1993; Sterman, 1989a, 1989b; see also Brehmer, 1992; Dörner, 1980, 1996; and Frensch & Funke, 1995). The usual explanation is that the complexity of the systems people are called on to manage overwhelms their cognitive capabilities. Implicit in this account is the assumption that although people are unable to correctly infer how a complex system will behave, they do understand the individual building blocks such as stocks and flows, feedback processes, and time delays. Our results challenge this view, suggesting the problems people have with dynamics are more basic and, perhaps, more difficult to overcome.

There are, of course, different schools of thought about the nature of complexity. Most scholars, however, agree that much of the art of systems thinking involves the ability to represent and assess dynamic complexity using mental and formal models, including qualitative models in which information is represented both textually and graphically (see, e.g., Bakken & Gould, 1992; Chandler & Boutilier, 1992; Forrester, 1971; Gould, 1993; Mandinach & Cline, 1994; Richardson, 1991; Richmond, 1993; Senge, 1990; Sterman, 2000; Vennix, 1996). Specific systems thinking skills include the ability to:

- Understand how behavior of the system arises from the interaction of its agents over time.
- Discover and represent feedback processes (both positive and negative) hypothesized to underlie observed patterns of system behavior.
- Identify stock and flow relations.
- Recognize delays and understand their impact.
- Identify nonlinearities.
- Recognize and challenge the boundaries of mental (and formal) models.

Effective systems thinking also requires more basic skills taught in high school, including scientific reasoning and the ability to understand data, which includes the creation and interpretation of graphs.

In this chapter, we describe results from a "systems thinking inventory" examining the ability of people to understand the basic elements of complex dynamic systems, primarily accumulations (stocks and flows), time delays, and feedbacks. The participants, students at the MIT Sloan School of Management, were highly educated in mathematics and the sciences compared to the public at large. Yet they performed poorly, exhibiting significant and systematic errors including violations of fundamental principles such as conservation of matter. Broad prevalence of such deficits poses significant challenges to decision makers, organizations, and educators.

We argue that for many important decision-making settings in public policy, business, and other domains, the type of tasks and mode of data presentation in our study is the naturalistic context. The long time delays and limited ability to conduct experiments in these systems simultaneously forces people to use numerical and graphical models and weakens the processes of learning and selection that might lead to the evolution of high-performing heuristics.

METHOD

We created several tasks to explore students' baseline systems thinking abilities. Each task consisted of a few paragraphs posing a problem. Participants were asked to respond by drawing a graph of the expected behavior over time. The items were designed to be done without use of mathematics beyond high school (primarily simple arithmetic).

Stocks and Flows: The Bathtub (BT) and Cash Flow (CF) Task

Stocks and flows are fundamental to the dynamics of systems (Forrester, 1961; Sterman, 2000). Stock and flow structures are pervasive in systems of all types, and the stock/flow concept is central in disciplines ranging from accounting to epidemiology. The BT/CF task tested participants' understanding of stock and flow relations by asking them to determine how the quantity in a stock varies over time given the rates of flow into and out of the stock. This ability, known as *graphical integration*, is basic to understanding the dynamics of complex systems.

To make the task as concrete as possible, we used two cover stories. The BT condition described a bathtub with water flowing in and draining out (Fig. 5.1); the CF condition described cash deposited into and withdrawn

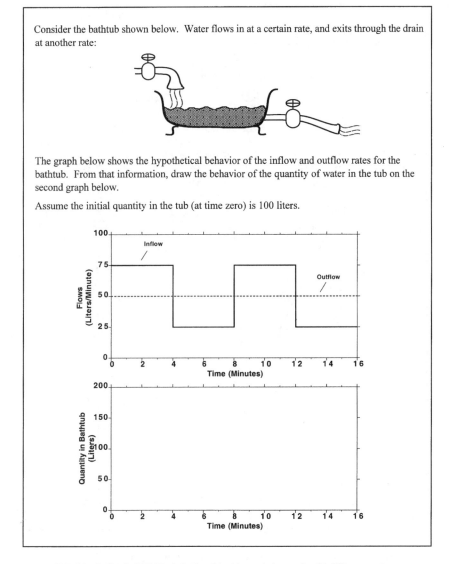

Consider the bathtub shown below. Water flows in at a certain rate, and exits through the drain at another rate:

The graph below shows the hypothetical behavior of the inflow and outflow rates for the bathtub. From that information, draw the behavior of the quantity of water in the tub on the second graph below.

Assume the initial quantity in the tub (at time zero) is 100 liters.

FIG. 5.1. Bathtub (BT) Task 1. Graphical integration task with BT cover story and square wave pattern for the inflow to the stock. The sawtooth pattern in Task 2 (shown in Fig. 5.2) was also used with the BT cover story.

from a firm's bank account (Fig. 5.2). Both cover stories describe everyday contexts quite familiar to the participants. Students were prompted to draw the time path for the quantity in the stock (the contents of the BT or the cash account). Note the extreme simplicity of the task. There are no feedback processes—the flows are exogenous. Round numbers were used, so it

Consider the cash balance of a company. Receipts flow in to the balance at a certain rate, and expenditures flow out at another rate:

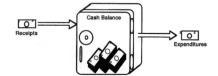

The graph below shows the hypothetical behavior of receipts and expenditures. From that information, draw the behavior of the firm's cash balance on the second graph below.

Assume the initial cash balance (at time zero) is $100.

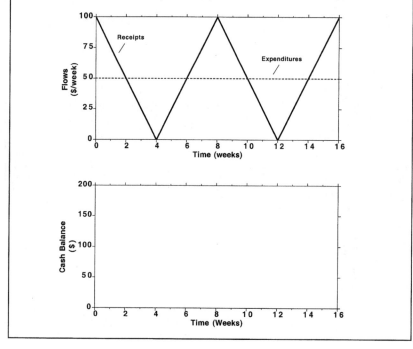

FIG. 5.2.   Cash flow (CF) cover story, Task 2. Graphical integration task with the CF cover story and sawtooth pattern for the inflow to the stock. The square wave pattern in Task 1 (shown in Fig. 5.1) was also used with the CF cover story.

was easy to calculate the net flow and quantity added to the stock. The form provided a blank graph for the stock on which participants could draw their answer. Note also that the numerical values of the rates and initial stock were the same in the BT and CF versions (the only difference was the time unit: seconds for the BT case, weeks for the CF case).

We also tested two different patterns for the flows: a square wave pattern (Task 1) and a sawtooth pattern (Task 2). Figure 5.1 shows the square wave pattern; Fig. 5.2 shows the sawtooth pattern. We tested all four combinations of cover story (BT/CF) and inflow pattern (Task 1/Task 2). Task 1 is among the simplest possible graphical integration tasks—during each segment of the behavior, the net flow is constant; therefore, the stock changes linearly. The different segments are symmetrical; therefore, solving the first (or, at most, first 2) gives the solution to the remaining segments. Task 2 tested participants' understanding that the net rate of flow into a stock equals the slope of the stock trajectory.

Solution to Task 1. Initial coding criteria were developed then tested on a subsample of results and revised to resolve ambiguities. A detailed coding guide is available from John D. Sterman and Linda Booth Sweeney. We coded each response for its conformance to seven criteria listed in Table 5.1. The first five items describe qualitative features of the behavior and did not require even the most rudimentary arithmetic. Indeed, the first three items (e.g., when the inflow exceeds the outflow, the stock is rising) are always true for any stock with any pattern of flows; they are fundamental to the concept of accumulation. Items 6 and 7 describe the behavior of the stock quantitatively, but the arithmetic required to answer them was trivial.

The solution to BT/CF Task 1 is shown in Fig. 5.3 (this is an actual participant response). First, note that the behavior divides into distinct segments

TABLE 5.1
Performance on the Bathtub (BT) Cash Flow (CF) Task 1

Item	Criterion	$Mean^a$	BT^b	CF^c	p
1	When the inflow exceeds the outflow, the stock is rising	.80	.87	.72	.83
2	When the outflow exceeds the inflow, the stock is falling	.80	.86	.73	.36
3	The peaks and troughs of the stock occur when the net flow crosses zero (i.e., at t = 4, 8, 12, 16)	.86	.89	.81	.52
4	The stock should not show any discontinuous jumps (it is continuous)	.89	.96	.82	.89
5	During each segment the net flow is constant so the stock must be rising (falling) linearly	.78	.84	.72	.21
6	The slope of the stock during each segment is the net rate (i.e., ±25 units per time period)	.66	.73	.58	.21
7	The quantity added to (removed from) the stock during each segment is the area enclosed by the net rate (i.e., 25 units per time period * 4 time periods = 100 units, so the stock peaks at 200 units and falls to a minimum of 100 units)	.63	.68	.56	.40
	Mean for all items	.77	.83	.69	.004

Note. The p values for individual items report χ^2 tests of H_0: Performance (BT) = Performance (CF).

[a]N = 182. [b]N = 95. [c]N = 87.

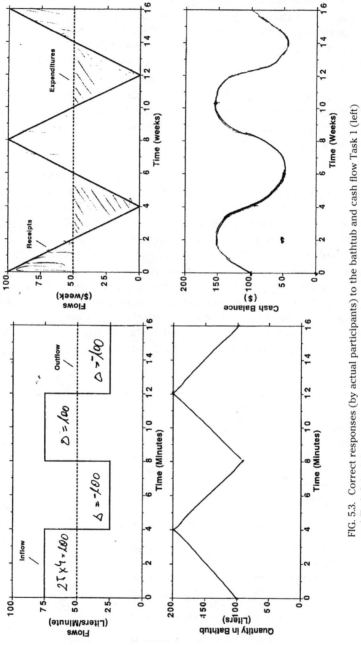

FIG. 5.3. Correct responses (by actual participants) to the bathtub and cash flow Task 1 (left) and Task 2 (right).

in which the inflow is constant (the outflow is always constant). During Segment 1 ($0 < t \leq 4$), the net inflow is $75 - 50 = 25$ liters per second (L/s). The total added to the stock over the first 4 s is therefore 25 L/s * 4 s = 100 L. Note that this quantity is given by the area bounded by the net rate curve between $0 < t \leq 4$. Because the net flow is constant, the stock rises at a constant rate (slope) of 25 L/s. In Segment 2 ($4 < t \leq 8$), the inflow drops to 25 L/s; therefore, the net flow is -25 L/s, and the stock loses the same quantity between Time 4 and Time 8 as it gained between Time 0 and 4. If the participant did not notice the symmetry, the same procedure used in Segment 1 can be used to determine that the stock loses 100 L by $t = 8$. Subsequent segments simply repeat the pattern of the first 2.

Solution to Task 2. Figure 5.3 shows the correct solution to Task 2. The solution must have the eight features listed in Table 5.2. As in Task 1, the first 5 items describe qualitative features of the behavior and did not require even the most rudimentary arithmetic. The last three items describe the behavior of the stock quantitatively, but the arithmetic required was trivial. The stock trajectory follows a parabola within each segment, but participants received full marks on Item 5 in Table 5.2 as long as they showed the slope for the stock changing qualitatively as shown in Fig. 5.3.

TABLE 5.2
Performance on the Bathtub (BT) Cash Flow (CF) Task 2

Item	Criterion	Mean[a]	BT[b]	CF[c]	p
1	When the inflow exceeds the outflow, the stock is rising	.47	.46	.48	.83
2	When the outflow exceeds the inflow, the stock is falling	.44	.41	.48	.36
3	The peaks and troughs of the stock occur when the net flow crosses zero (i.e., at t = 2, 6, 10, 14)	.40	.41	.39	.89
4	The stock should not show any discontinuous jumps (it is continuous)	.99	.99	.99	1.00
5	The slope of the stock at any time is the net rate	.28	.25	.30	.52
6	The slope of the stock when the net rate is at its maximum is 50 units per period (t = 0, 8, 16)	.47	.42	.52	.21
7	The slope of the stock when the net rate is at its minimum is -50 units per period (t = 4, 12)	.45	.41	.51	.21
8	The quantity added to (removed from) the stock during each segment of 2 periods is the area enclosed by the net rate	.37	.34	.41	.40
	Mean for all items	.48	.46	.51	.39

Note. The p values for individual items report χ^2 tests of H_0: Performance (BT) = Performance (CF).
[a]$N = 150$. [b]$N = 79$. [c]$N = 71$.

The Impact of Time Delays:
The Manufacturing Case (MC)

The BT/CF tasks addressed participants' understanding of the basic concepts of accumulation without any feedbacks or time delays. However, feedback and time delays are pervasive in complex systems and often have a significant effect on their dynamics. The MC assessed students' understanding of stock and flow relations in the presence of a time delay and a single negative feedback loop (Fig. 5.4).

The MC is a simple example of a stock management task (Sterman, 1989a, 1989b). Such tasks arise at many levels of analysis, from filling a glass of water, to regulating your alcohol consumption, to capital investment (see Sterman, 2000, chap. 17). The decision maker seeks to maintain a stock at a target or desired level in the face of disturbances, such as losses or usage, by regulating the inflow to the stock. Often there is a delay between a control action and its effect. Here the firm seeks to control its inventory in the face of variable customer demand and a lag in the production process. The task involves a simple negative feedback regulating the stock (boosting production when the stock is less than desired and cutting it when there is a surplus).

Solution to the MC.   There is no unique correct answer to the MC task. However, production and inventory must satisfy certain constraints, and their paths can be determined without any calculation. The production delay and unanticipated shift in customer orders mean shipments increase while production remains, for a time, at the original rate. Inventory therefore declines. The firm must not only boost output to the new order rate but also rebuild inventory to the desired level. Production must therefore overshoot and remain above shipments until inventory reaches the desired level, at which point it can drop back to equilibrium at the customer order rate. Furthermore, because the desired inventory level is constant, the area bounded by the production overshoot must equal the quantity of inventory lost when orders exceed production, which in turn is the area between orders and production as shown in Fig. 5.5. It is possible that production and inventory could fluctuate around their equilibrium values, but although such fluctuation is not inevitable, the overshoot of production is: The only way inventory can rise is for production to exceed orders in exactly the same way that the only way the level of water in a bathtub can rise is for the flow in from the tap to exceed the flow out through the drain.

A few modest assumptions allow the paths of production and inventory to be completely specified. When customer orders increase from 10,000 to 11,000 widgets per week, production remains constant at the initial rate due to the 4-week lag. Inventory, therefore, begins to decline at the rate of 1,000

Consider a manufacturing firm. The firm maintains an inventory of finished product. The firm uses this inventory to fill customer orders as they come in. Historically, orders have averaged 10,000 units per week. Because customer orders are quite variable, the firm strives to maintain an inventory of 50,000 units to provide excellent customer service (that is, to be able to fill essentially 100% of every order), and they adjust production schedules to close any gap between the desired and actual level. Although the firm has ample capacity to handle variations in demand, it takes time to adjust the production schedule, and to make the product – a total lag of four weeks.

Now imagine that the order rate for the firm's products suddenly and unexpectedly rises by 10%, and remains at the new, higher rate indefinitely, as shown in the graph below. Before the change in demand, production was equal to orders at 10,000 units/week, and inventory was equal to the desired level of 50,000 units.

Sketch the likely path of production and inventory on the graphs below. Provide an appropriate scale for the graph of inventory.

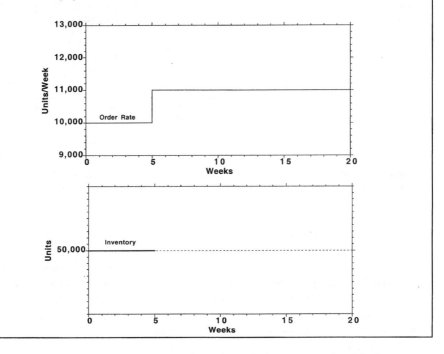

FIG. 5.4. The manufacturing case figure shows the inventory graph and feedback hint conditions. In the no inventory graph condition, the graph of inventory was not provided.

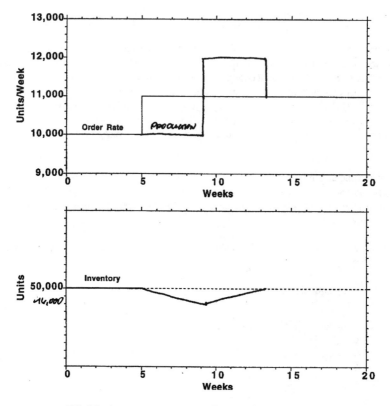

FIG. 5.5. A correct response to the manufacturing case.

widgets per week. What happens next depends on the distribution of the production lag. The simplest case, and the case most participants assumed, is a pipeline delay, that is,[1]

$$\text{Production}(t) = \text{Desired Production}(t - 4).$$

The 4-week delay means production continues at 10,000 widgets per week until Week 9. During this time, inventory drops by a total of 1,000 widgets per week * 4 weeks = 4,000 widgets, thus falling to 46,000. Assuming that management understands the delay and realizes that production will remain at its original level for 4 weeks, they will raise desired production above orders at Week 5, keep it above orders until an additional 4,000 widgets are scheduled

---

[1]We did not penalize participants if they selected other patterns for the delay, such as some adjustment before Week 9 and some after, as long as production did not begin to increase until after the step increase in orders.

for production, and then bring desired production back down to orders. Production then traces this pattern 4 weeks later. Assuming that production remains constant during the period of overshoot gives trajectories such as that shown in Fig. 5.5 in which production rises in Week 9 to 12,000 widgets per week and remains there for the next 4 weeks, giving a rectangle equal in shape to that for the period $5 < t \leq 9$ when shipments exceed production. Of course, the production overshoot can have any shape as long as the area equals 4,000 widgets. Few participants ($< 0.2\%$) drew a pattern with the duration of the overshoot other than 4 weeks while also maintaining the correct area relation.

In the basic version of the task, participants were asked only to sketch the trajectory of production. Doing so required them to infer correctly the behavior of the firm's inventory. Without a graph of inventory, it might be more difficult for participants to correctly trace the production overshoot. To test this hypothesis, we defined an inventory (I) graph treatment with two conditions. In the I condition, the page with the MC task included a blank graph for the firm's inventory, and participants were asked to provide trajectories for both production and inventory (as shown in Fig. 5.4). In the no I (~I) graph condition, participants were provided only with the graph showing customer orders and were not asked to sketch the trajectory of inventory.

In both conditions, performance was assessed by coding for the first four criteria listed in Table 5.3. Items 1 and 2 follow directly from the instructions, which specified that the system starts in equilibrium, that there is a 4-week production lag, and that the change in orders is unanticipated. Item 3 results from the firm's policy of adjusting production to correct any inventory imbalance and reflects the basic physics of stocks and flows, specifically, that a stock falls when outflow exceeds inflow and rises when inflow exceeds outflow. Item 4 tests conservation of material: Because desired inventory is constant, the quantity added to inventory during the production overshoot just replaces the quantity lost during the initial response when orders exceed production.

Responses to the inventory graph condition were also coded for Items 5 through 7 in Table 5.3.

Item 5 follows from Points 1 and 2: When orders increase, production must remain at the initial rate due to the adjustment delay. Until production increases, orders exceed output, and therefore, inventory must fall. Inventory should then rebound because the firm seeks to adjust inventory to its desired value (Point 6). Item 7 tests the consistency of the production and inventory trajectories and indicates whether participants understand that the slope of a stock at any moment is its net rate. Note that Item 7 does not require the production trajectory to be correct, only that the trajectory of

TABLE 5.3
Performance on the Manufacturing Case

Item	Criterion	Average[a]	~I[b]	I[c]	p
1	Production must start in equilibrium with orders	.53	.72	.33	.001
2	Production must be constant prior to Time 5 and indicate a lag of 4 weeks in the response to the step increase in orders	.44	.59	.29	.001
3	Production must overshoot orders to replenish the inventory lost during the initial period when orders exceed production; production should return to (or fluctuate around) the equilibrium order rate to keep inventory at or fluctuating around the desired level	.44	.63	.23	.001
4	Conservation of material: The area enclosed by production and orders during the overshoot of production (when production > orders) must equal the area enclosed by orders less production (when production < orders)	.11	.12	.10	.80
5	Inventory must initially decline (because production < orders)	.68	NA	.68	NA
6	Inventory must recover after dropping initially	.56	NA	.56	NA
7	Inventory must be consistent with the trajectory of production and orders	.10	NA	.10	NA
	Mean for all items	.41	.50	.32	.0001

Note. I = inventory graph; ~I = no inventory graph; NA = not applicable. Points 4 through 7 do not apply to the ~I treatment. The *p* values report χ^2 tests of H_0: Performance (I) = Performance (~I).
[a]$N = 225$. [b]$N = 116$. [c]$N = 109$.

inventory be consistent with the production path drawn by the participant, whatever it may be.

Participants and Procedure

We administered the preceding tasks to two groups of students enrolled in the introductory system dynamics course at the MIT Sloan School of Management. The first group received a background information sheet, the MC, and the "paper fold" case on the 1st day of class.[2] Two weeks later, the same class received BT/CF Task 1. On the 1st day of the next semester, a new set of students received the background information sheet and BT/CF

[2]The paper fold task tests understanding of positive feedback and exponential growth (see Sterman, 2000, chap. 8).

Task 2; the following week they received the global warming task discussed following. Students were given approximately 10 min in each session. They were told that the purpose of the questions was to illustrate important systems thinking concepts they were about to study and to develop a tool to assess systems thinking skills. Students were not paid or graded.

The two groups were similar and typical of the Sloan School's student body. About three fourths were master of business administration students; the rest were in other master's programs, PhD students, undergraduates, or students from graduate programs at other universities, primarily Harvard. About 75% were male, and about 80% were between 25 and 35 years of age. They came from 35 countries; English was a first language for about one half. More than one half had undergraduate degrees in engineering, computer science, mathematics, or the sciences; most of the rest had degrees in business or a social science (primarily economics). Fewer than 5% had degrees in the humanities. More than one third had a master's, doctoral, or other advanced degree, most in technical fields.

RESULTS

BT/CF Task I. Average performance on this simplest graphical integration task was 77%. Table 5.1 breaks performance down by the individual coding criteria and cover story. Participants did best showing the stock trajectory as a continuous curve with peaks and troughs at the correct times. They did worst on Items 6 and 7, which tested the basic concepts that the net rate is the slope of the stock and that the area enclosed by the net rate in any interval is the quantity added to the stock during the interval. One fifth did not correctly show the stock rising (falling) when the inflow was greater than (less than) the outflow. More than one fifth failed to show the stock rising and falling linearly during each segment, although the net rate was constant. Nearly two fifths failed to relate the net flow over each interval to the change in the stock. These concepts are the most basic and intuitive features of accumulation. Further, they are the fundamental concepts of calculus, a subject all MIT students are required to have.

Figure 5.6 illustrates typical errors for BT/CF Task 1. Figure 5.6a shows the stock changing discontinuously, jumping up and down in phase with the net rate (11% of the responses exhibited such discontinuities). The participant showed the stock as constant in each interval even though the net flow is nonzero. It appears the participant did not understand the basic concepts of accumulation, instead drawing a stock trajectory whose shape matched the shape of the net rate. Figure 5.6b shows a participant who understood something about the area swept out by the net rate (note the hash marks in the rectangle enclosed by the inflow and outflow up to Time

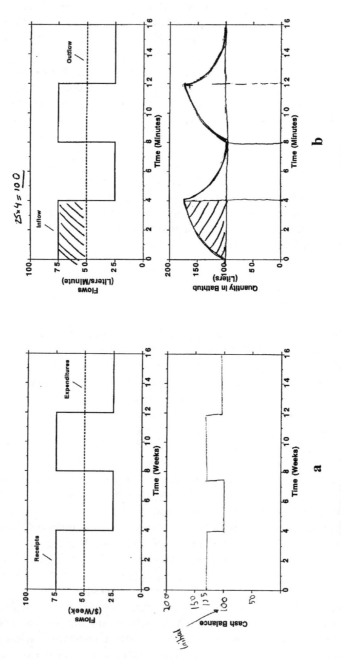

FIG. 5.6. Typical erroneous participant responses to the bathtub and cash flow Task 1.

4). However, the participant drew the stock in each segment as rising or falling at a diminishing rather than linear rate. The participant also drew hash marks in the area enclosed by the stock trajectory, an irrelevancy.

BT/CF Task 2. Average performance was just 48%. In general, participants did worse on comparable items than in Task 1 (Table 5.2). Fewer than one half correctly showed the stock rising (falling) when the inflow exceeds (is less than) the outflow compared to 80% in Task 1. Only 40% placed the peaks and troughs of the stock at the right times compared to 86% in Task 1. Only 37% correctly related the net rate over each interval to the change in the stock over the interval compared to 63% in Task 1. Only 28% correctly related the net rate to the slope of the stock. Fewer than one half correctly showed the maximum slope for the stock. The only item in which participants did better in Task 2 than Task 1 was showing the stock trajectory as continuous: Only 2 of 150 participants (1.3%) did this incorrectly compared to 11% in Task 1. Note that the net rate in Task 2 is continuous, whereas in Task 1 it is discontinuous, suggesting many participants drew stock trajectories that matched the pattern of the net rate.

Figure 5.7 illustrates typical errors in BT/CF Task 2. Figure 5.7a shows the most common errors. The participant correctly computed the quantity added to the stock during each interval of 2 periods (note the hash marks highlighting the area of the triangle enclosed by the net rate) and correctly places marks showing the value of the stock at t = 2, 4, 6, and so forth. However, the participant then drew straight lines between these points, insensitive to the fact that the net rate was not constant during each interval. The participant in Fig. 5.7b was one of a number who attempted to solve the problem analytically. The participant correctly wrote the equations for the system, $F = dQ/dt$, $Q = \int F dt$, and $F = \text{In} - \text{Out}$, yet the curve drawn bears no relation to the correct response. This participant has a PhD in physics.

The effects of the cover story are mixed. In Task 1, performance was significantly better in the BT condition, $t = 2.94$, $p < .004$. However, although participants with the BT cover story outperformed those with the CF cover story on every one of the individual categories, the individual differences were not significant. For Task 2, there was no significant difference between performance on the BT and CF treatments on any of the individual criteria or overall.

Manufacturing Case. Although more difficult than the BT/CF tasks, this task was still quite simple, involving only one stock, one time delay, and one negative feedback loop. Further, the make-to-stock system is a basic paradigm in manufacturing; most participants had previously taken operations management or had relevant real-world experience. Nevertheless, average performance was only 41% (Table 5.3). About one half of the partici-

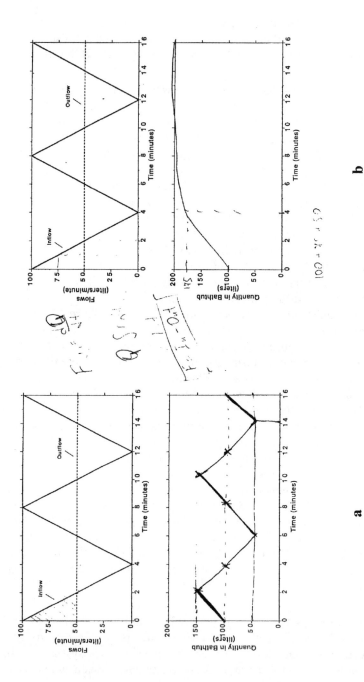

a

b

FIG. 5.7. Typical erroneous participant responses to the bathtub and cash flow Task 2.

pants failed to start in equilibrium or show the lag between the change in orders and the response of production. Only 44% showed production over-shooting orders. Instead, most showed production adjusting with a lag to the new customer order rate but not overshooting—they failed to under-stand that building inventory back up to its desired level requires produc-tion to exceed orders. Shockingly, 89% drew trajectories that violate the re-quired conservation of material, showing no production overshoot or an overshoot whose area does not equal the area of the production under-shoot they drew. Among those receiving the inventory graph condition, 68% correctly showed inventory initially declining, but only 56% showed it sub-sequently recovering. Also, 90% drew production paths inconsistent with their inventory trajectory.

Figures 5.8a and 5.8b show the most common error. Both participants showed production responding with a lag, but rising up only to the new level of orders. There was no production overshoot. Further, both drew inventory trajectories inconsistent with the production path they chose. In Fig. 5.8a, in-ventory immediately jumped to 55,000 and remained at that level. Actual in-ventory, given the participant's production path, would have fallen linearly to 46,000 and remained there. In Fig. 5.8b, production correctly lagged orders, but again there was no production overshoot. Inventory fell linearly through Week 10, whereas given the production path as drawn, it would have actually fallen at a diminishing rate. The participant then showed inventory rising, even though production equaled orders after Week 10.

The vast majority (90%) of participants drew trajectories for inventory in-consistent with the production path they chose. The participant in Fig. 5.8c correctly showed the production lag and the production overshoot (which, however, is far too large), but showed inventory immediately dropping in Week 5 to 46,000 units. Inventory then rose through about Week 10, even though production was less than orders. Inventory then stabilized, although the participant showed production exceeded orders. The participant in Fig. 5.8d also failed to show the time delay. In addition, the inventory trajectory was inconsistent with the production path. The participant showed inventory constant through Week 5, although production was drawn exceeding orders. Inventory then fell, although production equaled orders. This participant re-ported that he had a PhD in "nonlinear control theory."

We hypothesized that participants would do better in the inventory graph condition because the graph would prompt them to think about the trajectory of inventory and how it relates to production. Overall perform-ance in the inventory graph condition, however, was significantly worse than in the no graph condition, $t = 5.11$, $p < .0001$. Performance in the inven-tory graph condition was worse on all items, and these differences were highly significant (except for Item 4, conservation of material, in which per-formance was extremely poor for all). For example, 63% of those in the ~I

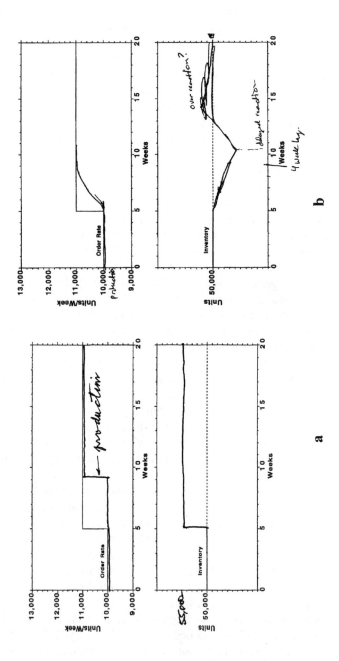

76

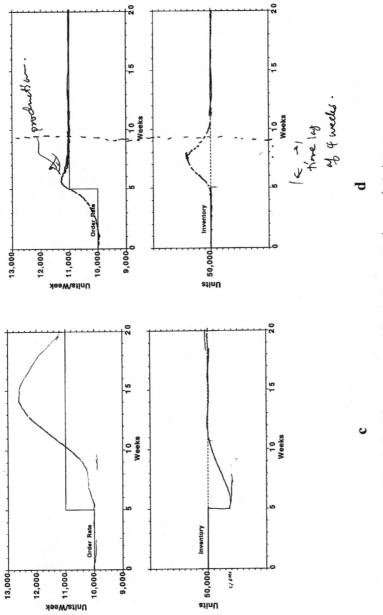

FIG. 5.8. Typical erroneous participant responses to the manufacturing case.

condition correctly showed production overshooting orders, compared to only 23% of those in the I condition. In contrast to our hypothesis, participants asked to sketch inventory did much worse specifying the production trajectory than those who were not.

Impact of Participant Demographics. We hypothesized that participants with more training in mathematics, the sciences, or engineering would outperform those with training in the social sciences or humanities. To test this hypothesis, we ran a variety of general linear models relating performance on the different tasks to participants' demographics including highest prior degree, major field, current academic program, gender, age, region of origin, and whether English was their native language. Although some items were significant, there was no consistent pattern. Prior academic field was significant for BT/CF Task 1 ($p = .006$), and highest prior degree was marginally significant ($p = .074$). As hypothesized, those with technical backgrounds did better than those in the social sciences, but these factors were far from significant in the other tasks. The degree program in which students were currently enrolled was not significant in any of the tasks. The results provide only limited support for the hypothesis that prior training in the sciences helps performance. It is possible that there simply was insufficient variation across participants to detect any effects. Other demographic factors also appeared to have only a weak impact. Age was not significant. Performance did not depend on whether English was the participant's first language, but region of origin was significant for BT/CF Task 1 ($p = .008$) and marginally significant in BT/CF Task 2 ($p = .078$) and the MC task ($p = .105$). Participants from North America generally did better. Men outperformed women on all three tasks, although the effect was only marginally significant ($p = .131, .062$, and $.055$ for BT/CF Task 1, BT/CF Task 2, and MC, respectively).

DISCUSSION

We found that highly educated participants with extensive training in mathematics and science had poor understanding of some of the most basic concepts of dynamics, specifically, stocks and flows, time delays, and feedback. The errors were highly systematic, and indicate violations of basic principles, including conservation of matter, and not mere calculation errors. This result was further reinforced by the significant deterioration in results between BT/CF Task 1 and BT/CF Task 2. Participants had poor understanding of the relation between the net flow into a stock and the slope of the stock trajectory. Many participants did not understand the relation between the area enclosed by the net rate over some interval and the change in the stock.

Many participants appeared to believe that the stock trajectory should have the same qualitative shape as the net rate. In BT/CF Task 1, the net rate was discontinuous, and 11% of the participants drew stock trajectories that were also discontinuous, similar to that shown in Fig. 5.6a. In BT/CF Task 2, the net rate was continuous, and only 2 of 150 participants (1.3%) drew discontinuous trajectories for the stock, the only criterion on which the participants did better on Task 2 than Task 1. However, 72 of 150 participants in Task 2 (48%) drew stock trajectories with discontinuous slopes similar to the net rate (as illustrated by Fig. 5.7a). We conjecture that participants with a weak grasp of stocks and flows relied on a heuristic that matches the shape of the output of the system to the shape of the input. That heuristic produces gross errors with first-order importance.

The two features that participants found problematic—the slope of the stock is the net flow, and the change in the stock over an interval is the area enclosed by the net rate in that interval—are the two fundamental concepts of the calculus. One might argue that calculus is rather advanced mathematics, so participants' errors are not too worrisome. Such a view, we believe, is erroneous. First, essentially every participant had taken calculus. Many had years of coursework; the majority had graduate or undergraduate degrees in mathematics, engineering, or the sciences. Nevertheless, for many participants, training and experience with calculus and mathematics did not translate into an intuitive appreciation of accumulations and of stocks and flows.

More important, these tasks did not require participants to use any of the analytic tools of calculus—no derivatives needed to be taken, no integrals needed to be written or evaluated. The tasks can be answered without use of any mathematics beyond simple arithmetic (and perhaps the formulae for the area of rectangles and triangles). The concepts of accumulation, although formalized in the calculus, are common and familiar to all of us through a host of everyday tasks, including filling a bathtub, managing a checking account, or controlling an inventory.

We turn now to consider alternative explanations for the results. One possibility is that participants did not put much effort into the tasks because there was insufficient incentive. In a review of more than 70 studies, Camerer and Hogarth (1999) found incentives sometimes improved performance. In other cases, even significant monetary incentives did not improve performance or eliminate judgmental errors, whereas in others, incentives worsened performance. It is possible that additional incentives in the form of grades or payment would improve the results. The resolution of this issue awaits future research. It is also possible that participants had insufficient time. This question also must be left for future research. We expect that more time would improve performance but suspect many of the same errors will persist, particularly violations of conservation laws and inconsistent net rate and stock trajectories. Given the importance and ubiquity of stock and flow structures, people should be able to infer their dynamics quickly and reliably; their failure to do so is a further indication of their poor grasp of these critical concepts. Further, time pressure is pervasive in many organizations in which decisions involving complex dynamics are made.

Advocates of the NDM movement argue that many of the apparent errors documented in decision-making research arise not because people have poor reasoning skills but as artifacts of unfamiliar and unrealistic laboratory tasks. They argue that people can often do well in complex settings because they have evolved "fast and frugal" heuristics that "are successful to the degree they are ecologically rational, that is, adapted to the structure of the information in the environment in which they are used" (Gigerenzer et al., 1999, vii). Following the naturalistic critique, perhaps people understand stocks, flows, delays, and feedback well and can use them in everyday tasks but did poorly here due to unfamiliar and unrealistic problem presentation. After all, people do manage to fill their bathtubs and manage their checking accounts. We agree that our decision-making capabilities evolved to function in particular environments; to the extent the heuristics we use in these environments are context specific, performance will not necessarily transfer to other situations even if their logical structure is the same.

The evolutionary perspective suggests the errors people exhibited in our tasks should be expected. It is not necessary to understand the relation between flows and stocks to fill a bathtub—nature accumulates the water "automatically." It is far more efficient to simply monitor the level of water in the tub and shut off the tap when the water reaches the desired level—a simple, effectively first-order negative feedback process, which, in experiments such as Diehl and Sterman's (1995) show, people can do well. As Laplace remarked, "Nature laughs at the difficulties of integration" (quoted in Krutch, 1959, p. 510). That is, stocks in nature always properly accumulate their flows even when mathematicians cannot solve the equations of motion for the system. For a wide range of everyday tasks, people have no need to infer how the flows relate to the stocks—it is better to simply wait and see how the state of the system changes and then take corrective action.

Unfortunately, the wait-and-see strategy can fail spectacularly in systems with high dynamic complexity (Diehl & Sterman, 1995; Sterman, 1989a, 1989b). More and more of the pressing problems facing us as managers and citizens alike involve long delays. The long time scale means there is little opportunity for learning through outcome feedback. Instead, we must rely on models of various types to help us project the likely dynamics of the system. These models typically present information in the form of spreadsheets, graphs, or text—the same type of data presentation used in our experiments. Managers are called on to evaluate spreadsheets and graphs projecting revenue and expenditure, bookings and shipments, and hiring and attrition. These modes of data presentation are not unique to business. Epidemiologists must understand the relation between the incidence and prevalence of disease, urban planners need to know how migration and population are related, and everyone, not only climatologists, needs to understand how greenhouse gas emissions alter global temperatures. For many of the most pressing issues in business and public policy, the mode of data presentation in our tasks is the naturalistic context.

There is abundant evidence that sophisticated policymakers suffer from the same errors we observed in our experiments. Sterman (2000) documented many examples, including boom and bust cycles in the real estate, shipbuilding, semiconductor, machine tool, and other industries.

Consider global warming (GW). The climate is a quintessential complex dynamic system. The spatial scale is global; the time scale dwarfs normal human concerns. The dynamics of the climate are exquisitely complex and remain imperfectly understood. Nevertheless, the essentials are simple and can be easily understood with basic knowledge of stocks and flows. The temperature at the earth's surface—the land, lower atmosphere, and surface layer of the ocean (the top 50 to 100 m where most sea life exists)—is determined primarily by the balance of the incoming solar radiation and the outgoing reradiated energy. Incoming solar energy warms the earth. The warmer the earth, the greater the flow of energy radiated back into the cold of space. The temperature rises until the earth is just warm enough for the energy radiated back to space to balance the incoming solar energy in a straightforward negative feedback process.

The amount of energy radiated back into space depends on the composition of the atmosphere. Greenhouse gases (GHGs) such as carbon dioxide and methane absorb some of the energy radiated by the earth instead of allowing it to escape into space. Thus, an increase in GHGs causes the earth to warm. The earth heats up until the energy escaping through the atmosphere to space rises enough to again balance the incoming solar energy. GHGs reduce the emissivity of the atmosphere enough to warm the surface of the earth (including the oceans) to a life-sustaining average of about $15°C$ (59 °F). Without GHGs in the atmosphere, the mean global temperature would be about -17 °C (0 °F), and a blanket of ice would perpetually cover the earth.

Natural biogeochemical processes cause the concentration of carbon dioxide in the atmosphere to fluctuate over geological time, and surface temperatures have fluctuated with it. Human activity has now reached a scale in which it affects these processes significantly. Anthropogenic GHG emissions have been growing exponentially since the beginning of the industrial age. Consequently, atmospheric concentrations of CO_2 and other GHGs including nitrous oxide (N_2O), methane (CH_4), and others have been growing exponentially, with concentrations of CO_2, N_2O, and CH_4 up by 31%, 17%, and

151%, respectively, since 1750. The United Nations sponsored Intergovern-
mental Panel on Climate Change (IPCC; 2001) noted that "The present CO_2
concentration has not been exceeded in the last 420,000 years and likely not
during the past 20 million years. The current rate of increase is unprecedented during at least the past 20,000 years" (p. 7).

Current GHG concentrations contribute about 2.4 watts per square meter
of net radiative forcing, that is, incoming solar radiation exceeds outgoing
radiation by 2.4 W/m^2. Consequently, mean global surface temperatures are
rising. Mean temperatures rose in the 20th century by 0.6 ± 0.2 °C. The
warming has been accompanied by glacier retreat and a decline in winter
snow cover, a 40% decline in summer sea-ice thickness in the arctic, an increase in average precipitation and in extreme weather events, and a rise of
0.1 to 0.2 m in sea level, among other effects (IPCC, 2001).

Debate continues about the dynamics of the global climate system and
the consequences of warming. The public discussion has been polarized by
well-financed campaigns to discount the science. Nevertheless, consensus
is emerging. The most recent report of the IPCC (2001) concluded that GW
is real and that "most of the warming observed over the last 50 years is attributable to human activities" (p. 10).

To examine people's ability to understand the basic properties of the
global climate system, we administered the GW task shown in Fig. 5.9 to a
subset of the participants discussed previously. These participants had previously done BT/CF Task 2. Participants were asked to imagine that
anthropogenic CO_2 emissions fell instantly to zero in the year 2000. The task
provided a brief description of the GW issue, with graphs showing the historical data. Two treatments were used. In Treatment 1, participants were
provided with graphs showing the historic data for CO_2 emissions, CO_2 in
the atmosphere, and global mean temperature and asked to sketch the
likely path of atmospheric CO_2 and mean temperature from 2000 to 2050, assuming the hypothesized drop to zero in human CO_2 emissions. In Treatment 2, the graph of atmospheric CO_2 was omitted, and participants were
asked only to sketch the global mean temperature. We hypothesized that
performance in Treatment 1 would be better than in Treatment 2 because
explicitly showing the stock of atmospheric CO_2, which determines net radiative forcing and thus the change in global mean temperature, should help
participants think about the stock and flow relations.

Of course no one knows exactly how the climate would respond to such
a shock. However, the stock/flow structure of the climate system and basic
laws of physics sharply constrain the possible trajectories. Simulation models of various types are the primary research tools used to explore these issues. The detailed general circulation models calculate climate at finely
spaced intervals covering the entire surface of the earth but take GHG emissions as exogenous inputs. At the other extreme, so-called integrated cli-

Consider the problem of global warming. Carbon dioxide (CO_2) is a greenhouse gas that traps heat and contributes to warming. CO_2 emissions from combustion of fossil fuels like oil, gas, and coal have been increasing since the start of the industrial revolution. The curve labelled "Anthropogenic CO_2 Emissions" in Figure 1 shows the worldwide emission rate of CO_2 from fossil fuel combustion since 1950. Figure 2 shows the stock of carbon dioxide in the atmosphere, along with a trend line generated by a global climate simulation model. Figure 3 shows data on average global temperatures since 1950, along with a trend line.

In 1995, a UN scientific panel concluded that these emissions were contributing to global warming, stating that "The balance of evidence suggests a discernible human influence on climate." In 1997 the industrialized nations agreed to stabilize their CO_2 emissions near mid 1990 rates. Implementation, however, remains elusive.

Now let's do a mental exercise we call an extreme conditions test. What do you think would happen to the average global temperature if anthropogenic CO_2 emissions suddenly stopped completely, so that annual emissions were zero? This imaginary scenario is shown in Figure 1. In the year 2000 anthropogenic CO_2 emissions drop instantaneously to zero and remain there forever.

Assume anthropogenic CO_2 emissions follow this scenario. Sketch the likely path (the continuation of the simulation) for atmospheric CO_2 for the next 50 years using the space provided in the right half of Figure 2. Then sketch the likely path (the continuation of the simulation) for the average global temperature for the next 50 years using the space provided in the right half of Figure 3.

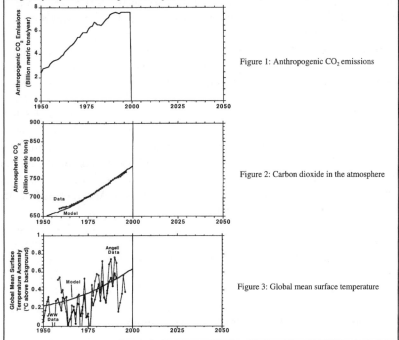

Figure 1: Anthropogenic CO_2 emissions

Figure 2: Carbon dioxide in the atmosphere

Figure 3: Global mean surface temperature

FIG. 5.9. The global warming task.

mate-economy models close some of the feedbacks among the human economy, carbon emissions, and global climate but treat the carbon cycle and climate as global aggregates with a small number of stocks. Fiddaman (1997) analyzed many of the most widely used climate-economy models, identifying a number of problems and inconsistencies in them, including many of the errors in stock-flow representation discussed in this chapter. For example, in his widely cited climate-economy model, Nordhaus (1992a,

1992b) violated the law of conservation of mass by assuming a significant fraction of carbon emissions simply disappear (flowing into a limitless sink outside the model boundary). Fiddaman (1997, 2002) developed a model that corrects these and other defects and linked it to a model of the economy and energy system. The model sectors were based on the relevant scientific knowledge of the global carbon cycle and climate system and carefully calibrated to the available data.

Despite the differences among the models, all show the climate to possess enormous inertia. Changes in GHG emissions only slowly affect temperature and climate, and their impact persists for many decades. Figure 5.10 shows the response of Fiddaman's (1997) model to the extreme conditions test in the GW task, along with a typical participant response. Although the rate of CO_2 emissions falls to zero in the year 2000, simulated mean global temperature continues to rise for about three more decades. It then falls very slowly.

Simple stock and flow considerations explain how it is possible for the global temperature to rise even after human GHG emissions fall to zero. When emissions fall to zero, the inflows to the stock of atmospheric carbon fall below the outflows (absorption of carbon by biomass and by the oceans). Therefore, the stock of CO_2 in the atmosphere peaks and begins to fall. The concentration of CO_2 in the atmosphere falls only slowly, however, because carbon previously taken up by terrestrial and aquatic biomass and dissolved in the ocean is recycled to the atmosphere. Global mean temperature, in turn, rises as long as incoming solar radiation exceeds the heat radiated back to space or transferred to the deep ocean. Although falling after the year 2000, high global atmospheric CO_2 concentrations still cause net radiative forcing. Global mean temperature continues to grow but at a diminishing rate as the net forcing slowly declines. By about 2030, the surface has warmed enough and the concentration of CO_2 in the atmosphere has fallen enough for insulation to be balanced again by radiation of heat to space and the rate of heat transfer to the deep ocean. Temperature peaks and begins to fall. The decline is slow because CO_2 concentrations remain high and because heat previously stored in the ocean begins to flow back to the atmosphere.

Although the numerical values in the simulation shown in Fig. 5.10 are obviously uncertain, the key features of the behavior are not. We coded each response as correct if it qualitatively approximated the patterns shown in Fig. 5.10. Specifically, the CO_2 trajectory was considered correct if it peaked at or shortly after the year 2000 and then declined. The temperature trajectory was considered correct if it continued to rise, peaked sometime after the year 2000, and then declined. We were generous in the timing and magnitudes, marking a response as wrong only when it clearly violated one of the basic physical criteria dictated by the stock and flow structure.

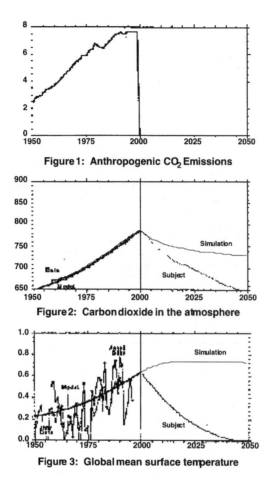

Figure 1: Anthropogenic CO_2 Emissions

Figure 2: Carbon dioxide in the atmosphere

Figure 3: Global mean surface temperature

FIG. 5.10. Simulated response to the global warming task with a typical participant response. In the simulation (Fiddaman, 1997), CO_2 in the atmosphere
peaks when emissions fall to zero, whereas simulated mean surface temperature must continue to rise for several decades because net radiative forcing, although falling, is still positive.

Even with this generous coding, only 22% of the CO_2 trajectories were
correct (Table 5.4). The most common error showed atmospheric CO_2 stabilizing in 2000 and remaining constant thereafter. Such trajectories would be
correct if the flux of CO_2 into the atmosphere from natural sources was exactly balanced by the absorption of CO_2 out of the atmosphere, or if, as
seems more likely, participants ignored the natural flows and assumed
anthropogenic emissions are the only flow.

Across both treatments, only 36% of the participants correctly approximated the trajectory for global temperature (Table 5.5). The most common

TABLE 5.4

Performance on the Global Warming Task: CO_2 Trajectory

Totals May Not Add to 1.00 Due to Rounding

Item	CO_2 Trajectory	%
	Correct	
1	CO_2 peaks at or very shortly after the year 2000 then declines at a diminishing rate	22
	Incorrect	78
2	CO_2 stabilizes in or after 2000 and never drops	31
3	CO_2 keeps rising forever	8
4	CO_2 immediately drops and continues to go down (shows a sudden, discontinuous jump down at or very shortly after 2000)	4
5	CO_2 stabilizes then decreases	7
6	CO_2 increases then decreases	16
7	The CO_2 trajectory is discontinuous (has a sudden jump up or down at some other time than at or very shortly after 2000)	1
8	The CO_2 trajectory follows some other path	10

Note. $N = 97$.

TABLE 5.5

Performance on the Global Warming Task: Global Mean Temperature

Item	Global Mean Temperature Trajectory	Total[a]	CO_2 Graph (%)[b]	No CO_2 Graph (%)[c]
	Correct			
1	Temperature continuing to rise for about 20 to 30 years, then falls (slowly)	36	**28**	**46**
	Incorrect	64	**72**	**54**
2	Immediate peak and drop in temperature in or very shortly after the year 2000	22	23	20
3	Temperature rising forever	11	13	9
4	Temperature stabilizing in or after 2000 and never declining	18	**25**	**9**
5	A fluctuation in temperature	1	0	3
6	A discontinuous path (temperature has a sudden jump up or down)	3	4	1
7	Temperature stabilizing then decreasing	2	**3**	**0**
8	Temperature decreasing then increasing	2	2	3
9	Some other path for temperature	6	3	10

Note. Bold figures indicate significant differences between the CO_2 and no CO_2 graph conditions at $p < .05$ by the Fisher exact test. Totals may not add to 1.00 due to rounding. The number of responses for the CO_2 graph in the CO_2 graph condition (97) was less than the 109 who provided usable temperature graphs in that condition.

[a]$N = 186$. [b]$N = 106$. [c]$N = 80$.

error was to show temperature peaking in the year 2000 and then falling, as seen in Fig. 5.10. The participant's trajectory for CO_2 overestimated the rate of decline but was still coded as correct. The participant, however, also showed temperature peaking in 2000, although simulated mean surface temperature must continue to rise for several decades because net radiative forcing, although falling, is still positive. We had hypothesized that asking participants to sketch the trajectory for the stock of atmospheric CO_2 would improve performance on the trajectory for temperature, but just the opposite was observed: 46% of those asked to sketch temperature only were correct, whereas only 28% of the temperature trajectories were correct for those asked to sketch both. The difference was highly significant ($p = .014$ by the Fisher exact test). It is possible that performance was worse when both graphs were requested because participants had more to do in the time available. More likely, consistent with our hypothesis, people tend to assume the inputs and outputs in a system should be highly correlated: 76% drew CO_2 and temperature trajectories with the same pattern (e.g., both peaking in 2000 and then falling).

The results suggest that highly educated people have extremely poor understanding of GW. In a simple scenario for future emissions, about two thirds drew trajectories for CO_2 and global temperature inconsistent with the most basic stock and flow considerations. There are several lessons. First, many people drew trajectories in which CO_2 and temperature followed the same pattern, suggesting they intuitively felt CO_2 and temperature should be correlated. However, the stock/flow structure means climate dynamics are fundamentally incompatible with such naive "common sense" approaches. Second, the inertia of the system means further warming and climate change are already underway. Actions to halt warming must be taken decades before the consequences of warming will be known. Yet many people drew trajectories in which global temperature responded immediately to changes in emissions of GHGs, significantly underestimating the time delays and inertia of the system. Most important, the stock and flow structure of the global climate means stabilizing emissions near current rates will not stabilize the climate. Stabilizing emissions means atmospheric CO_2 continues to rise, leading to significantly higher global surface temperatures. Yet policy proposals such as the Kyoto accord seek to stabilize the emission rate, not the stocks of GHGs that drive the climate.

Here lies both the challenge and the opportunity for NDM. The challenge is that policymakers and the public do not understand the immensely complicated models developed by climate scientists. People judge the plausibility of model-based projections such as those of the IPCC by whether the projections "make sense" relative to their intuitive understanding of the system and their intuitive ability to relate flows and stocks, understand time delays, and account for feedbacks. The results of such models are presented, even in nontechnical reports such as those of the IPCC, in the form of charts and graphs—the same mode of data presentation used in our tasks. As we have seen, people's intuitive understanding of even the simplest dynamic systems is poor. As long as "common sense" tells people that stabilizing emissions is sufficient, there can be little political will or public pressure for policies that could prevent additional climate change. As long as people believe the delays in the response of the system are short, they will conclude it is best to wait and see if further warming will be harmful before taking action. Such heuristics may work well in everyday tasks with low dynamic complexity in which delays are short, outcome feedback is unambiguous and timely, the opportunities for corrective action frequent, and the costs of error are modest. None of these conditions hold in systems with high dynamic complexity—delays between actions and impacts are long, outcome feedback is ambiguous and delayed, many actions have irreversible consequences, and the costs of error are often immense. Decision-making heuristics that work well in simple systems may lead to disaster in complex dynamic systems such as the climate.

The opportunity is that although the complexities of the global climate are daunting, the essence of the problem is as simple as filling a bathtub. Elementary education about stocks and flows may go a long way to overcome people's poor intuitive understanding of dynamics. The higher the concentration of GHGs in the atmosphere, the higher the global temperature will eventually become. GHG concentrations are already at levels higher than any in the past 20 million years and are rising at rates without any precedent in human history. GHG concentrations must continue to rise as long as the emission rate exceeds the absorption rate, just as a bathtub continues to fill as long as the inflow exceeds the outflow. To stabilize GHG concentrations at even the record high levels they have now attained requires deep cuts in emissions. To reduce CO_2 concentrations, emissions must fall below the absorption of carbon into terrestrial and oceanic sinks. Neither the public nor many policymakers understand these simple facts. So far, the nations of the world have been unable to agree even to stabilize emissions near current record rates, even though such policies guarantee further warming. The world has yet to face up to the inexorable logic of the stocks and flows of the global climate system.

ACKNOWLEDGMENTS

Financial support was provided by the MIT Sloan School of Management Organizational Learning Fund. We thank Berndt Brehmer, Jim Doyle, Henry Montgomery, Michael Radzicki, Nelson Repenning, Terry Tivnan, and seminar participants at the University of Chicago and MIT for helpful suggestions. Christopher Hunter assisted with data entry.

REFERENCES

Bakken, B., & Gould, J. (1992). Experimentation in Learning Organizations: A Management Flight Simulator Approach. *European Journal of Operations Research, 59*, 167–182.

Brehmer, B. (1992). Dynamic decision making: Human control of complex systems. *Acta Psychologica, 81*, 211–241.

Camerer, C., & Hogarth, R. (1999). The effects of financial incentives in experiments: A review and capital-labor-production framework. *Journal of Risk and Uncertainty, 19*, 7–42.

Chandler, M., & Boutilier, R. (1992). The development of dynamic system reasoning. *Contributions to Human Development, 21*, 121–137.

Diehl, E., & Sterman, J. (1995). Effects of feedback complexity on dynamic decision making. *Organizational Behavior and Human Decision Processes, 62*, 198–215.

Dörner, D. (1980). On the difficulties people have in dealing with complexity. *Simulations and Games, 11*, 87–106.

Dörner, D. (1996). *The logic of failure*. New York: Metropolitan Books/Henry Holt.

Fiddaman, T. (1997). *Feedback complexity in integrated climate-economy models*. Unpublished doctoral dissertation, MIT Sloan School of Management, Cambridge, MA. Retrieved http://www.sd3.info

Fiddaman, T. (2002). Exploring policy options with a behavioral climate-economy model. *System Dynamics Review, 18*(2), 243–267.

Forrester, J. W. (1961). *Industrial dynamics*. Waltham, MA: Pegasus Communications.

Forrester, J. W. (1971). Counterintuitive behavior of social systems. *Technology Review, 73*, 52–68.

Frensch, P., & Funke, J. (Eds.). (1995). *Complex problem solving—The European perspective*. Mahwah, NJ: Lawrence Erlbaum Associates.

Gigerenzer, G., & Goldstein, D. (1996). Reasoning the fast and frugal way: Models of bounded rationality. *Psychological Review, 103*, 650–669.

Gigerenzer, G. P., Todd, P. M., & ABC Research Group. (1999). *Simple heuristics that make us smart*. New York: Oxford University Press.

Gould, J. (Ed.). (1993). Systems thinking in education [Special issue]. *System Dynamics Review, 9*(2).

IPCC. (2001). Climate Change 2001: *The Scientific Basis*. Cambridge, UK: Cambridge University Press.

Klein, G. A., Orasanu, J., Calderwood, R., & Zsambok, C. E. (Eds.). (1993). *Decision making in action: Models and methods*. Norwood, NJ: Ablex.

Lipshitz, R., Klein, G., Orasanu, J., & Salas, E. (2000). Taking stock of NDM. In H. Friman (Ed.), *Proceedings of the 5th Naturalistic Decision Making Conference, Tammsvik, Sweden, 26–28 May* (CD-ROM).

Lipshitz, R., & Strauss, O. (1997). Coping with uncertainty: A naturalistic decision making analysis. *Organizational Behavior and Human Decision Processes, 69*, 149–163.

Mandinach, E., & Cline, H. (1994). *Classroom dynamics: Implementing a technology-based learning environment*. Hillsdale, NJ: Lawrence Erlbaum Associates.

Nordhaus, W. (1992a). *The "DICE" Model: Background and structure of a dynamic integrated climate-economy model of the economics of global warming* [Cowles Foundation Discussion Paper No. 1009]. New Haven, CT: Cowles Foundation for Research in Economics.

Nordhaus, W. (1992b, November 20). An optimal transition path for controlling greenhouse gases. *Science, 258*, 1315–1319.

Omodei, M., & Wearing, A. (1995a). Decision making in complex dynamic settings—A theoretical model incorporating motivation, intention, affect, and cognitive performance. *Sprache & Kognition, 4*, 75–90.

Omodei, M., & Wearing, A. (1995b). The fire-chief microworld generating program—An illustration of computer-simulated microworlds as an experimental paradigm for studying complex decision making behavior. *Behavior Research Methods Instruments & Computers, 27*, 303–316.

Paich, M., & Sterman, J. (1993). Boom, bust, and failures to learn in experimental markets. *Management Science, 39*, 1439–1458.

Richardson, G. (1991). *Feedback thought in social science and systems theory.* Philadelphia: University of Pennsylvania Press.

Richmond, B. (1993). Systems thinking: Critical thinking skills for the 1990s and beyond. *System Dynamics Review, 9*, 113–134.

Senge, P. (1990). *The fifth discipline: The art and practice of the learning organization.* New York: Doubleday.

Sterman, J. (1989a). Misperceptions of feedback in dynamic decision making. *Organizational Behavior and Human Decision Processes, 43*, 301–335.

Sterman, J. (1989b). Modeling managerial behavior: Misperceptions of feedback in a dynamic decision making experiment. *Management Science, 35*, 321–339.

Sterman, J. (1994). Learning in and about complex systems. *System Dynamics Review, 10*, 291–330.

Sterman, J. (2000). *Business dynamics: Systems thinking and modeling for a complex world.* New York: Irwin/McGraw-Hill.

Vennix, J. (1996). *Group model building: Facilitating team learning using system dynamics.* Chichester, England: Wiley.

Zsambok, C., & Klein, G. (1997). *Naturalistic decision making.* Mahwah, NJ: Lawrence Erlbaum Associates.

6

Not Only for Experts: Recognition-Primed Decisions in the Laboratory

Raanan Lipshitz
Adi Adar Pras
University of Haifa

According to Bales and Strodtbeck's (1951) phase theorem, "[problem solvers] go through certain stages or phases in the process of solving problems, [and] . . . problem-solving would somehow be more effective if some prescribed order were followed" (p. 485). Summarizing the studies that tested the validity of the phase theorem, Nutt (1984) concluded that "the sequence of problem definition, alternative generation, refinement, and selection, called for by nearly every theorist, seems rooted in rational arguments, not behavior" (p. 446). This conclusion is consistent with studies of naturalistic decision making that show that decision making is driven by situation assessment and recursive information search and option generation (Klein, Orasanu, Calderwood, & Zsambok, 1993).

Lipshitz and Bar Ilan (1996) tested the descriptive and prescriptive validities of a six-phase model adapted from Brim, Glass, Lavin, and Goodman (1962): identification (becoming aware that something is "out of order"), definition (formulating the problem as a discrepancy between a desired and an actual state of affairs requiring remedial action), diagnosis (identifying the cause of the problem), generation of alternatives (identifying which measures can counteract the causes of the problem), evaluation (weighing the pros and cons of alternative solutions), and choice (choosing and implementing the preferred alternative). Analyzing retrospective case reports of success and failure in problem solving of low to middle rank managers, Lipshitz and Bar Ilan found that the probability that a certain theoretical phase will occupy a certain location in the observed process or follow another phase was consistent with the model (accompanied by a marked but statistically nonsignificant tendency to leap forward to choice).

Because Lipshitz and Bar Ilan (1996) analyzed retrospective reports, it is not clear if their findings pertain to how problems are actually solved or to a cognitive schema that drives the reconstruction of problem-solving processes from long-term memory. We designed this study to clarify this question and to test the phase theorem with respect to ill-defined versus well-defined problems.

The difference between ill-defined and well-defined problems has been formulated in various ways and various terminologies. For example

> Problems that are well-structured vs. problems that are ill-structured are problems that vary in terms of both the completeness which the problem can be specified and the certainty with which a solution can be recognized as being correct or incorrect. (Arlin, 1989, p. 233)
>
> A problem is structured if the solver is readily able to identify a promising solution strategy. (Smith, 1988, p. 1499)
>
> Well-structured problems are those in which the initial state, goal state, constraints, and operators are precisely defined. Ill-structured problems, on the other hand, are those in which one or more of the above are not precisely defined. (Voss, 1990, p. 315)

Problems are ill defined because of missing information or inadequate understanding of (a) the nature of the problem (i.e., current vs. goal states) and (b) solution strategies and the causal relations in the problem's domain. Problems are not as ill defined or well defined in an absolute sense. These terms, rather, constitute a continuum on which problems are ill defined or well defined relative to one another (as are the problems studied in this research) and relative to problem solvers' competence. A problem that is ill defined (hence intractable) to a novice is well defined (hence routine) to the expert (Kahney, 1986; Reitman, 1965; Taylor, 1974).

Lack of structure in problem definition leads to reduced structure in the solution process (Fredriksen, 1984; Gettys, Pliske, Manning, & Casey, 1987; Sweller, 1983). For example, Newell and Simon (1972) suggested that well-structured problems are problems for which there exist "strong" (i.e., essentially algorithmic) solution methods that ensure smooth transition from problem to solution states, whereas ill-structured problems are those for which only "weak" (i.e., essentially heuristic) solution methods exist, such as means–end analysis, which entail at least a certain amount of trial and error. Some of the process differences associated with the difference between well-defined and ill-defined problems are expressed in terms of compatibility with some problem-solving phase model. Kochen and Badre (1974) suggested that the solution process for ill-defined problems is basically similar to that for well-defined problems except for the greater attention paid to problem definition and diagnosis. By contrast, Klein and Weitzenfeld (1978) suggested that the two processes are dissimilar inasmuch as solving well-defined problems progresses linearly from definition to action basically in the fashion found by Lipshitz and Bar Ilan (1996), whereas ill-defined problems are solved in cyclical process in which definition, option generation, and option evaluation are intertwined with one another.

In conclusion, the purpose of this study was to examine how problem formulation (ill defined vs. well defined) affects the ensuing problem-solving process. To this end, we used Voss' (1990) definition mentioned previously that enables us to represent identical problems in relatively ill- or relatively well-defined forms. Our methodology enabled us both to test Kochen and Badre's (1974) and Klein and Weitzenfeld's (1978) hypotheses and to find out if the solution processes of well-defined and ill-defined problems differ in ways not hypothesized by them.

METHOD

Participants

Twenty-two undergraduate students (17 women and 3 men) in the Department of Psychology at the University of Haifa participated in the study for required credit hours. Participants had not taken courses in problem solving or decision making prior to the experiment.

Design and Manipulation

Participants were required to think aloud as they solved one relatively well-defined and one relatively ill-defined problem in randomized order. To examine the effects of problem presentation on problem definition, the manipulation of a well-defined versus ill-defined problem had to allow the problem solver, rather than the experimenter, to define the problem. This was accomplished by use of Thematic Apperception Test (TAT) pictures to present the problems visually with minimal verbal instructions that established the problem as well defined or as ill defined (see Geiser & Stein, 1999). The TAT pictures used in the study were 3GF (a young woman standing before an open door apparently in grief) and 6BM (an old woman and a young person, typically perceived as "mother and son" in TAT protocols, with grave demeanor standing side by side seemingly avoiding one another). These pictures were selected because they depict ambiguous, apparently problematic naturalistic situations to which participants could readily relate.

Each picture was presented as a relatively well-defined or a relatively ill-defined problem by variation of the amount of information on two of the three parameters that define the two problem types (Reitman, 1965): the current state and the goal state. (The third parameter, admissible operators, is irrelevant in this context in which there are no well-defined operators.) In addition, pretests showed that it was necessary to define the participants' role in the situation to help them produce think-aloud protocols. For example, the instructions for the relatively well-defined version of the "Young woman" read as follows:

> The young woman has just emerged from the room in which her seriously ill sister lies in bed. She feels that the situation is desperate, and that she does not know how to cope with it. You are a close friend who went to visit and found her in this situation. Your task is to help your friend to feel that she can cope with the situation.

The corresponding instructions for the relatively ill-defined version simply read as follows:

> The picture presents a young woman in a problematic situation. You are a close friend. Your task is to help her in her situation.

Procedure

Data collection included general instructions followed by three phases of data collection with each picture:

1. General instructions. Participants were told that they would be shown pictures of people in problematic situations and that their task was to imagine the situation of these people and think aloud as they deliberated on how to help them in any way that they thought was appropriate.

2. Think-aloud training. Participants were shown a TAT picture that was not used in the study and trained to "report anything that went on in their heads from the moment that the picture was presented to the point that they reached a solution." Half the participants were presented with a relatively ill-defined version of the training picture and half were presented with a relatively well-defined version of the picture in random order.

3. Instructions for the first problem (see previously).

4. Thinking aloud. At this phase, participants proceeded to solve the problem without any experimenter's intervention. The experimenter audio-taped and took long-hand notes of each participant's think-aloud protocol.

5. Nondirective probes. At this phase, the experimenter used reflection (reading back from his notes and trailing off, e.g., "You said that the young

woman was sad . . .") and used nonverbal cues (e.g., head nodding) and mini-mal verbal cues (e.g., "aha . . .") to encourage the participant to elaborate his or her unaided protocol.

6. Repeat Phases 3 to 5 for the second problem.

Coding

Participants' verbal protocols of the three data collection phases were tran-scribed and arranged in two versions, one the verbatim record of partici-pants' think-aloud protocols, the other the think-aloud protocol with the ad-ditional information obtained by the nondirective probing inserted as appropriate (e.g., references to definition produced by probing were at-tached to references to this phase obtained through thinking aloud). Both protocols were coded by an 11-item coding scheme adapted from Lipshitz and Bar Ilan (1996) and Cummings, Murray, and Martin (1989) and extended to capture more fully the information contained in our own protocols:

1. Situation description. The problem solver provides factual details of the problem at hand and the internal and external conditions of the people involved (e.g., "She feels deeply for her sick sister").

2. Definition. The problem solver refers to a desirable situation, an unde-sirable situation, or both (e.g., "The mother is apparently afraid to be left alone").

3. Diagnosis. The problem solver refers to causes of the gap between de-sirable and undesirable situations (e.g., "She is old and sick and her only son is going abroad") for the preceding problem.

4. Choice/action. The problem solver refers to a potential course of ac-tion open to him or her (e.g., "I'll try to explain to the mother why her son has to go abroad and promise to call and visit her periodically"; "I won't say any-thing unless they ask me for my opinion"). Note that because participants de-liberated single options and did not implement their solutions, this phase in-corporates the phase of option generation in the model used by Lipshitz and Bar Ilan (1996).

5. Option evaluation. The problem solver refers to reasons for or pros and cons of potential courses of action (e.g., "This [i.e., potential solution] will keep the mother occupied while her son is away").

6. Reflection. The problem solver reflects on the problem, the persons in-volved, his or her own options, or potential outcomes (e.g., "I think that the mother is wrong to object to her son going away"; "Doing this [the contem-plated action] is very difficult").

7. Self-reference. The problem solvers refer the problem or problem-solving process to themselves via their feelings and personal experiences

(e.g., "I remember how hard it was for me to leave my old parents"; "This is not my style"; "It is really difficult for me to deal with this").

8. Elaboration of definition. The problem solver elaborates (i.e., does not merely repeat a previously stated definition, e.g., "You can see that the mother is really very worried").

9. Elaboration of diagnosis. The problem solver elaborates (i.e., does not merely repeat a previously stated diagnosis, e.g., "She had a heart attack not long ago and has been particularly apprehensive since then").

10. Elaboration of choice. The problem solver elaborates (i.e., does not merely repeat a previously stated choice, e.g., "I can also tell her how easy it is to call abroad these days").

11. Elaboration of option evaluation. The problem solver elaborates (i.e., does not merely repeat a previously stated option evaluation, e.g., "This usually works, but she may object, who knows?").

Two problem-solving phases in the model employed by Lipshitz and Bar Ilan (1996) were not found in participants' protocols: identification (because they were informed that a problem exists), and option generation (which was merged with action in participants' protocols). Codes were attached to text segments (of varying lengths) that corresponded to them. We refer to the three additional phases of situation description, reflection, and self-reference as metacognitive phases because the first seems to prepare for the problem solving process, and the latter two control it.

RESULTS

We did not calculate the reliability of coding; however, Lipshitz and Bar Ilan (1996), using identical methodology, obtained a Cohen kappa coefficient of ($r = .83$). Figure 6.1 presents the distribution of the 11 coded phases in the think-aloud protocols of "Mother and Son." As the figure shows, all phases appeared, albeit with different frequencies, in both protocols, and the addition of data obtained by nondirective probes did not change the relative distribution of phases in the think-aloud protocol: a clear domination of the choice phase, a nearly rectangular distribution of all other nonelaboration phases, and a negligible appearance of the latter except for action elaboration. These conclusions held, with slight nonsystematic variation, in the remaining three data sets (nondirective probes of "Mother and Son" and think-aloud and nondirective probes of "The Young Woman"). Considering their negligible appearance, elaboration phases (except for action elaboration) were joined with the corresponding phases, leaving 8 phases to be used in further analyses. The results from the four data sets (two pictures ×

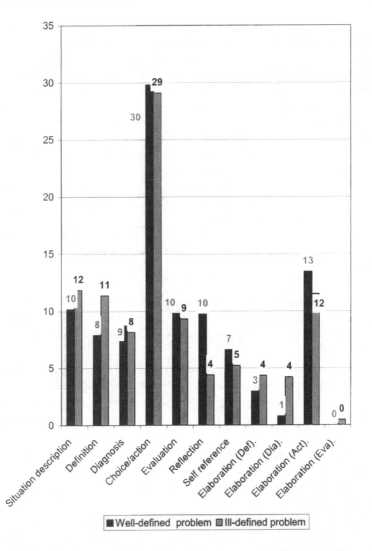

FIG. 6.1. Mean frequencies of problem-solving elements, think aloud, "Mother and Son." Def = definition; Dia = diagnosis; Act = action; Eva = evaluation.

two elicitation methods) were likewise similar. Only 3 of the 44 pair-wise statistical comparisons testing the mean frequency of appearance of the 11 phases in the four data sets were significant below the p = .05 level, leading to the conclusion that the four data sets were identical except for chance fluctuations. We therefore present the remaining findings from one data set, the think-aloud protocols of "Mother and Son," and a summary of findings from all four sets.

The relative location of the theoretical phases in the observed problem-solving processes was determined by examination of their cumulative distribution along the observed problem-solving process (Fig. 6.2). Because cumulative distributions of all phases tapered off unevenly after 15 steps (a *step* is defined as any reported observed datum that can be mapped onto one of the elements of the coding scheme), Step 15+ in Fig. 6.2 contains all observations beyond Step 14. Figure 6.2-a presents the cumulative distributions of five phases traditionally associated with problem solving (definition, diagnosis, choice, evaluation, and elaboration of choice) in the relatively well-defined version of "Mother and Son." Figure 6.2-b presents the corresponding results for the relatively ill-defined version. Figure 6.3 presents the cumulative distributions of the three metacognition phases for the two versions of "Mother and Son."

The general conclusions regarding the location of the five problem-solving phases and three metacognition phases in all four data sets were as follows:

1. Definition is the first phase in the solution process of both relatively well-defined and relatively ill-defined problems. Its frequency of appearance in the early phases of solving relatively ill-defined problems is higher.
2. Diagnosis is the second phase in the problem-solving process, appearing earlier in the solution of relatively ill-defined problems.
3. The third phase is choice (or action) that appears midway through the solution of both types of problems intertwined with elaboration of choice.
4. Option evaluation appears next in the latter third of the solution processes of both types of problems. This finding and the absence of a distinct phase of option generation indicate a cyclical process of serial option generation and evaluation.
5. Self-reference appears early on, particularly in solving relatively well-defined problems, and disappears around Step 7 in both solution processes.
6. Situation description appears early in the solution of relatively ill-defined problems and midway in the solution of relatively well-defined problems.
7. Reflection appears midway through the process of solving both types of problems.

The findings concerning the location of the phases on the problem-solving process served to fix their order in the search for systematic patterns in the observed sequences of phases using lag analysis (Bakeman &

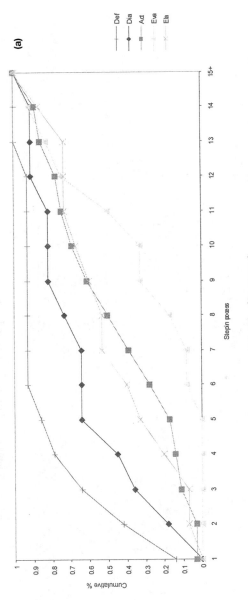

FIG. 6.2. (Continued).

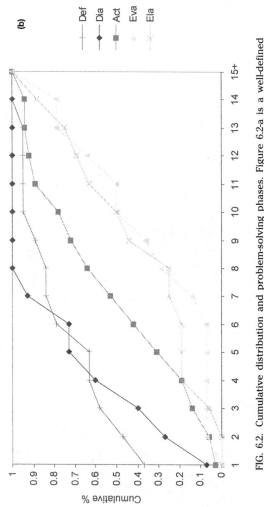

FIG. 6.2. Cumulative distribution and problem-solving phases. Figure 6.2-a is a well-defined problem; Fig. 6.2-b is an ill-defined problem.

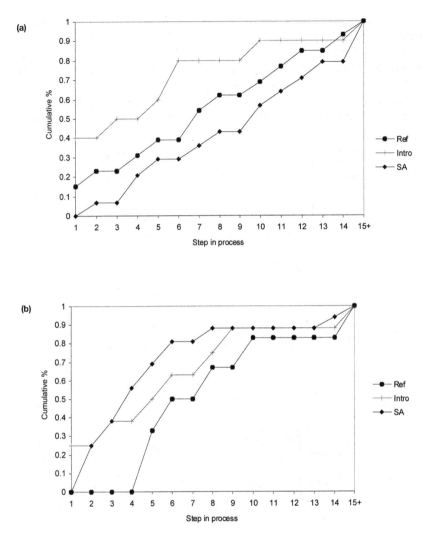

FIG. 6.3. Cumulative distribution of metacognition phases. Figure 6.3-a is a well-defined problem; Fig. 6.3-b is an ill-defined problem.

Gottman, 1986), a method for detecting stable dependencies in sequential data. Lag analysis is performed on transition probabilities such as the probabilities that a person who exhibits behavior i (e.g., problem identification) at one step in the process will exhibit behaviors j, k, and so forth (e.g., definition, diagnosis, etc.) in the following step in the process. Figure 6.4, which presents the results of lag analysis of the "Mother and Son" problem, reveals a pattern that was found in all four data sets (numbers within rectangles denote the probabilities of phase repetition):

1. Phases gathered in two clusters—one consisting of definition, situation description, and diagnosis (in relatively ill-defined problems), and the other consisting of action, option evaluation, and elaboration. Choice/action, which was linked to most elements in both clusters, served as the linchpin connecting them. The two clusters are defined mostly by the backward paths that create feedback loops that are contained within them (e.g., evaluation → action → elaboration → evaluation in Fig. 6.4-a; see also Point 4 following).

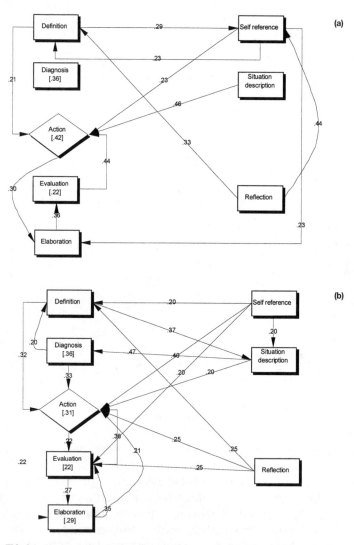

FIG. 6.4. Lag analysis of "Mother and Son"; a. is a well-defined problem; b. is an ill-defined problem.

2. Problem solvers tended to move forward from early to late phases, albeit with frequent leaps forward and loops backward. In particular, the four relatively well-defined problems included five moves forward between consecutive problem-solving phases, eight leaps forward, and seven loops backward from late to earlier phases. The corresponding statistics for the four relatively ill-defined problems were 7, 7, and 9.

3. The solution processes of relatively ill-defined and relatively well-defined problems differed in two respects: the solution of ill-defined problems involved more phases ($M = 6.75$ vs. $M = 5.25$ for relatively ill-defined and relatively well-defined problems, respectively), and metacognitive activity was far more intense in the solution processes of relatively ill-defined than of relatively well-defined problems ($M = 7.75$ vs. $M = 5.5$ for links between problem-solving and metacognition phases for relatively ill-defined and relatively well-defined problems, respectively; it is impossible to test these differences statistically owing to the sample size of four).

4. The basic two-cluster structure of the solution processes of both relatively well-defined and relatively ill-defined problems was consistent with the recognition-primed decisions (RPD) model (Klein, 1993) in which the first cluster corresponds to the situation recognition stage (which roughly corresponds to problem definition up to choice-action), and the second cluster corresponds to the second stage of mental simulation (in which options are critically evaluated one at a time, roughly corresponding to choice-evaluation-elaboration). This relation is confirmed by the fact all 7 paths from definition and diagnosis in the four data sets led either to one of these phases or to action (links to metacognitive phases are irrelevant to the RPD model), and all 14 paths from choice-action, evaluation, and elaboration in the four data sets led to one another.

DISCUSSION

Our findings reveal certain similarities coupled with significant differences between reconstructed problem-solving processes and problem solving in vivo. These results partially confirm our hypotheses regarding differences between solving well-defined and relatively ill-defined problems and show that the problem-solving process observed in this laboratory study is essentially identical with the decision-making processes observed by Klein and others (Klein, 1998; Lipshitz & Ben Shaul, 1997) in naturalistic settings. We elaborate on these three issues in turn.

Solving Problems in Vivo and in Retrospect

The underlying phase structure of concurrent and retrospective problem solving case reports are similar in two respects: Both are dominated by choice-action, and both show a preference for moving forward from early to late phases, including leaping directly to choice/action. The dissimilarities between these processes are more noteworthy. First, similar to the RPD model (Klein, 1993), option evaluation does not precede but follows choice. Second, the process is less orderly, that is, instead of the smooth linear progression found by Lipshitz and Bar Ilan (1996), it includes numerous leaps forward and loops backward. Third, similar to Cummings et al. (1989), the process includes metacognitive activity that is totally missing from the retrospective reports. These findings are consistent with the notion that retrieval from long-term memory is schema driven and with Brewer and Dupree's (1983) model of plan schemata that suggests that elements in a story that are in a purposeful (in-order-to) relation, such as consecutive elements in an action plan, are remembered better than elements not related in this fashion. The problem-solving phases follow this logic (e.g., problems must be defined to be diagnosed). In contrast, the metacognitive activities reported in the think-aloud protocols, which serve to monitor the process, do not follow this logic and hence are not retrieved from long-term memory.

Solving Well-Defined Versus Ill-Defined Problems

Findings provided support to Kochen and Badre's (1974) hypothesis and disconfirmed Klein and Weitzenfeld's (1978) hypothesis in that the sequences of problem-solving phases differed in the greater attention to early phases prompted by ill-defined problems. Neither hypothesis pertained to the second major difference between the two types of processes, namely, the greater intensity of metacognitive activity and the greater complexity of the links between problem-solving and metacognition phases engendered by solution of ill-defined problems. The disconfirmation of Klein and Weitzenfeld's proposition should be treated cautiously given the minimal manipulation of problem structuredness and the fact that the analyzed processes were truncated on the end side (solutions were neither implemented nor subjected to vagaries of dynamic situations). Nevertheless, both findings seem reasonable: The greater amount of information contained in relatively well-defined problem presentations reduces both the amount of attention that problem solvers must pay to defining and diagnosing the problem and to monitoring their progress throughout the problem-solving process.

Problem Solving and RPDs

It is fair to say that the principal, and unanticipated, finding of this study is that the solution processes of both relatively well-defined and ill-defined problems conform to the RPD model. This finding, which illustrates the role of serendipity in science, has implications for both the distinction between problem solving and decision making and the RPD model.

The distinction between problem solving and decision making is artificial inasmuch as problem solvers and decision makers in the "real world" typically do both. One suspects that the distinction endured so long because researchers have tended to congregate in specialized communities; Newell and Simon (1972) chose to title their seminal book, which partly dealt with chess players' decision making, *Human Problem-Solving*. The distinction began to gradually give way when Klein (1989) found that firefighters, like chess players, made matching-mode decisions (Lipshitz, 1994). He refrained from using Newell and Simon's model and terminology, titling his own model RPDs, forcing the community of decision researchers to confront the fact that decisions are not made by concurrent choice unless presented this way, for example, by experimental economists and supermarket managers.

Because Klein (1993) and other researchers who validated his model tended to work in naturalistic settings and with proficient decision makers, the RPD model is typically associated with high stakes, pressure of time, and expert decision makers. None of these conditions hold in this research with the possible partial exception of the latter condition, assuming that freshmen students in psychology possess some knowledge relevant to interpersonal problems. Thus, we hypothesize that the principal factor that determines the use of RPDs is the mode of problem presentation. Decision makers will use RPDs if decisions are not presented to them as choice among alternatives and if the decision problem has a significant perceptual component. Under these conditions, the difference between expert (or experienced) decision makers and novice (or journeyman) decision makers should be sought in the ways in which they execute recognition-primed decision making as much as in the frequency in which they prefer it to concurrent choice.

REFERENCES

Arlin, P. K. (1989). The problem of the problem. In J. D. Sinnott (Ed.), *Everyday problem-solving* (pp. 229–237). New York: Praeger.

Bakeman, R., & Gottman, J. R. (1986). *Observing interaction*. Cambridge, England: Cambridge University Press.

Bales, R. F., & Strodtbeck, F. L. (1951). Phases in group problem-solving. *Journal of Abnormal and Social Psychology, 46*, 485–495.

Brewer, W. F., & Dupree, D. A. (1983). Use of plan schemata in recall and recognition of goal-directed action. *Journal of Experimental Psychology: Learning, Memory, and Cognition, 9*, 117–129.

Brim, O. G., Glass, D. C., Lavin, D. E., & Goodman, N. (1962). *Personality and decision processes: Studies in the social psychology of thinking*. Stanford, CA: Stanford University Press.

Cummings, A. L., Murray, H. G., & Martin, J. (1989). Protocol analysis of the social problem solving of teachers. *American Educational Research Journal, 26*, 25–43.

Fredriksen, N. (1984). Implications of cognitive theory for instruction in problem-solving. *Review of Educational Research, 54*, 363–407.

Geiser, L., & Stein, M. I. (1999). *Evocative images: The thematic apperception test and the art of projection.* Washington, DC: American Psychological Association.

Gettys, C., Pliske, R. M., Manning, C., & Casey, J. T. (1987). An evaluation of human act generation performance. *Organizational Behavior and Human Decision Processes, 39*, 23–51.

Kahney, H. (1986). *Problem-solving: A cognitive approach.* Philadelphia, PA: Open Press.

Klein, G. A. (1989). Recognition primed decisions. In W. R. Rouse (Ed.), *Advances in man-machine systems research, 5* (pp. 47–92). Greenwich, CT: JAI Press.

Klein, G. A. (1993). Recognition-primed decision (RPD) model of rapid decision making. In G. A. Klein, J. Orasanu, R. Calderwood, & C. E. Zsambok (Eds.), *Decision making in action: Models and methods* (pp. 138–147). Norwood, NJ: Ablex.

Klein, G. (1998). *Sources of power: How people make decisions.* Cambridge, MA: MIT Press.

Klein, G. A., Orasanu, J., Calderwood, R., & Zsambok, C. (Eds.). (1993). *Decision making in action: Models and methods.* Norwood, NJ: Ablex.

Klein, G. A., & Weitzenfeld, J. (1978). Improvement of skills for solving ill defined problems. *Educational Psychologist, 13*, 31–41.

Kochen, M., & Badre, A. N. (1974). Questions and shifts of representation in problem-solving. *American Journal of Psychology, 87*, 369–383.

Lipshitz, R. (1994). Decision making in three modes. *Journal for the Theory of Social Behavior, 24*, 47–66.

Lipshitz, R., & Bar Ilan, O. (1996). How problems are solved: Reconsidering the phase theorem. *Organizational Behavior and Human Decision Processes, 65*, 48–60.

Lipshitz, R., & Ben Shaul, O. (1997). Schemata and mental models in recognition-primed decision making. In C. Zsambok & G. A. Klein (Eds.), *Naturalistic decision making* (pp. 293–304). Mahwah, NJ: Lawrence Erlbaum Associates.

Newell, A., & Simon, H. A. (1972). *Human problem-solving.* Englewood Cliffs, NJ: Prentice Hall.

Nutt, P. C. (1984). Types of organizational decision processes. *Administrative Science Quarterly, 29*, 414–450.

Reitman, W. R. (1965). *Cognition and thought: An information processing approach.* New York: Wiley.

Smith, G. H. (1988). Towards a heuristic theory of problem structuring. *Management Science, 34*, 1489–1506.

Sweller, J. (1983). Control mechanisms in problem-solving. *Memory and Cognition, 11*, 31–40.

Taylor, R. N. (1974). Nature of problem ill-structuredness: Implications for problem formulation and solution. *Decision Sciences, 5*, 632–643.

Voss, J. (1990). On the solving of ill-structured problems: A review. *Unterrichts Wissenschfat, 18*, 313–337.

7

Using a Prediction Paradigm to Compare Levels of Expertise and Decision Making Among Critical Care Nurses

Monique Cesna
Kathleen Mosier
San Francisco State University

Naturalistic decision making (NDM) was described by Lipshitz (1997) as the process of making decisions under time pressure in a dynamic or changing environment. The decisions made in these situations often involve poorly structured problems, uncertain information, and shifting goals. These types of decisions call for immediate action, with sometimes-dire consequences for bad choices. It is in these settings that the level of expertise among operators is seen as vital in the decision process. The importance of expertise is evident in such domains as fire fighting, aviation, medicine, and nursing (Klein, Calderwood, & Clinton-Cirocco, 1986; Buch & Diehl, 1982; Bogner, 1997; Crandall & Getchell-Reiter, 1993). In describing NDM further, four facets identify experience/expertise as having a critical influence on the quality of the operator's decision-making process.

The first characteristic of NDM impacted by expertise is how experts see and process information. The more experience one has the quicker they are able to identify the most important information first (Orasanu & Connolly, 1997). Second, experience gives operators mental models that allow them to adapt to constantly changing values of information (Waag & Bell, 1997). The quality of NDM also depends on the level of experience an operator has with similar situations. Situations which are familiar to allow experts to comprehend what may be at the core of a problem and come to a conclusion about what actions should be taken (Cellier, Eyrolle, & Marine, 1997). Last, NDM occurs within time constraints. As the operator gains experience/expertise, prior decisions and outcomes can more quickly be recalled from similar past situations, which can lead to a keen sense of timing (Klein, 1997). This sense of timing gives the operator flexibility to change course in the middle of a situation when a decision is not working effectively.

Understanding the role of experience/expertise in NDM is important not only because of the complexity of decisions made in dynamic environments but also because of the high cognitive demands placed on the operators. Studying how people practically apply their knowledge and experience in making critical decisions has highlighted the importance of intuitiveness as closely tied to expertise in NDM. The role of intuitiveness relative to NDM can be more easily understood by examining the recognition primed decision making (RPD) model (Klein, 1997).

RECOGNITION PRIMED DECISION-MAKING MODEL

The RPD model outlines one framework of NDM and identifies three aspects that are at the core of RPD: the quality of the decision maker's situation assessment, his or her level of experience/expertise, and the use of intuitive rather than analytical decision processes.

In situation assessment, the degree to which the operator can recognize that a situation is similar to one previously encountered comes into play. They can gain this recognition through the evaluation of relevant cues, expectancies, and actions. In familiar situations, operators look for relevant clues, which confirm that the situation has been experienced before. They will recall mental pictures of what worked to solve the problem (expectancies). Situation assessment is at the core of the RPD model and is highly dependent on level of experience/expertise. The RPD model supports expertise/experience as relevant to intuitive decision making. Experts are able to utilize an intuitive rather than analytical processes to make decisions more quickly and accurately than novices.

Expert Versus Novice Decision Making

Studies comparing expert and novice decision making cover a wide variety of domains, ranging from aviation (Buch & Diehl, 1982), to chess playing (Klein & Peio, 1989), fire fighting (Klein et al., 1986), fighter pilots (Waag & Bell, 1994), medicine (Nyssen & De Keyser, 1998) and nursing (Tabak, Bar-Tal, & Cohen-Mansfield, 1996).

The consistent findings concerning expertise and decision making are summarized in four points:

1. experts are more accurate at inference. They are able to see causality, that is, that X would lead to Y, and how that affects a situation (Spence & Brucks, 1997);
2. experts are better at anticipating problems. While novice or less experienced operators are able to solve a problem, they are not as good at preventing them. (Cellier et al., 1997);
3. experts have a better functional view of the decision process, that is, they know which decisions require immediate action (Benner, Hooper-Kyria, & Stannard, 1999);
4. experts and novices differ with respect to domain-related decision processes.

Less experienced people often used analytical decision model, looking for the best answer to a problem. Experts are not as concerned about coming up with the best solution, only a "good" one. They often describe their processes as intuitive involving little or no conscious deliberation (Klein, 1997).

Expert/Novice Research in Medicine

Benner, Tanner, and Chesla (1992) studied 105 nurses in eight hospitals working in newborn, adult and pediatric Intensive care units (ICU). The participants were interviewed by a clinical expert and were classified novices or experts according to their skill levels. The nurses were asked to describe cases in which they had to make decisions leading to diagnosis. The experts identified several decision points which illustrated the differences between the skill levels relevant to decision making and their quality of situation assessment. Benner et al. (1992) found that novice, or less experienced nurses focused on tasks according to a mental "checklist" based on a specific protocol for certain symptoms of illness. Moreover, novice nurses could not draw on mental models as they had no prior experiences, to inform their level of situation assessment, and would fall back on a more analytical approach to decision making. They were able to match information but were not able to identify causality for their patient's symptoms; they could identify problems but seemed unable to prevent or anticipate them. In contrast, experts had a very complex grasp of perceived changes in their patients' status. They were able to deviate from the checklist and identify important information, which called for specific immediate actions.

Tabak et al. (1996) focused on nurses' confidence and accuracy in their diagnosis comparing experienced and novice nurses. They studied 92 experienced and 65 novice (still in training) hospital nurses. The investigators gave the participants two clinical scenarios; both required the nurses to make a clinical diagnosis. The nurses then rated how difficult they found their decisions and how confident they were with their diagnosis. The results show that experts were more accurate at initial diagnosis with a higher degree of confidence than the novice group. It has been shown that confidence and insight are closely linked. Confident decision makers have greater insight into their own decision making processes (Christensen-Szalanski & Bushyhead, 1981; Mahajan, 1992). Therefore, experts who are better decision makers display a higher level of confidence.

A third study of expertise in nursing conducted by Benner et al. (1999) combine data collected from two separate studies which explore expertise and skill acquisition. The purpose of this study was to trace the knowledge and development of expertise as articulated by interventions in critical care. Benner et al. (1999) identified two habits of thought and action that lead to clinical expertise: clinical grasp, which is the quality of situation assessment, and clinical forethought, or perceptions of what may be causing a problem. These habits which are closely related to the RPD elements of situation assessment are described by Benner et al. (1999) as elements of good clinical practice.

One of these elements of good clinical practice tied to situation assessment is described as thinking in action. This concept is tied to expertise and learning under pressure as well as to thinking linked with action-based decision in ongoing situations. Through subjective interviews, Benner et al. (1999) elicited information on decision making from critical care nurses who described decision points that articulated their levels of situation assessment. The nurses recalled accrual critical incidents, which formed the bases of mental models for expert knowledge and could be used to create clinical scenarios for training less experienced nurses.

The other element of good clinical practice identified by Benner et al. (1999) was the idea of reasoning in transition (or reasoning about the changes in a situation). Reasoning in transition describes an operator's ability to evaluate, in a dynamic environment, whether or not a decision will work as is or if it needs modification prior to implementation.

Although the previously mentioned studies have examined nurses' decision-making processes, only one investigation used an objective measure (Tabak et al., 1996). The study we describe in this chapter extended this research by creating additional objective measures of expertise and decision making among critical care nurses.

These measures are modeled, in part, after a prediction paradigm tool used by Klein and Peio (1989). Their research studied 34 chess players with varying degrees of expertise and asked them to predict moves an expert would make at different points within a given chess game. At each decision point, Klein and Peio found that the more experienced players had a higher degree of accuracy in predicting the correct sequence of moves than less experienced players. In addition the researchers also found that the more experienced players often generated the correct move as their first option.

For this study, two clinical scenarios with defined decision points were created. Nurse participants were asked at each decision point to predict what an expert would do and made these predictions under a time constraint. They were asked to rate their confidence and difficulty of their decisions. Each nurse was scored for accuracy at each decision point. In addition they were scored according to how many answers they generated at each decision point.

It was hypothesized that more experienced nurses would be more accurate and confident in predicting what the experts would do than the less experienced nurses. It was also anticipated that the less experienced nurses would find decision points to be more difficult in more complex clinical situations and generate more options than the more experienced nurses would.

METHOD

Participants

Participants were critical care nurses at the University of California, San Francisco medical center, with varying degrees of experience ($n = 30$; 6 men and 24 women). All participants worked in an adult critical care area and ranged in age from 27 years to 51 years. The average level of college education for all nurses was 4 years.

Nurses were placed into their respective groups based on levels of experience. These levels were determined by the number of years working in the ICU (Benner et al., 1992):

Advanced beginner—up to 2 years experience in the ICU

Intermediate—at least 2 years experience in the ICU

Expert—at least 5 years experience in the ICU

Group 1 ($n = 6$) consisted of advanced beginners who ranged in years of critical care experience from 1 to 2. Group 2 ($n = 6$), were intermediate nurses who ranged in years of critical care from 2 to 18. Group 3 ($n = 8$) were expert nurses who ranged years of critical care from 9 to 27.

Instruments

Two clinical scenarios were drafted by M. Cesna, RN (an expert critical care nurse) and reviewed by a panel of four expert nurses. These expert nurses were identified as such based in part on the Dreyfus model of expert skill acquisition (1986) using the following set of criteria:

Expert—Nurses who are recognized by their peers and supervisors as expert practitioners in the clinical environment. These nurses provide direction to less experienced nurses and other healthcare professionals. They easily triage care of multiple patients and assign direction to additional personnel in emergency situations. They are highly skilled in clinical tasks and are often called on to consult on clinical situations as needed.

Two different clinical scenarios were presented to each nurse. These scenarios were based on a typical emergency situation in the ICU that requires quick decisions and actions. The first scenario was based on a more complex clinical situation; the second contained more straightforward information. The expert panel also identified decision points based on Klein et al.'s (1989) RPD model. After the panel agreed on the actions that should be taken for each decision point, a correct answer key was created (Klein & Peio, 1989). The experts were able to agree that the first two actions of each decision point were the most important ones, and that the subsequent action options were not as important, but did contribute to the treatment plan of the patient.

Procedures

One to three nurses met together in a quiet environment, and a written scenario was distributed. Each part of the scenario (to a decision point) was read aloud by the research assistant, as the participants read along silently. After the excerpt was read, the nurses were given 2.5 minutes to list options and prioritize actions they thought the experts would take at that decision point and wrote down their answers. (When two or more nurses participated together, the nurses were told not to discuss what they were thinking.) After making their decisions, participants were asked to rate the confidence level of their prediction as well as how difficult they found the decision. Following this, the expert panel responses at that decision point were told to the group, as well as what happened in the actual clinical situation. The same procedure was followed until all the decision points were completed within the given scenario.

The data were scored as follows:

Difficulty of decision: At the end of each decision point, the nurse was asked to rate the difficulty of the decision on a scale ranging from 1 to 5.

Confidence Scale: At the end of each decision point, the nurses were asked to rate their level of confidence, on a scale ranging from 1 to 10, in correctly predicting what the experts would do.

Options generated: Options generated and prioritized at each decision point were recorded and were added together for a total number of options for each scenario.

Accuracy points: The decision accuracy score for each participant was based on the total number of correct predictions of the experts' action at each decision point. The scores were measured against the total number of possible correct answers.

The decisions were made under time pressure (a stopwatch was used to time decisions and nurses were stopped after 2.5 min). Each decision point was scored for a possible total of 6 points. The nurses received the full 6 points if they correctly predicted what the experts would do as a first and second choice. They scored 3 points if they predicted one of the top two correct actions. They got no points if they were unable to predict either of the two most important decisions the experts made. It was felt that using a score of 0, 3, or 6 would make the differences in scores more obvious.

RESULTS

According to Klein and Peio (1989), a more complex scenario will differentiate expertise and decision making better than a simple scenario. For this chapter, therefore, the complex scenario was used to analyze expertise and decision making among critical care nurses. The panel of experts felt that correct treatment was most important at the first decision point in the complex scenario and that measures at this point would further highlight differences in decision making among the levels of expertise. The independent variable was the level of expertise of the nurses. The dependent variables were overall prediction accuracy, prediction accuracy at decision point 1, confidence rating, difficulty of decision rating, and number of options generated.

Mean accuracy scores for the first decision point as well as the whole scenario, confidence levels, and difficulty ratings are displayed in Table 7.1. As shown, Group 3 (experts) had the highest accuracy and confidence levels. The mean difficulty of decision ratings were nearly the same for every

TABLE 7.1
Mean Scores

Group	Accuracy— Decision Pt #1		Accuracy— Entire Scenario		Confidence		Difficulty of Decision		Total Options Generated	
	M	SD	M	SD	M	SD	M	SD	M	SD
1. Advanced Beginners	2.33	1.03	3.00	1.05	5.40	0.98	2.80	0.65	28.3	6.12
2. Intermediates	2.25	1.00	2.93	0.54	6.82	1.41	2.87	0.87	22.9	5.18
3. Experts	4.25	1.98	3.80	1.00	7.57	0.81	2.86	0.86	20.0	3.07

group. In comparing means for the total number of options generated, Group 1 was highest.

Analyses of variance (ANOVAs) were used to compare all the groups on the dependent variables. At the first decision point, analyses of accuracy revealed significant differences among all the groups of nurses. A main effect was found for accuracy of decision at the first decision point, $F(2, 27) = 6.72$, $p < .05$, as well as the total overall prediction accuracy, $F(2, 27) = 3.37$, $p < .05$. A main effect was also found for confidence of decision rating, $F(2, 27) = 7.43$, $p < .05$. Main effects were also found in total options generated, $F(2, 27) = 4.96$, $p < .05$.

One-way ANOVAs were used to compare Groups 1 and 2, 2 and 3, and 1 and 3 on accuracy, confidence, difficulty, and number of options generated. Significant differences were found between Groups 2 and 3 (intermediates vs. experts) in the area of total overall decision accuracy, $F(1, 22) = 7.82$, $p < .05$. Comparing Groups 2 and 3 in accuracy at the first decision point, significant differences were also found, $F(1, 22) = 10.00$, $p < .05$. Although the mean accuracy was higher for Group 3 than Group 2, there were no significant differences found in levels of confidence or difficulty of decision rating. Comparing Groups 1 and 3 (advanced beginners vs. experts), there were significant differences found in number of options generated, $F(1, 12) = 6.31$, $p < .05$. Group 1 was less confident than Group 3 in their decisions, $F(1, 12) = 38.18$, $p < .05$. Groups 1 and 3 also showed significant differences in accuracy at the first decision point, $(1, 12) = 6.21$, $p < .05$, which was the most important decision point as determined by the expert panel.

DISCUSSION

The results of this study support similar findings in studies on NDM. More experienced nurses were found to be more accurate at decision making, predicting the two most important options each decision point. This finding was important as the experts who created the answer key to the scenario were from the same work environment as the nurse participants. The expert panel establishes a standard of practice creating a good evaluation tool of clinical judgment. Establishing a standard of practice would imply that it would be expected that less experienced nurses would not match as many of the answers as the more experienced nurses. The less experienced nurses are still in the process of gaining experience in their domain. This supports the findings in Klein and Peio's (1989) study of chess players in several ways: 1) experienced nurses were more accurate in predicting experts' would do next, than those who were less experience. 2) experienced nurses were able to consider the best option earlier than less experienced nurses. 3) experienced nurses generated significantly fewer options than less experienced nurses at each decision point. These findings in the number of options generated were also consistent with the RPD descriptions of expert intuitive decision making. This description supports the notion that experts generate fewer options in any given decision point. Those with less experience would generate more options, looking for the best answer, using a more analytical approach to their decisions.

A finding that was consistent with the Tabak et al. (1996) study of nurses was in the area of confidence. Tabak et al. found that nurses with more experience were more confident overall in their decisions than the less experienced nurses. This is consistent with the findings of Christensen-Szalanski and Bushyhead (1981) and Mahajan (1992) as well who found confidence levels higher in experts than non experts in domains other than nursing. It was also found that there were no significant differences in confidence in decisions between the intermediate and the more experienced nurses. This is interesting, as the intermediate group was found to have the lowest score in accuracy of decisions among all the groups. This again is consistent with Tabak et al., who found that less experienced nurses were less accurate at decision their decision of diagnosis.

Using a prediction paradigm as an objective measure to investigate NDM among critical care nurses does not capture all of the components of RPD. However, there are important ways to use a scenario-based prediction paradigm in clinical training and evaluation of critical care nurses (Klein & Peio, 1989; Cannon-Bowers, Burns, Salas, & Pruitt, 1998). For example the scenario based tool could be used as a think out loud exercise, with the nurses talking through what they might do in a given situation, using the more experienced nurses as consultants. This process may help the less experienced nurses identify areas in which more experience and education is needed. It will also aid in forming theoretical mental models to help them in situations they have yet to experience. The limitations to this approach would be that determination of the standard of practice might be based on how things have always been done as opposed to other options, which may be just as good but are not as familiar or done as often. This could stifle creativity and individuality in the workplace. However, tapping into the skills of a few experts in a particular area may capture the nuances of each hospital's practice in clinical care and treatments for patients. A scenario that is based on actual clinical situations, including outcomes of expert decisions, can be a powerful teaching tool. Benner et al. (1999) suggested that this is a good way to teach less experienced nurses to make better decisions as they are gaining experience and expertise.

Identifying experts within the work environment is essential when considering the importance of decision making in dynamic environments. Understanding of the RPD model and the impact of levels of situation assessment and awareness in intuitive decision making in critical care nursing is vital in mentoring and training less experienced nurses. It is also important to understand the differences in the levels of situation awareness present at differing levels of expertise. This knowledge should be incorporated in developing training and evaluations tools in clinical performance. It is frequently found that novice nurses are expected to make decisions at a much more proficient level than they are able to accomplish. Utilizing experts more in the clinical setting to support less experienced nurses in decision making will not only improve patient care but also increase nurses' level of job satisfaction.

A possible area of future study in NDM in critical care environments would be in collaborative decision making among nurses and physicians. Health care providers need to know whether the level of expertise possessed by the team of relevant decision makers will delay or hasten urgent treatment for patients in the ICU.

REFERENCES

Buch, G., & Diehl, A. (1982). An investigation of effectiveness of pilot judgment training. *Human Factors, 26,* 557–564.

Benner, P., Hooper-Kyria, P., & Stannard, D. (1999). Thinking-in-action and Reasoning-in-transition: An overview. In T. Eoyang (Ed.), *Clinical wisdom and interventions in critical care: A thinking-in-action approach* (pp. 1–22). Philadelphia: Saunders.

Benner, P., Tanner, C., & Chesla, C. (1992). From beginner to expert: Gaining a differentiated clinical world in critical care Nursing. *Advances in Nursing Science, 14,* 13–28.

Bogner, M. (1997). Naturalistic decision making in health care. In C. E. Zsambok & G. Klein (Eds.), *Naturalistic decision making* (pp. 61–69), Mahwah, NJ: Lawrence Erlbaum Associates.

Cannon-Bowers, J., Burns, J., Salas, E., & Pruitt, J. (1998). Advanced technology in scenario-based training. In J. Cannon-Bowers (Ed.), *Making decisions under stress: Implications for individual and team training* (pp. 365–374). Washington, DC: American Psychological Association.

Cellier, J., Eyrolle, H., & Marine, C. (1997). Expertise in dynamic environments. *Ergonomics, 40,* 28–50.

Christensen-Szalanski, J., & Bushyhead, J. (1981). Physician's use of probabilistic information in a real clinical Setting. *Journal of Experimental Psychology: Human Perception and Performance, 7,* 928–935.

Crandall, B., & Getchell-Reiter, K. (1993). Critical decision method: A technique for eliciting concrete assessment of indicators from the intuition of NICU nurses. *Advanced Nursing Science, 16,* 42–51.

Dreyfus, H., & Dreyfus, S. (1986). Five steps from novice to expert. *Mind over machine.* New York: The Free Press.

Klein, G., Calderwood, R., & Clinton-Cirocco, A. (1986). Rapid decision-making on the fireground. *Proceedings of the 30th Annual meeting of the Human Factors Society, 1,* 576–580.

Klein, G., & Peio, K. (1989). Use of a prediction paradigm to evaluate proficient decision-making. *American Journal of Psychology, 102,* 321–331.

Klein, G. (1997). A Recognition-primed decision (RPD) model of rapid decision-making. In C. E. Zsambok & G. Klein (Eds.), *Naturalistic decision making* (pp. 138–147). Mahwah, NJ: Lawrence Erlbaum Associates.

Lipshitz, R. (1997). Converging themes in the study of decision-making in realistic Settings. In C. E. Zsambok & G. Klein (Eds.), *Naturalistic decision making* (pp. 107–111). Mahwah, NJ: Lawrence Erlbaum Associates.

Mahajan, J. (1992). The overconfidence effect of marketing management predictions. *Journal of Marketing Research, 29*, 329–342.

Nyssen, A., & De Keyser, V. (1998). Improving training in problem solving skills: Analysis of anesthetists' performance in simulated problem situations. *Le Travail Humain, 61*, 387–401.

Orasanu, J., & Connolly, T. (1997). The reinvention of decision making. In C. E. Zsambok & G. Klein (Eds.), *Naturalistic decision making* (pp. 3–22). Mahwah, NJ: Lawrence Erlbaum Associates.

Spence, M., & Brucks, M. (1997). The moderating effects of problem characteristics on experts' and novices' judgments. *Journal of Marketing Research, 34*, 233–247.

Tabak, N., Bar-Tal, Y., & Cohen-Mansfield, J. (1996). Clinical decision making of experienced and novice nurses. *Western Journal of Nursing Research, 18*, 534–547.

Waag, W., & Bell, H. (1997). Situation assessment and decision-making in skilled fighter pilots. In C. E. Zsambok & G. Klein (Eds.), *Naturalistic decision making* (pp. 247–354). Mahwah, NJ: Lawrence Erlbaum Associates.

Wickens, C., & Flach, J. (1989). Information Processing. In E. Wiener & D. Nagel (Eds.), *Human factors in aviation* (pp. 127–145). San Diego, CA: Academic Press.

8

The Psychology of Economic Forecasting: A Possibility for Cooperation Between Judgment and Decision Making and Naturalistic Decision-Making Theories?

Henry Montgomery
Stockholm University

The prototypical investigation within the naturalistic decision-making (NDM) paradigm focuses on decision making in a fastly moving dynamic environment (see, e.g., Klein, Orasanu, Calderwood, & Zsambok, 1993). The time scale of these decisions is typically seconds or minutes.

This study concerns decision making and judgment in an environment in which long changes are in focus. The time scale in this case is years.

Also in difference to many other NDM researchers, I utilize notions from the so called judgment and decision making (JDM) tradition to understand how professionals make decisions in a naturalistic environment. More precisely, I am interested in how expertise (typically studied in NDM research) is combined with being a victim to cognitive biases and limitations (typically studied in NDM research). The experts in this study are economists whose task is to make forecasts of the economy of a country.

DATA

The Swedish Ministry of Finance delivers every year a bill—the so called budget bill—concerning the state budget of the coming fiscal year. The budget bill presents an analysis of the Swedish economy including various forecasts. The data used in this study are estimates of the Swedish gross national product (GNP) given in the budget bill during the period 1970 to 1998. I analyzed the data by comparing predicted GNPs with data on the actual

GNP presented in later budget bills. The predicted GNPs included predictions 1 year ahead (during the entire time period) and predictions 2 years ahead (for 1988–2000). The estimates of the actual GNP for a given year include estimates made at the end of the same year (implying that this estimate actually is a prediction but a safer prediction than those made in preceding years) and estimates made 1 year after the given year.

The forecasts rely partly on mathematical modeling but also, to an unknown extent, on a subjective integration of available data (Persson, 1995).

RESULTS

Figure 8.1 presents estimates of GNP plotted against year. An analysis of the data in Fig. 8.1 is given in Table 8.1.

The following observations can be made from Fig. 8.1 and Table 8.1.

1. Predicted GNPs tended to be higher than actual GNPs (compare filled and empty symbols in Fig. 8.1, and see col. Mx – My in Table 8.1). On the average, predicted GNPs are approximately 0.5 percentage units higher than actual GNPs. See D2B in Table 8.1, which shows sums of squared deviations between actual GNPs and predicted GNPs after correction for the mean

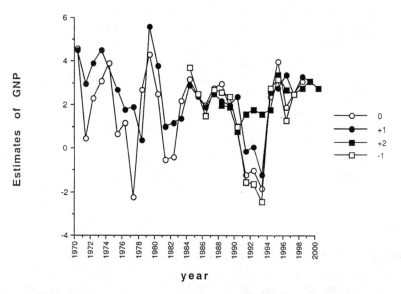

FIG. 8.1. Estimates of the gross national product (GNP) of Sweden made by the Swedish Ministry of Finance (0 = GNP estimated at the end of Year 0; +1 = GNP predicted 1 year ahead; +2 = GNP predicted 2 years ahead; –1 = GNP estimated 1 year later).

TABLE 8.1
Accuracy of Forecasts of the Swedish Gross National Product (GNP)

Period	x	y	Rxy	SDx/Sdy	Mx − My	D2A	D2B	D2C
1970 to 1998	y0	y + 1	.72	.82	−.66	2.00	1.54	1.48
1980 to 1998	y0	y + 1	.86	.74	−.39	0.90	0.75	0.72
1985 to 1997	y − 1	y + 1	.90	.67	−.41	1.00	0.84	0.67
1988 to 1997	y − 1	y + 2	.48	.42	−.53	4.08	3.32	3.28

Note. x = criterion (estimate of actual GNP); y = forecast (predicted GNP); y0 = estimate of GNP in year n made at the end of year n; y + 1 = estimate of GNP in year n made at the end of year $n − 1$; y + 2: estimate of GNP in year n made at the end of year $n − 2$; y − 1 = estimate of GNP in year n made at the end of year $n + 1$; rxy = correlation between x and y; SDx, SDy = standard deviation of x and y, respectively; Mx, My = mean of x and y, respectively; D2A = sum of $(x − y)**2$; D2B = sum of $(x − (y + C))**2$, where C = $Mx − My$; D2C: sum of $(x − (b*y + c))**2$, where b and c have been estimated by a linear regression analysis of y as a function of x.

difference between actual and predicted GNPs. As can be seen by comparing D2A with D2B in Table 8.1, the accuracy in the predictions would improve by approximately 20% by simply deducting a constant value from the predictions.

2. The correlation between actual GNPs and predicted GNPs 1 year ahead is rather high, particularly for 1980 to 1998. This means that the forecasts between actual and predicted are sensitive to ups and downs in the economy.

3. For predictions 2 years ahead, the correlation is moderately high ($r = .48$). It is interesting to note that if the 2-year forecasts had been the real world, then the deepest economic crisis in Sweden since the 1930s would not have existed, and instead there would be quite good economic times during this period (see the black squares in Fig. 8.1 for years 1991–1993).

4. There is a tendency for predicted GNPs to vary less than actual GNPs. This reduction in the variation of the predicted GNPs tends to be greater than would be expected from the regression of predicted against empirical values, particularly when estimates made 1 year later (Y − 1 in Table 8.1) are compared with predictions made 1 year ahead (Y + 1 in Table 8.1). If the reduced variation only was a result of taking regression into account, then SDx/Sdy should equal rxy (cf. Magnusson, 1967), but as can be seen in Table 8.1, this ratio is lower than rxy for estimates from 1980 and later. See also D2C in Table 8.1 that shows sum of squared deviation of predicted GNPs that have been corrected for nonoptimal change in variability and change in general level. The data in Table 8.1 suggest that when the most accurate GNP estimates (made 1 year later than the year of the GNP) are used as criterion, then an additional 20% improvement could be obtained as compared to correcting only for differences in mean level (D2B).

What kind of data is used as basis for the forecasts? The correlation between predicted GNPs and the actual GNPs of the preceding year is .59, which suggests that the GNP of the year preceding the year of the predicted GNP may be an important determinant of the predictions. However, in reality, the correlation between actual GNPs of successive years is only .29.

DISCUSSION

The results indicate that economical forecasts may rely on professional knowledge (cf. the relatively high correlations between actual and predicted GNPs). However, in addition, forecasts may be victim to psychological biases, such as (a) an optimism bias (cf. the difference in mean level of actual and predicted GNPs), (b) an anchoring and adjustment bias shown by insufficient adjustments of a prototypical value (cf. the relatively small variation of predicted GNPs), and (c) an availability bias (cf. the strong relation between predicted GNPs and actual GNP of the preceding year, which also can be interpreted as an anchoring and adjustment bias).

To the extent that systematic biases can be identified in economic forecasts, there will be a possibility to correct the forecasts for these biases. These results suggest two simple corrections: (a) subtracting a constant to compensate for an optimist bias and (b) multiplying a constant to compensate for an anchoring and adjustment bias. However, data from several countries are needed before any conclusions can be drawn with respect to general biases in economic forecasts on a national level. Nevertheless, these results suggest a possibility of using cognitive psychology for improving economic forecasts: a new challenge for economic psychology!

It may be concluded that studies of how professionals make decisions and judgments would benefit by taking into account how professional knowledge on one hand and psychological heuristics and biases on the other hand determine the experts' achievement. In other words, a happy marriage between NDM and JDM research.

REFERENCES

Klein, G. A., Orasanu, J., Calderwood, R., & Zsambok, C. E. (Eds.). (1993). *Decision making in action: Models and methods*. Norwood, NJ: Ablex.

Magnusson, D. (1967). *Test theory*. Reading, MA: Addison-Wesley.

Persson, M. (1995). Vad har vi för glädje av den nationalekonomiska vetenskapen? [To what use is the science of economics?]. *Ekonomisk Debatt, 23*, 610–623.

9

Modes of Effective Managerial Decision Making: Assessment of a Typology

Bernard Goitein
Edward U. Bond, III
Bradley University

Economics provides a normative model for business decision making: subjective expected utility (SEU) maximization (von Neumann & Morgenstern, 1947). SEU maximization requires specification of the possible alternative actions, identification of all consequences of each action, valuation of each of these consequences for the decision maker—that is, its utility—and estimation of the probability of each consequence. Computation of the SEU of each action is based on the sum of the utilities of the action's consequences weighted by their probabilities, permitting selection of the action with the maximum SEU.

The real-world context of managerial decision making makes it difficult to implement the SEU maximization model. Workplace decisions involve multiple players and require functioning within the framework of organizational goals and norms, two of eight factors that inhibit SEU maximization in field settings (Orasanu & Connolly, 1993). Managers must often contend with the other complicating factors discussed by Orasanu and Connolly: ill-structured problems; uncertain, dynamic environments; shifting, ill-defined, or competing goals; action/feedback loops; time stress; and high stakes.

Managerial decision makers with bounded rationality (Simon, 1957) confronted with these complicating factors are unlikely to conform to the normative model. Rather, experienced managers would be expected to develop modes of decision making that are effective in dealing with the challenges of particular complicating factors. Researchers of naturalistic decision making (NDM) in field settings have concluded that these deci-

TABLE 9.1
Decision-Making Modes

Mode	Emphasis
Pragmatic	Sensing and seizing opportunities
Systems	Modeling organizational effectiveness to enable intervention
Empiricist	Finding evidence-based solutions
Value focused	Articulating and pursuing common goals
Structuralist	Determining responsibility and ensuring procedural control
Multiparty	Negotiating agreements
Imaginative	Fostering creativity

sions are made in diverse and qualitatively distinct ways (Lipshitz, 1993). Kinston and Algie (1988) identified seven management decision-making modes that they encountered in their consulting practice: seven "distinctive and formally coherent approaches to decision and action" (p. 118) that they viewed as effective modes for managerial decision making (see Table 9.1).

We discuss these seven modes from the perspectives of the managerial decision making and the NDM empirical literatures and consider how these address the complicating factors enumerated by Orasanu and Connolly (1993). We collected and analyzed data from 325 decision makers and assessed the extent to which they recognized these seven managerial decision-making modes in their workplaces.

MODES OF DECISION MAKING

Pragmatist

The pragmatist mode's focus on immediate action is particularly useful for decisions that must be addressed under time stress (Orasanu & Connolly, 1993). For the pragmatist decision mode, decision making is an opportunity to act. Effective decision making is about motivated individuals who sense and seize opportunities because delayed action may mean that opportunities are lost. Brunsson (1985) is an articulate proponent of this mode for managerial decision making. Brunsson questioned several components of traditional decision-making prescriptions, for example, specification of all possible courses of action because they delay necessary action and can erode commitment among adherents to nonselected courses of action.

The pragmatist mode recommends quick action "from the gut," responding to opportunities as they are sensed. Such sensing is apparent in a study of decision making with risky gambles (Bechara, Damasio, Tranel, & Damasio, 1997) in which participants responded to the gambles with changes in skin conductance, "feelings," and "hunches" before they could articulate the risks that the gamble posed. Quick action is found in the behavior of senior managers given a business case by Isenberg (1986), who began action planning, on average, 40% of the way into presentation of the case. In a questionnaire study of managerial decision making, Hodgkinson and Sadler-Smith (2000) identified a distinct management decision-making mode that they termed "intuition," measured by items such as "I prefer chaotic action to orderly inaction," items that are pragmatic in character. The pragmatist emphasis on the value of improvisation over inaction is a position that has been supported by Eisenhardt and Tabrizi (1995) and Moorman and Miner (1998).

Entrepreneurs thrive on identifying and seizing opportunities; therefore, they would be expected to use the pragmatist decision-making mode. Kinston and Algie (1988) emphasized that the pragmatist concern for immediate action means that the Pragmatist will look for actions that build on existing strengths and available or easily acquired resources (p. 122). Consistent with this characterization are the entrepreneurs studied by Shane (2000) who rely on their own resources and knowledge (e.g., of a market segment) to identify unique opportunities from the same technological advance.

The NDM literature offers models comparable to the pragmatist that are used by decision makers under time stress. Firefighters, for example, are concerned with "finding actions that were workable, timely, and cost effective" (Klein, 1993, p. 139). Montgomery's (2001) perspective model in which evaluative judgments are made quickly and Klein's (1997) recognition-primed decision (RPD) model's "simple match" are comparable to the pragmatic decision mode. The simple match is used when the decision maker can match the decision situation to a prototype in terms of its key features (plausible goals, relevant cues, expectancies, and typical actions) and reacts accordingly (Klein, 1993, 1997).

Systems

The presence of "action feedback loops" (Orasanu & Connolly, 1993) presents a more complex case than that addressed by the pragmatist mode, and Klein's (1997) RPD model allows for a decision process that goes beyond the simple match in which "the course of action is deliberately assessed by conducting a mental simulation to see if the course of action runs into any difficulties" (p. 285). Other NDM researchers (e.g., Serfaty, MacMilan, Entin, & Entin, 1997) have also highlighted the importance of mental models, with Endsley's (1997) situation awareness model "Level 3" requiring a mental model to project "future states of the system" (p. 274).

To forestall unintended consequences of managerial action, Kinston (1994; Kinston & Algie, 1988) distinguished the systems decision mode that avoids looking at actions in isolation and constructed a model of how these actions fit together within the larger organizational system and its relations with the environment. The systems mode models organizational effectiveness to enable appropriate intervention. The organizational literature has long advocated open systems thinking (e.g., Katz & Kahn, 1966). Senge (1990) discussed managerial decision making from a systems perspective, one that looks for interrelations and processes and avoids solutions that only treat symptoms. Dean and Sharfman (1996) found that successful organizational decisions depend on "implementation quality," that is, consideration of the many related decisions and tasks that must be coordinated with the chosen course of action (p. 378).

Empiricist

For workplace decision makers confronted with uncertain, dynamic environments (Orasanu & Connolly, 1993), Kinston and Algie (1988) distinguished the empiricist mode, which relies on comprehensive analysis of data to allow sufficient understanding of the problem to eliminate uncertainty and reveal a solution. The empiricist mode approaches decision making as a problem-solving task. The empiricist values knowledge, necessary for the generation of evidence-based solutions. The empiricist advises that the solutions be pilot tested so that data can be collected about the inevitable glitches that accompany implementation, and uncertainties can be reduced before widespread introduction.

In the managerial decision-making literature, March and Heath (1994) discussed a decision engineering effort to improve the quality of decision making "to exploit what is known" (p. 238) which analyzes data to identify successful solutions. Hodgkinson and Sadler-Smith (2000) found an analytical cognitive style for managerial decisions that is comparable to the empiricist approach. Dean and Sharfman (1996) described the empiricist mode under the rubric of "procedural rationality," defined as "the extent to which the decision process involves the collection of information relevant to the decision, and the reliance upon analysis of this information in making the choice" (p. 373), and found that use of this mode is associated with successful organizational decision outcomes.

The NDM literature reveals decision-maker concern for critical unknowns in the hourglass model of Serfaty et al. (1997) that is empiricist in nature, as is the data-driven situation awareness described by Endsley (1997) and the diagnostic activities "initiated in response to uncertainty" of the revised RPD model described by Klein (1997, p. 290).

Value-Focused

A workplace decision may be characterized by shifting, ill-defined, or competing goals (Orasanu & Connolly, 1993). The value-focused (or "rationalist"; Kinston & Algie, 1989, p. 118) management decision-making mode focuses on identifying and achieving those objectives valued by the manager. The value-focused decision mode finds that goal clarification is essential, asking, "If you don't know what you want to achieve, how can you ever achieve it?" The value-focused mode emphasizes articulating and pursuing common goals for the organization, unifying staff in their pursuit. The value-focused mode finds that once objectives are set and their relative priorities specified, it is straightforward to work out a plan for achievement of the desired outcomes.

Decision analyst Keeney (1992, 1994) shares the value-focused concerns in what he termed "value-focused" thinking for effective management decision making. Keeney (1992, 1994) argued that one should begin the decision process by clarifying the full range of values pertinent to the decision, asking questions such as the following: What are our goals? What do we care about? What should we care about? Which goals have a high priority, which low? Once the goals are understood, the means for their achievement can be specified.

The organizational theory literature has long emphasized the importance of goals, with goal direction usually cited as one of the defining characteristics of organizations (e.g., Daft, 2001). Dean and Sharfman (1996) found that organizational decisions that focus on organizational objectives and member openness about their preferences are more successful, on average, than decisions that emerge from internal politics.

In the NDM literature, Brehmer (1990) viewed dynamic decision making as "a matter of providing direction ... and monitoring progress towards some goal" (p. 277). Endsley (1997) spoke of dynamic goal selection and goal-driven situation awareness, and Montgomery's (1993) "dominant alternative" is the option that is as good or better on all criteria than are the competing options.

Structuralist

Workplace decisions require functioning within the framework of organizational goals and norms (Orasanu & Connolly, 1993). The structuralist decision mode relies on organizational rules and consistently applied procedures and standards. The structuralist mode is concerned that actions conform to standards developed by competent authorities. The structuralist mode stresses assignment of responsibility to a position filled by a properly qualified expert. Levine, Higgins, and Choi (2000) identified a comparable decision-making mode termed a "prevention focused" strategic orientation focused on "security, safety, and responsibility" (p. 91).

Fischhoff (1984) described a bureaucratic setting standards decision-making mode that he contrasted with case-by-case decision making. Standard setting invokes a noncompensatory decision rule in which certain actions are ruled out if they exceed a standard (e.g., building a facility with less than a 25-ft setback from the property line), regardless of any compensating benefits derived from the action (e.g., jobs at the facility or return to investors). In the organizational decision-making literature, Hickson, Butler, Cray, Mallory, and Wilson (1986) praised this mode as "reasonable" and referred to it as the "rationality of control" that follows the "rules of the game" (p. 250). March and Health (1994) identified this mode as the "rule following" (p. 57) approach to organizational decision making in which obedience to rules and conformance to standards is central.

The NDM literature provides examples of rule-based decision making in which experts deliberately apply rules acquired from professional training or experience of the type, "If condition C holds, then take Action A" (Yates, 2001, p. 22). Structuralist deference to experts is the basis for the secondary decision modes identified by Yates (2001) in which decision makers rely on others (experts) to help make the decision.

Multiparty

Workplace decisions involve multiple players (Orasanu & Connolly, 1993). In large organizations, the units frequently pursue their own departmental agenda without concern for working effectively with the other departments in the bureaucracy (Hickson et al., 1986). Handling the conflicts between units in a multiparty decision context requires negotiation, as advocated by the multiparty (or "dialectical"; Kinston & Algie, 1989, p. 123) decision-making mode. The multiparty mode resolves conflicts by encouraging negotiations between the parties in an agreement that satisfies both sides' needs. It emphasizes confronting divisive issues rather than ignoring them or "sweeping them under the rug" so that a negotiated agreement may be reached. In the organizational decision-making literature, Hickson et al. (1986) praised this mode as reasonable, representing an "interest-accommodating rationality" (p. 250); and March and Heath (1994) offered extensive discussion of multiparty decision making. A similar logic supports the bargaining literature's emphasis on win–win negotiations (Fisher & Ury, 1991).

Imaginative

When a task is ill-structured (Orasanu & Connolly, 1993), creative new options that have not previously been conceived may be needed. The imaginative decision-making mode values personal growth. Personal growth encourages imagination and creativity, enabling the imaginative decision maker to generate new and better options (Kinston, 1994; Kinston & Algie, 1988). Inspiration and inner commitment to develop one's potential in a high-trust organizational setting are sources of innovation for the imaginative decision-making mode.

In the organizational decision-making literature, Langley, Mintzberg, Pitcher, Posada, and Saint-Macary (1995) advocated looking at the decision maker with the emphasis on the "making" and referred to an "insightful" approach to organizational decision making in which the decision maker relies on "intuitive sensibilities" to restructure thinking and create entirely new options (pp. 268–269). March and Heath (1994) discussed a decision engineering effort to improve the quality of decision making that argues for exploration of the unknown, not just the empiricist's exploitation of the known, because the empiricist fails to develop new directions and capabilities.

METHOD AND RESULTS

Data for this study were collected from working managers, engineers, and accountants enrolled in evening MBA classes in 1998 and 1999. These students solicited further participation in the study from their supervisors. The 1998 sample totaled 164 respondents: 108 men, 34 women, and 22 for whom gender data were not available. The 1999 sample of 161 respondents included 100 men, 46 women, and 15 for whom gender data were not available.

Participants in the study were asked to read a two-page summary of the seven modes of decision making derived from the work of Kinston (1994). Respondents rated the appeal of each mode as an ideal for decision making on a scale ranging from 1 (*Does not appeal to me*) to 7 (*Very strongly appeals to me*).

On a separate page, respondents were asked to indicate the relation between the seven decision-making modes and their own decision behavior. The scale for responses ranged from 1 (*Does not describe me at all*) to 7 (*Describes me very well*). Managers in the 1999 sample in addition used a comparable scale to rate how well the decision-making modes described the decision behavior of their supervisors and of the firms for which they worked.

The self-rating responses for each of the two samples were subjected to principal components analyses extracting factors with eigenvalues greater than 1.0. The solutions revealed seven factors in each of the two samples. Table 9.2 reports factor loadings for the seven factors after varimax rotation.

To explore the modes used by these workplace decision makers, we computed dummy codes from the 1999 data to represent whether a decision-making mode characterized the respondent, supervisor, and firm.

TABLE 9.2

Factor Analysis of Decision-Making Modes

Item	Pragmatist	Systems	Empiricist	Value Focused	Structuralist	Multiparty	Imaginative
Pragmatist (ideal)	.926 ***.919***						
Pragmatist (actual)	.927 ***.901***				***-.110*** -.127		
Systems (ideal)	***-.134***	.925 ***.897***					
Systems (actual)		.922 ***.908***					
Empiricist (ideal)	***.123***	.112	.913 ***.899***				-.104
Empiricist (actual)			.924 ***.905***				
Value focused (ideal)				.910 ***.896***		***-.118***	
Value focused (actual)		***.107***		.915 ***.886***		***.150***	
Structuralist (ideal)					.918 ***.903***	–	
Structuralist (actual)	-.106	-.121			.901 ***.882***	***-.144***	
Multiparty (ideal)				.129	.189	.858 ***.903***	***.189***
Multiparty (actual)		.115		***.103***		.912 ***.894***	***.144***
Imaginative (ideal)		.137					.911 ***.910***
Imaginative (actual)							.916 ***.927***

Note. 1998 loadings are reported in plain text; 1999 loadings are reported in bold italic. Cross-loadings less than .100 suppressed.

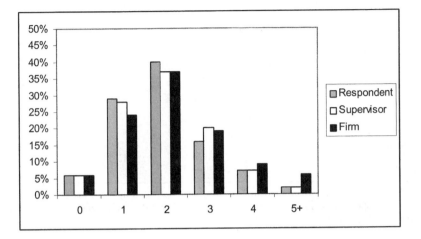

FIG. 9.1. Number of modes characterizing respondents, supervisors, and firms.

Using responses to the question of the degree to which each decision-making mode described decision behavior, we coded responses of 6 or 7 on the 7-point scales as 1 to represent a decision mode that characterized the respondent, supervisor, or firm. Responses of 1 through 5 were coded as 0.

Figure 9.1 reveals that more than 90% of the workplace decision makers rated themselves, their supervisors, and their firms as conforming to at least one of the seven modes. It also shows that fewer than 30% were characterized by only one mode, with a majority relying on two or three modes. Fewer than 10% perceived themselves to be characterized by many (i.e., five or more) modes.

Table 9.3 reports the frequency with which managers cited the specific decision modes as characteristic of themselves, their supervisors, and their firms. Each decision mode characterized the decision behavior of at least 10% of workplace decision makers, supervisors, and firms. A majority of firms were characterized as relying on the value-focused and the structuralist decision-making modes.

DISCUSSION

Our results show that workplace decision makers use each of the seven qualitatively distinct decision-making modes. Each mode emphasizes a different element of effective managerial decision making: quickly seizing opportunities, systems thinking, use of data, objectives focus, using the formal organization, negotiating agreements, and creativity. The factor analyses

TABLE 9.3
Distribution of Characteristic Decision-Making Modes

Decision-Making Mode	Respondents	Supervisors	Firms
Pragmatic	20	23	25
Systems	19	19	22
Empiricist	33	31	26
Value focused	45	48	66
Structuralist	43	39	52
Multiparty	11	15	12
Imaginative	24	27	16

Note. $N = 161$. Numbers represent percentage characterized by each mode. Columns total more than 100 because most respondents, supervisors, and firms were characterized by more than one decision making mode.

revealed these seven distinct management decision-making modes in two samples of workplace decision makers. These decision-making modes are recognized as effective by management scholars, with proponents of each management decision-making mode to be found (e.g., the value-focused mode of strategic planning proponents, the systems mode among open systems theorists), and empirical research of management decision-making modes (e.g., Dean & Sharfman, 1996) has found positive impacts of the modes on decision outcomes.

The ratings reveal that although some workplace decision makers rely on just one mode, a majority of decision makers, their supervisors, and their firms use two or three modes. Several NDM models of decision making in field settings incorporate more than one mode: Klein's (1997) RPD model, for example, includes three (pragmatic, systems, and empiricist) decision modes that may be used depending on the task. Some research (e.g., Khatri & Ng, 2000; Orasanu & Fischer, 1997) has shown that particular decision modes may not work equally well in all conditions. Future research is needed to elaborate the indications and counterindications for each mode.

A limitation of this study is that its examination of workplace decision-making modes is restricted to one culture. Yates and Lee (1996) reported that preferred decision-making mode varies by culture; therefore, future research will need to study cultural variation in the prevalence and popularity of these seven workplace decision-making modes.

REFERENCES

Bechara, A., Damasio, H., Tranel, D., & Damasio, A. (1997). Deciding advantageously before knowing the advantageous strategy. *Science, 275,* 1293–1295.

Brehmer, B. (1990). Strategies in real-time, dynamic decision making. In R. Hogarth (Ed.), *Insights in decision making: A tribute to Hillel J. Einhorn* (pp. 262–279). Chicago: University of Chicago Press.

Brunsson, N. (1985). *The irrational organization: Irrationality as a basis for organizational action and change.* Chichester, England: Wiley.

Daft, R. L. (2001). *Organization theory and design.* Cincinnati, OH: South-Western College Publishing.

Dean, J. W., & Sharfman, M. P. (1996). Does decision process matter? A study of strategic decision making effectiveness. *Academy of Management Journal, 39,* 368–396.

Eisenhardt, K. M., & Tabrizi, B. N. (1995). Accelerating adaptive processes: Product innovation in the global computer industry. *Administrative Science Quarterly, 40,* 84–110.

Endsley, M. R. (1997). The role of situation awareness in naturalistic decision making. In C. E. Zsambok & G. Klein (Eds.), *Naturalistic decision making* (pp. 269–283). Mahwah, NJ: Lawrence Erlbaum Associates.

Fischhoff, B. (1984). Setting standards: A systematic approach to managing public health and safety risks. *Management Science, 30,* 823–843.

Fisher, R., & Ury, W. (1991). *Getting to yes.* New York: Penguin.

Hickson, D. J., Butler, R. J., Cray, D., Mallory, G. R., & Wilson, D. C. (1986). *Top decisions.* San Francisco: Jossey-Bass.

Hodgkinson, G. P., & Sadler-Smith, E. (2000, August). *Complex or unitary? A critique and empirical reassessment of the Allison-Hayes Cognitive Style Index.* Paper presented at the Annual Meeting of the Academy of Management, Toronto.

Isenberg, D. (1986). Thinking and managing: A verbal protocol analysis of managerial problem solving. *Academy of Management Journal, 29,* 775–788.

Katz, D., & Kahn, R. L. (1966). *The social psychology of organizations.* New York: Wiley.

Keeney, R. L. (1992). *Value-focused thinking.* Cambridge, MA: Harvard University Press.

Keeney, R. L. (1994). Creativity in decision making with value-focused thinking. *Sloan Management Review, 35*(4), 33–41.

Khatri, N., & Ng, H. A. (2000). The role of intuition in strategic decision making. *Human Relations, 53,* 57–86.

Kinston, W. (1994). *Strengthening the management culture.* London: Sigma Centre.

Kinston, W., & Algie, J. (1988). Seven distinctive paths of decision and action. *Systems Research, 6,* 117–122.

Klein, G. (1993). A recognition-primed decision (RPD) model of rapid decision making. In G. A. Klein, J. Orasanu, R. Calderwood, & C. E. Zsambok (Eds.), *Decision making in action: Models and methods* (pp. 138–147). Norwood, NJ: Ablex.

Klein, G. (1997). The recognition-primed decision (RPD) model: Looking back, looking forward. In C. E. Zsambok & G. Klein (Eds.), *Naturalistic decision making* (pp. 285–292). Mahwah, NJ: Lawrence Erlbaum Associates.

Langley, A., Mintzberg, H., Pitcher, P., Posada, E., & Saint-Macary, J. (1995). Opening up decision making: The view from the black stool. *Organization Science, 6,* 260–279.

Levine, J. M., Higgins, E. T., & Choi (2000). Development of strategic norms in groups. *Organizational Behavior and Human Decision Processes, 82,* 88–101.

Lipshitz, R. (1993). Converging themes in the study of decision making in realistic settings. In G. A. Klein, J. Orasanu, R. Calderwood, & C. E. Zsambok (Eds.), *Decision making in action: Models and methods* (pp. 103–137). Norwood, NJ: Ablex.

March, J. G., & Heath, C. (1994). *A Primer on decision making: How decisions happen.* New York: Free Press.

Montgomery, H. (1993). The search for a dominance structure in decision making: Examining the evidence. In G. A. Klein, J. Orasanu, R. Calderwood, & C. E. Zsambok (Eds.), *Decision making in action: Models and methods* (pp. 182–187). Norwood, NJ: Ablex.

Montgomery, H. (2001). Reflective versus nonreflective thinking: Motivated cognition in naturalistic decision making. In E. Salas & G. Klein (Eds.), *Linking expertise and decision making* (pp.
159–170). Mahwah, NJ: Lawrence Erlbaum Associates.

Moorman, C., & Miner, A. S. (1998). Organizational improvisation and organizational memory.
Academy of Management Review, 23, 698–723.

Orasanu, J., & Connolly, T. (1993). The reinvention of decision making. In G. A. Klein, J. Orasanu,
R. Calderwood, & C. E. Zsambok (Eds.), *Decision making in action: Models and methods* (pp.
3–20). Norwood, NJ: Ablex.

Orasanu, J., & Fischer, U. (1997). Finding decisions in natural environments: The view from the
cockpit. In C. E. Zsambok & G. Klein (Eds.), *Naturalistic decision making* (pp. 343–357).
Mahwah, NJ: Lawrence Erlbaum Associates.

Senge, P. (1990). The leader's new work: Building learning organizations. *Sloan Management Review, 32*(1), 1–17.

Serfaty, D., MacMillan, J., Entin, E. E., & Entin, E. B. (1997). The decision making expertise of battle
commanders. In C. E. Zsambok & G. Klein (Eds.), *Naturalistic decision making* (pp. 233–246).
Mahwah, NJ: Lawrence Erlbaum Associates.

Shane, S. (2000). Prior knowledge and the discovery of entrepreneurial opportunities. *Organization Science, 11*, 448–469.

Simon, H. A. (1957). *Models of man.* New York: Wiley.

von Neumann, J., & Morgenstern, O. (1947). *Theory of games and economic behavior.* Princeton,
NJ: Princeton University Press.

Yates, J. F. (2001). "Outsider:" Impressions of naturalistic decision making. In E. Salas & G. Klein
(Eds.), *Linking expertise and decision making* (pp. 9–33). Mahwah, NJ: Lawrence Erlbaum Associates.

Yates, J. F., & Lee, J. W. (1996). Chinese decision making. In M. H. Bond (Ed.), *Handbook of Chinese
psychology* (pp. 338–351). Hong Kong: Oxford University Press.

10

Superior Decision Making as an Integral Quality of Expert Performance: Insights Into the Mediating Mechanisms and Their Acquisition Through Deliberate Practice

K. Anders Ericsson

Florida State University

Knowledge within the academic discipline of general psychology to this day remains organized around a number of human activities and abilities, including perception, memory, concept formation, decision making, reasoning, and problem solving. Most psychological researchers choose to focus on one of these broad areas and design research programs that search for general mechanisms and laws within the area's associated systems.

Most people studying naturalistic decision making (NDM) would likely describe themselves as researchers of judgment and/or decision making. In the beginning of my career, I described myself as a researcher of problem solving and memory. However, subsequent studies on the superior performance of experts in chess, music, and sports has led me to question the merits of such traditional approaches, that is, attempts to study psychological activities in their purest, most abstract and generalized form. In this chapter, I explore the salient issues surrounding my study of expertise in many divergent domains and the implications this research has on how we as researchers view and conduct our psychological research efforts.

In the last decades, numerous influential researchers in psychology have criticized the search for basic capacities of memory and attention. These researchers have subsequently proposed a return to the study of memory phenomena in everyday life (cf. Cohen, 1996; Gibson & Pick, 2000; Neisser, 1976, 1982). In the domain of decision making, this influential approach has led to NDM. In this presentation, I proceed even further with this line of thought and question whether human performance in everyday life—such as memory, problem solving, reasoning and decision making—permits researchers to distinguish different independent aspects of activities in a useful manner. In skilled and expert performance, these activities are so tightly interrelated that the principal challenge is to attain optimal integration of different activities and in particular, to maintain this integration over time during the development of vastly improved expert performance.

OUTLINE OF CHAPTER

I begin this chapter by sketching the historical motivation for the original scientific approach in psychology that focused on uncovering and describing basic nonmodifiable human capacities and processes. I briefly discuss the emergence of information-processing models of cognitive processes and how these models were based on declarative and procedural knowledge, fixed limits of memory capacities, and elementary information processes. I then briefly review recent empirical evidence on the effects of practice and adaptation and demonstrate that under certain external conditions, biological systems—including humans—are capable of dramatic change, including changes all the way down to the cellular level.

The remarkable modifiability of human performance raises issues for finding basic invariant phenomena that can inform researchers about the potential (and limits) of human achievement. One approach, that is, the expert performance approach (Ericsson & Smith, 1991), searches for reliably superior performance by experts and then attempts to capture and reproduce that performance with representative tasks in the laboratory. The structure of the intact performance is assessed and examined using standard process tracing techniques such as protocol analysis of verbal reports of thinking. Given that most forms of expert performance involve the rapid generation and/or selection of superior courses of action, researchers can learn about the cognitive mechanisms underlying decision making and determine how these mechanisms are integrated within the overall structure that mediates expert performance. I conclude with a discussion of how these integrated structures and mechanisms are acquired through the application of deliberate practice.

HISTORICAL BACKGROUND TO THE TRADITIONAL APPROACH SEARCHING FOR BASIC ABILITIES

Pioneering psychologists in the 19th century were inspired by the successful approach of the natural sciences during the 17th through 19th centuries. Motivated by such examples as Newton's discovery and mathematical description of his three laws in mechanics, these psychologists searched for general laws and the simplest observable mental phenomena that could be elicited in a controlled laboratory context. The first attempts to study decision making in a laboratory were conducted at the beginning of the 20th century and involved collecting detailed introspective reports on the cognitive processes mediating simple judgment tasks such as determining which of two weights is heavier. Unfortunately, these introspective investigations did not always produce results on which researchers could agree, and when such disputes over introspection could not be settled empirically, scientists rejected the validity of introspective data in the study of higher level cognition (see Ericsson & Simon, 1993, for the important distinction between this type of analytic introspection and think-aloud reports).

During the behaviorist reign that dominated the first half of the 20th century, researchers rejected the earlier focus on introspective analysis of complex experience and focused on discovering general laws of learning. (In a similar shift, economists also turned away from empirical studies of descriptions of decision-making processes and turned to formal analysis of choice under uncertainty; Goldstein & Hogarth, 1997). Researchers of decision making developed and optimized mathematical analyses for discovering the best alternative or strategy that was general and independent of particular decision-making situations. In the 1950s, it became increasingly clear that humans did not perform "rationally" on these abstract decision-making tasks and that the proposed mathematical models did not provide plausible mechanisms for how humans made their decisions (Simon, 1955).

With the emergence of cognitive psychology and human information-processing models in the second half of the 20th century, researchers began collecting data on the cognitive processes that mediated performance on tasks—such as decision making involving choices between well-defined alternatives. Pioneering studies using protocol analysis of decision making (cf. Montgomery & Svensson, 1976; Payne, 1976) have shown that the challenge with these types of tasks involved the integration of information about each choice option to find the alternative with the highest overall appeal to the individual making the decision. More specifically, participants in these tasks had problems comparing unfamiliar alternatives with many aspects within the constraints of their limited capacity of short-term memory (STM).

The difficulty of making decisions among unfamiliar choices in the laboratory, however, did not seem to characterize decision making in everyday life where memory problems are rarely experienced. In their influential theory of expertise, Simon and Chase (1973) reconciled this discrepancy between laboratory and everyday life and described how skilled and expert performance could be developed without violating invariant memory constraints and the fixed speed of basic processing. With experience, individu-

als were shown to build associations in memory between previously encountered situations (represented as combinations of chunks) as well as
the appropriate actions that should be taken in these situations. During
work-related performance and competitions, these experienced individuals
would simply use the situational cues to retrieve the associated action from
memory and bypass the complex generation and decision-making processes and thus relieve the demand for memory storage in STM.

Simon and Chase (1973) argued that the acquisition of expertise was
closely linked to the gradual accumulation of patterns and knowledge gained
from much extended experience in the domain. For example, Simon and
Chase found that chess experts had to spend at least ten years playing—
roughly comparable to the time taken to master a language with a vocabulary of 50,000 to 100,000 words—to attain international levels of performance.

Contrary to Simon and Chase's (1973) predictions of the benefits of extended experience, the accuracy of many expert decision makers does not
exceed the accuracy of simple decision rules based on statistics (Dawes,
Faust, & Meehl, 1989), nor are experts typically clearly superior to less experienced individuals (Camerer & Johnson, 1991; Shanteau, 1988). Hence, a
vast amount of knowledge, experience, and education is not always closely
associated with accuracy of judgments. In sum, the number of years of
work and experience in a domain has been found to be a poor predictor of
attained performance after the first few months to a year of initial experience in the domain (Ericsson & Lehmann, 1996).

In the domain of decision-making research, investigators (Klein, Orasanu, Calderwood, & Zsambok, 1993) started to question whether the selection among alternatives in well-defined choice tasks, such as those necessary for the application of mathematical and statistical decision models,
captures the essence of decision-making expertise. In everyday life, it is
rare to encounter decision-making situations in which all alternatives are
immediately available and described by the same set of dimensions. NDM
typically starts with a problem (such as finding an apartment or buying a
computer) and then proceeds into a more intense phase of attempts to generate alternatives as well as collect valid information about the most promising options. Everyday decisions frequently have to be made with incomplete and biased information under a time pressure constraint.

A number of theoretical models have been proposed to account for NDM
(Lipshitz, 1993). Most of these models, such as image theory (Beach, 1993)
and explanation-based decisions (Pennington & Hastie, 1993), are based on
the description of the complex constructive processes involved in comprehension (Kintsch, 1998). As an alternative, Montgomery (1993) proposed
that individuals—even experts—try to avoid making detailed comparisons of
alternatives in working memory and seek a representation of the alternatives that allow the superior alternative to emerge. Perhaps the only new

model that is closely related to the traditional information-processing models is Klein's (1993) recognition-primed decision (RPD) model of rapid decision making. This model is explicitly related to Chase and Simon's (1973) model for cued retrieval of actions (Calderwood, Klein, & Crandall, 1988) and Simon's (1955) notion of "satisficing," that is, finding an acceptable rather than best alternative.

In sum, during a century of laboratory research, the focus of decision-making research has moved toward the study of the processes mediating the selection of the most appropriate action in a given situation. However, the selection of the most appropriate options appears to be an attribute of virtually all goal-directed activities induced by laboratory tasks. Hence, laboratory studies on decision making focused for some time on the special strategies involved in selecting the best option among multiattribute choices within the limited memory capacity of STM and working memory. However, these selection tasks may not capture the essence of decision making in everyday life, and the memory problems may be an artifact of the unfamiliar alternatives. In natural decision-making settings, individuals reach an acceptable and seemingly effortless level of performance (typically within a few months to a year), at which point any increases in knowledge and experience are no longer reliably associated with further improvements in decision-making performance. This lack of predictable benefit from further experience and instruction has led some investigators to argue that individual differences in performance at the expert level must be determined by individual differences in fixed abilities and capacities that cannot be influenced by training and experience.

In the next section, I question the general claim that adults cannot change their basic capacities. I describe evidence for the dramatic modifiability of virtually every aspect of the human ability and bodily characteristics. I then describe an approach for studying reproducibly superior performance in everyday life, its structure and its mediating acquired mechanisms. When we as researchers fully understand the mechanisms that allow experts to reliably produce their superior performance, we will also understand how experts can consistently select the best actions (i.e., decision making).

THE POWERFUL EFFECTS OF SOME TYPES OF PRACTICE

Some 20 years ago, Chase and I (see Ericsson, Chase, & Faloon, 1980) were interested in whether it was possible to increase what was at that time commonly believed to be the primary constraint on information processing, namely the limited capacity of STM. The standard test of STM involves presenting a series of digits and asking for immediate recall of the digits. The average performance of college students for such an activity was found to be about seven digits, the equivalent of a local phone number.

Ericsson et al.'s (1980) study began by establishing the digit memory of several college students before the start of the training and verifying that recall performance was normal (i.e., limited to around seven digits). After 50 hr of practice, all of the trained students increased their memory performance by 200% to over 20 digits. After 200 to 400 hr, two of the students improved their recall by more than 1,000% to over 80 digits. Many other investigators have replicated these dramatic improvements as a result of practice. It was discovered that the observed 1,000% improvement in digit recall relied on mechanisms similar to those that expert performers in other domains acquire to expand their ability to store and process relevant information in working memory (Ericsson & Lehmann, 1996). Adults attaining skilled performance in reading and text comprehension, for example, acquire similar mechanisms for expanding their working memory through storage in long-term memory (Ericsson & Kintsch, 1995).

The effects of specialized training are not limited to memory and other cognitive capacities. With specific practice, speed of performance in many performance domains considerably increases, and even physiological capacities can be shown to markedly increase (Ericsson, Krampe, & Tesch-Römer, 1993). The record for the number of consecutive push-ups illustrates one striking example of the far-reaching effects of practice. High-school students in physical education classes can perform, on the average, about 20 push-ups in a row (Knapp, 1947). However, after specialized practice, one individual was able to complete over 6,000 push-ups and set the first official record in 1965. The Guinness Book of Records (McWhirter & McWhirter, 1975) has shown that this record has been regularly broken and now the event has been slightly changed, and all push-ups have to be completed within 24 hours. The current record (Young, 1997) is over 26,000 push-ups—this value represents an average of nearly a completed push-up every 3 sec for 24 hr straight.

These examples, as well as numerous other findings, raise doubt about fixed innate capacities that limit an individual's ability to reach the highest levels of performance. Reviews of expert performance (Ericsson, 1996; Ericsson & Lehmann, 1996) have uncovered no evidence of innate talents that are critical to expert performance that cannot also be altered or circumvented with extended practice. There is, however, at least one well-documented exception to this rule: height and body size. Height and body size are generally related to success in many types of sports activities. It is obviously an advantage for a player to be taller in basketball and shorter in gymnastics, and additionally, there is no known practice activity that can increase the height of humans. When one excludes height-related characteristics, recent reviews (Ericsson & Lehmann, 1996; Howe, Davidson, & Sloboda, 1998) have not found any accepted evidence that innate characteristics are required for healthy adults to attain elite performance. When appropriately designed training is maintained with full concentration on a regular basis for weeks, months, or even years, there appears to be no firm empirical evidence for innate capacities besides physical size that limits the attainment of high-level performance.

Reconciling Great Modifiability With Stable Adult Characteristics

The fact that adults are capable of dramatically improving their performance in various domains is not inconsistent with the finding that adult performance is normally very stable across time. For adults to change, they must alter their behavior. To greatly improve their performance, adults must engage in extended deliberate practice (Ericsson et al., 1993). Most adults reach a stable life situation in which their behavior and activity achieve equilibrium. Research has shown that adults' physical fitness level closely matches the intensity and duration of their habitual physical activity. When adults merely maintain their current level of activity, there are no further improvements or changes in their physical fitness. To improve physical aerobic fitness, adults have to dramatically increase physical activity. To see significant performance gains, adults must engage in training and exercise that pushes them beyond the their current level of adaptation to their activity.

The improvement of aerobic fitness by young adults is a good case in point. It is well documented that adults have to engage in intense aerobic exercise to improve their aerobic fitness. Specifically, young adults have to exercise at least a few times each week for a minimum of 30 min per session and with a sustained heart rate that is 70% of their maximal level (i.e., around 140 beats per minute for a maximal heart rate of 200; Roberts & Roberts, 1997). Similarly, improvements of strength and endurance require that individuals strain themselves on a weekly basis, and each training session pushes the associated physiological systems outside the comfort zone to stimulate physiological growth and adaptation. By straining the metabolic system, biochemical waste products are generated that serve as signals to cells to adapt or proliferate (Ericsson, 2001, 2002, 2003).

When the human body is put under exceptional strain, a whole range of extraordinary biochemical processes are activated resulting in physiological responses and anatomical changes. For example, adults' bodies can recover from surgery and broken bones and spontaneously heal over time. When an adult donates a kidney, the remaining kidney automatically grows in size by around 70% to handle the increased load on the endocrine system.

In sum, when adults encounter demanding, exceptional situations or challenge themselves to improve by choosing to engage in deliberate practice, they are able to transcend the stable structure of abilities and capacities so characteristic of most adults in everyday life. When researchers include all the evidence for training-induced changes in the size of hearts, thickness of bones, and allocation of cortical areas in the brain, then virtually all aspects of human's bodies and nervous systems are modifiable, with the exception of height and body size (Ericsson, 2001, 2002).

Theoretical Implications of Far-Reaching Adaptability of the Body and the Nervous System

If physiological systems—including systems down to the level of individual cells—can be modified by engaging in various strenuous forms of physical or mental activity, then the existence of fixed limits or invariant capacities are no longer tenable. For individuals who are sufficiently motivated to engage in the training activities necessary to induce physiological and anatomical changes, the stability of traits, abilities, and inferred capacities cannot be assumed to remain stable during adulthood. This modifiability raises doubt about the popular analogy between computers and human adults, shown in Fig. 10.1. Computer performance of some task depends on many different factors. Part of the computer's performance is attributable to its hardware, such as the central processing unit, speed of access to secondary memory, and capacity of primary memory. These characteristics of computer hardware are fixed for a given computer. The computer's performance of a given task, however, also depends on another factor—the computer's software, including application programs and the operating system. A computer's software, unlike the computer's hardware, can be easily exchanged and improved. However, unlike computers, the hardware of humans—cells and organs—are modifiable in response to demands to preserve homeostasis. These same biochemical mechanisms that are activated to heal wounds and broken bones in the human body can also grow organs and assist the body in adapting to repeated strain induced by practice and performance. Humans' bodies do not wear out from repeatedly performing the same action (as long as the actions do not injure tissue and provide enough recuperation to avoid repeated stress injuries). The human body is able to adapt over time and will perform the same action more and more efficiently (Bernstein, 1996).

This remarkable adaptability at the level of physiological, behavioral, and cognitive systems raises a major challenge to scientists interested in the highest levels of achievement and performance. What is the nature of the limits of performance for motivated individuals searching for their highest attainable level of performance in a domain?

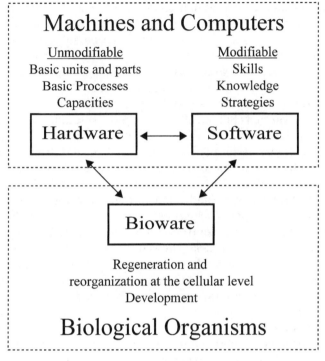

Hardwired Components of the Mind

Machines and Computers

Unmodifiable	Modifiable
Basic units and parts	Skills
Basic Processes	Knowledge
Capacities	Strategies

Hardware ⟷ **Software**

Bioware

Regeneration and
reorganization at the cellular level
Development

Biological Organisms

FIG. 10.1. The limits of applicability of the hardware and software distinction in computers to biological organisms that can repair injuries to tissue and adapt to physiological systems to the repeated demands of extended performance.

Smith and I (Ericsson & Smith, 1991) proposed that researchers should search for the highest levels of reproducible performance that has ever been attained in any domain of expertise or any aspect of everyday life. Once such individuals are identified, researchers should capture and reproduce these phenomena in the laboratory to enable the application of standard process-tracing techniques in assessing the structure of the performance's mediating mechanisms. This type of focused inquiry raises some important questions. To what extent will these types of exceptional performances reflect mechanisms that have been gradually acquired, or adapted, in response to extended practice and training? Can we identify types of practice and training that are particularly effective in improving performance?

THE SCIENTIFIC STUDY OF EXPERT
PERFORMANCE AND ITS ACQUISITION

Everyone has heard impressive anecdotes about the sometimes unbelievable achievements of athletes, musicians, and scientists. When I (Ericsson, 1996) examined hard scientific evidence for the most amazing achievements, most of these incidents could not even be substantiated by independent and unbiased sources. The only source was often the exceptional persons themselves who told stories about their childhood. For example, the famous stories of the brilliant mathematician Gauss as a prodigious child were recounted by Gauss himself as an old man to his closest friends and cannot be verified (Bühler, 1981). In other cases, individuals observing the event may have misinterpreted what actually happened. To build a science of exceptional performance, researchers must restrict the scientific evidence to phenomena that can be repeatedly and reliably observed then reproduced and analyzed under standardized conditions in the laboratory.

Smith and I (Ericsson & Smith, 1991) proposed methods for objectively measuring different types of expert performance and reproducing such superior performance under controlled laboratory conditions. A recent review (Ericsson & Lehmann, 1996) showed that performance of experts has been successfully reproduced in the laboratory where methods of process tracing (such as analysis of think-aloud protocols and eye movements) have been applied to assess mechanisms that mediate experts' superior performance.

Procedures for fair measurement of superior performance in many domains have been developed in our western industrialized culture. Methods of measuring performance have become extremely precise. In many sports, the conditions of competition are so highly standardized that it is common to use an individual's best performance at local and regional competitions to assess their qualification to participate in national and international competitions. Sport organizations also keep historical archives that record the best performances on particular events in the nation and the world.

Competitions in music, dance, and chess have a similar long history of attempting to design standardized situations that allow fair competition between individuals. In all of these domains, elite individuals reliably outperform less accomplished individuals. In most individual sporting events, the athletes' performance is even objectively measured, such as the speed of swimming and the accuracy of throwing darts, with a precision and rigor that matches measurement in the psychological laboratories. Expert performers can reliably reproduce their performance at any time when required during competitions and training, and thus, experts can also exhibit their superior performance, when required, under controlled laboratory conditions.

In the next section, I identify several claims about expertise that generalize across different domains for superior reproducible performance of experts. First, I review the development of the reliably superior performance of experts and discuss the evidence for its acquired nature. I then focus in more depth on the characteristics of activities that appear necessary for the acquisition of expert performance, particularly the essential role of deliberate activities designed to optimize improvement of performance.

The Acquired Nature of Expert Performance

Engagement in music, chess, and sports generally starts early in a person's life, and most children come into some type of contact with activities in these domains. This diverse and varied exposure to activities in these domains makes it difficult to measure the amount of experience and relate the amount and quality of that experience to a specific individual's attained level of performance. The available evidence, however, allows researchers to support a weaker claim: Extended engagement in domain-related activities is necessary to attain expert performance in a domain (Ericsson, 1996; Ericsson & Lehmann, 1996).

Longitudinal assessments reveal that performance levels within individuals increase gradually over time. There is no evidence for abrupt increases in performance from one time to the next, as is shown in Fig. 10.2.

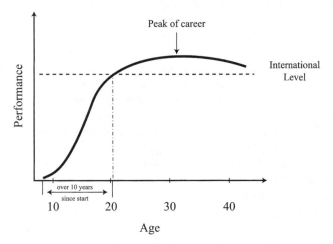

FIG. 10.2. An illustration of the gradual increases in expert performance in domains such as chess as a function of age. The "international level," which is attained after more than 10 years of involvement in the domain, is indicated by the horizontal dashed line. *Note.* From "Expertise," by K. A. Ericsson and Andreas C. Lehmann, 1999, in M. A. Runco & S. Pritzer (Eds.), *Encyclopedia of Creativity.* New York: Academic. Copyright 1999 by Academic Press. Reprinted with permission.

Even child prodigies in music and chess whose performance is vastly superior to children of the same age show a gradual, steady performance increase over time. Furthermore, when adults and children are measured using the same standards by playing in the same chess tournaments or music competitions as adults, without special considerations for their age, it is clear how much the child prodigies have to improve to reach the level of elite adults.

The age of peak performance differs across domains of expertise. However, expert performers continue to improve their performance beyond the age of physical maturation (the late teens in industrialized countries) for many years and even decades, as shown in Fig. 10.2. Performers in many vigorous sports typically reach their peak career performance is the middle to late 20s; in the arts and sciences, the performance peak is a decade later, in the person's 30s and 40s. This continued and often extended performance development beyond the age of physical maturity implies that the highest performing individuals are able to engage in activities that lead to continued improvements of performance for decades. It is important to recognize though that in many domains such as music, golf, and chess, performance decreases in older adults are often hard to detect and not uniform across performers. Expert performers in their 50s and 60s may remain competitive with younger individuals even at the highest levels of performance in these domains.

Finally, the most compelling evidence for the need to engage in domain-related activities prior to attaining high levels of performance is that even the most "talented" performers require some 10 years of intense involvement in a specific skill domain before reaching a level of performance equivalent to an international level, and for most individuals, this period is considerably longer than 10 years. Simon and Chase (1973) originally proposed the 10-year rule, showing that no modern chess master had reached the international level in less than approximately 10 years of chess playing. Subsequent reviews show that the 10-year rule extends to music composition, as well as to a wide range of sports, sciences, and arts (Ericsson et al., 1993). In sum, the need to actively engage in domain-related activities to improve domain expertise is a well-established criterion for expert levels of performance. Few individuals sustain that commitment for more than a few months; therefore, many individuals will never discover their true potential in a given domain.

The Need for Particular Types of Experience, Such as Instruction, and for Designed Training Environments

Extensive involvement in a domain is clearly necessary for the select group of elite individuals who steadily increase their performance and reach very high levels. However, extensive involvement does not benefit amateurs in the same way as experts. After first being introduced to a domain, amateurs improve performance for a short period until reaching an acceptable level of performance, at which point performance remains relatively stable. Recreational golfers, tennis players, and skiers are easily recognizable examples of individuals who have not improved after years, or even decades, of regular experience. The striking difference between elite and average performance seems to result not just from the duration of an individual's activity, but from the particular types of domain-related activities they choose.

From retrospective interviews of international-level performers in several domains, Bloom (1985) showed that elite performers are typically introduced to their future domain in a playful manner. As soon as these individuals are found to enjoy the activity and show promise compared to peers in the neighborhood, they are encouraged to seek out a teacher and begin regular practice. Bloom (1985) showed the importance of access to the best training environments and the most qualified teachers. The parents of the future elite performers spend large sums of money for teachers and equipment and devote considerable time to escorting their child to training and weekend competitions. In some cases, the performer (as well as the performer's family) relocates to be closer to the teacher and training facilities. Based on their interviews, Bloom (1985) argued that access to the best training resources is necessary to reach the highest levels.

The best evidence for the value of current training methods and practice schedules comes from historical comparisons (Ericsson et al., 1993; Lehmann & Ericsson, 1998). Historically, very dramatic improvements in the level of performance are found in sports. In competitions such as the marathon and swimming events, many serious amateurs of today could easily beat the gold medal winners of the early Olympic Games. For example, after the fourth Olympic Games in 1908, the Olympic Committee almost prohibited double somersault dives because these dives were thought to be dangerous and uncontrollable. In competitive domains such as baseball, it is sometimes difficult to demonstrate the increased level of today's performers because both the level of pitching and batting has improved substantially (Gould, 1996).

The Insufficiency of Experience, Instruction, and Opportunities for Practice: The Role of Deliberate Practice

It is generally accepted that experience, opportunities for practice, and instruction are all necessary for improving performance. It is often proposed that when performers are given the appropriate opportunities, learning and skill acquisition are essentially automatic. Individuals have been traditionally assumed to reach an asymptote for their performance that is set by innate talent. Further, according to this view, an individual's performance is not substantially modified by further training. Although this view appears to be correct for the majority of individuals mastering skills in everyday life, it does not hold true for expert performers who continue developing their specific expertise over extended periods of time. This raises an important question: Could the type of training activity and the type of cognitive processes mediating practice be critical for allowing future expert performers to keep improving? Krampe, Tesch-Römer, and I (Ericsson et al., 1993) tried to identify those training activities most closely associated with optimal improvement and classified them as "deliberate practice." To continue the investigation into this notion, I first consider some characteristics of deliberate practice.

The laboratory studies of learning and skill acquisition during the last century (such as the memory training studies discussed earlier) have demonstrated continued improvement as a function of practice. Ericsson et al. (1993) found that performance uniformly improved when motivated individuals were given well-defined tasks, were provided with immediate feedback on their performance, and had ample opportunities for repeated performance on the same or similar tasks. Individuals were often able to keep improving during a series of training sessions as long as the sessions were limited to about 1 hr—the time that college students could maintain sufficient concentration to make active efforts at performance improvement. These deliberate efforts to improve performance beyond a current level involve solving problems and finding better strategies. Engaging in an activity with the primary goal of improving some aspect of performance is a prerequisite of deliberate practice.

The vast majority of individuals active in popular domains (such as tennis and golf) spend little, if any, time on deliberate practice. Most of the time is spent on "playful interaction" in which the primary goal is inherent enjoyment of the activity. For example, consider someone who plays tennis in a tournament and misses a backhand volley. Several tennis matches later, this individual may encounter another similar opportunity for this shot, but without intervening practice, the chance for another miss nevertheless remains high. In contrast, a tennis coach working with a player could give the individual several hundred opportunities to gradually perfect this shot in a single tennis lesson.

Professionals, due to the nature of the work environment in which they interact, gain a great deal of on-the-job experience. The goal of such activity is to generate a quality product in a reliable fashion. To produce optimal performance in work activities, individuals rely on previously established, well-entrenched methods rather than exploring new methods with unknown reliability. Hence, work activities offer opportunities for making established methods more efficient but not for changing methods to improve future performance. It is difficult to acquire new and better methods while conducting one regular job in an efficient and reliable fashion. In most work situations, individuals are thus frequently encouraged to attend training courses where they may focus on performance improvement in protected learning environments.

All major domains of expertise that train individual performers, including music, ballet, and sports, have accepted methods for effective teaching based on accumulated knowledge and skills. Over the last 200 years, teachers and coaches have gained insight into how students master a sequence of increasingly difficult training tasks. Teachers know to what extent simpler tasks have to be mastered to serve as building blocks of more complex skills. Unlike the beginners themselves, the teacher can foresee future performance demands and avoid the need for complete relearning of previously attained skills. By gradually improving performance through acquisition of these basic skills, students eventually master complex tasks that might have seemed virtually unattainable when they were beginners. The core assumption of deliberate practice (Ericsson et al., 1993) is that expert performance is acquired gradually and that effective improvement of students' performance depends on the teachers' ability to isolate sequences of simple training tasks that the student can master sequentially.

A very similar development of the quantity and quality of practice was observed in most domains for which Bloom (1985) discovered individuals who exhibited an international level of performance. Promising children started training with teachers at very young ages after only a brief period of playful interaction. Because of the need for sustained concentration, the duration of training is initially quite short—typically no more than around 15 to 20 min per day, leaving no room for many other domain-related activities (including play, as illustrated in the first phase of Fig. 10.3). Many parents help their children to concentrate during practice, help establish regular practice patterns, and encourage their children by pointing out practice-related improvements in their performance (Bloom, 1985; Lehmann, 1997). Future expert performers become increasingly involved within their domain of choice as they mature. Toward the end of adolescence, their commitment to the domain-related activities becomes essentially a full-time pursuit.

During the third and final phase of Bloom's (1985) framework for the acquisition of expert performance (shown in Fig. 10.3), students assimilate the available knowledge and skills in a domain until they have been fully mastered. The very best performers attempt to go even further, when they initiate a fourth and final stage (Ericsson et al., 1993) in which they focus on making creative innovations that extend the current knowledge and achievements in their domain (Ericsson, 1998, 1999).

The importance of deliberate practice in attaining expert performance was first demonstrated by Ericsson et al. (1993) by using diaries and other

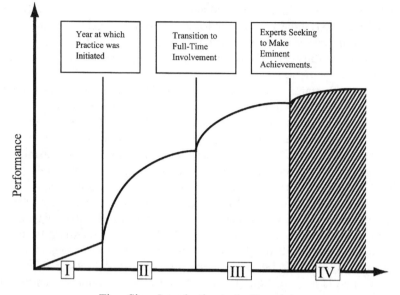

Time Since Introduction to the Domain

FIG. 10.3. Three phases of acquisition of expert performance followed by a qualitatively different fourth phase when, to make a creative contribution, experts attempt to go beyond the available knowledge in the domain. *Note.* From "Can We Create Gifted People?" by K. A. Ericsson, R. T. Krampe, and S. Heizmann in *The Origins and Development of High Ability* (p.), 1993, Chichester, England: Wiley. Copyright 1993 by CIBA Foundation. Adapted with permission.

methods to investigate how three groups of expert musicians who differed in their level of attained music skills spent their time during a typical day. Although each musician spent about the same amount of time on all types of music-related activities, the better musicians spent more time in deliberate practice. The best musicians, who were judged to have the potential for an international level of performance, spent around 4 hr every day (including weekends) in solitary practice. From retrospective estimates of practice, Ericsson et al. (1993) calculated the number of hours of deliberate practice that five groups of musicians at different performance levels had accumulated by a given age, shown in Fig. 10.4. By the age of 20, the best musicians had spent over 10,000 hr of practice, which is respectively some 2,500 and 5,000 hr more each than the two less accomplished groups of expert musicians and 8,000 hr more than amateur pianists of the same age (Krampe & Ericsson, 1996).

Several studies and reviews have found a consistent relation between performance and the amount and quality of deliberate practice in chess (Charness, Krampe, & Mayr, 1996), in music (Krampe & Ericsson, 1996;

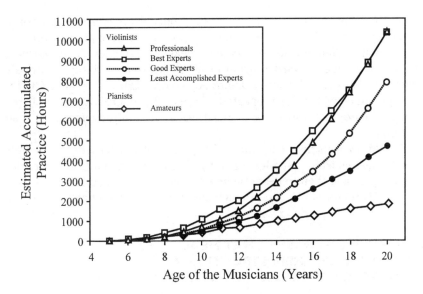

FIG. 10.4. Estimated amount of time for solitary practice as a function of age for the middle-aged professional violinists (triangles), the best expert violinists (squares), the good expert violinists (empty circles), the least accomplished expert violinists (filled circles), and amateur pianists (diamonds). *Note.* From "The Role of Deliberate Practice in the Acquisition of Expert Performance," by K. A. Ericsson, R. T. Krampe, and C. Tesch-Römer, 1993, *Psychological Review, 100*, p. 379 and p. 384. Copyright 1993 by American Psychological Association. Adapted with permission.

Lehmann & Ericsson, 1996; Sloboda, 1996), and in different types of sports (Helsen, Starkes, & Hodges, 1998; Hodges & Starkes, 1996; Starkes, Deakin, Allard, Hodges, & Hayes, 1996). The concept of deliberate practice also accounts for many earlier findings in other domains (Ericsson & Lehmann, 1996), as well as for the results from the rare longitudinal studies of elite athletic performers (Schneider, 1993).

From Correlations to Causal Mechanisms

The evidence reviewed so far indicates that instruction and access to training environments are necessary—but not sufficient—to attain elite levels of performance. The highest correlate of attained performance observed thus far is the amount of time individuals have spent in activities that meet requirements for deliberate practice. Deliberate practice requires that individuals (or their teachers) design training to improve specific aspects of performance. In essence, this means that the individual is consciously attempting to exceed their current ability and attain a higher performance level. This type of practice requires full concentration, and the desired higher level of performance is achieved only through the gradual process of experiencing failures and overcoming weaknesses. When the performers stretch themselves toward a higher level of performance during deliberate practice, mistakes and failures cannot be avoided (Ericsson et al., 1993). Experts do not consider these repeated failures to be inherently enjoyable but rather focus on the subsequent breakthrough and improvements of their performance.

To fully explain the acquisition of expertise, it is necessary to identify specific mechanisms that allow expert performers to exhibit their superior performance and then to describe how these mechanisms can be gradually acquired through deliberate practice and relevant experience.

SPECIFIC MECHANISMS THAT MEDIATE EXPERT PERFORMANCE: THEIR STRUCTURE AND ACQUISITION THROUGH DELIBERATE PRACTICE

Any complete scientific account of how expertise develops must explain the underlying changes that enable future experts to continue improving. Improvement by children and adolescents can be partially explained by maturation. However, improvement in the performance of mature adults represents change in the mechanisms mediating individual performance. Isolating these processes that enable experts in various domains to continue modifying these mechanisms is a daunting task. Even explaining the performance of aspiring experts in a single domain is very difficult because the required changes in mediating mechanisms will differ as a function of the expert's current level of performance. For example, when chess players at the club level improve their performance during a season, the required changes in their mediating mechanisms will be quite different from those necessary to improve the performance of a chess master. Consequently, the deliberate practice activities that induce these changes must also differ, at least at their most detailed level.

I first address the fundamental issue, namely, how may expert performers continue to improve their performance? Traditional theories of skill acquisition explain how most amateurs reach an acceptable level of performance that they then maintain from that point onward. The same theories cannot explain the extended improvement of experts for specific tasks within a given domain.

When individuals are first introduced to an activity (such as golf and tennis), their primary goal is to reach a level of mastery that will allow them to engage in recreational activities with their friends. During the first phase of learning (Fitts & Posner, 1967), individuals need to concentrate on what

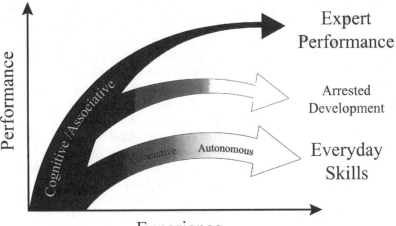

FIG. 10.5. An illustration of the qualitative different path leading to expert performance and to mastery of everyday activities. The goal for everyday activities is to reach a satisfactory level that is stable and automatic by rapidly passing through the cognitive and associative phases (see the lowest curve in the graph). In contrast, expert performers' development (see the top curve in the graph) involves extended cognitive and associative phases when the expert performers acquire increasingly complex mental representations to attain higher levels of control of their performance. *Note.* From "The Scientific Study of Expert Levels of Performance: General Implications For Optimal Learning and Creativity," by K. A. Ericsson, 1998, *High Ability Studies, 9*, p. 90. Copyright 1998 by European Council for High Ability. Adapted with permission.

they are going to do to reduce embarrassing mistakes, as is illustrated in the lowest curve in Fig. 10.5. With more experience, gross mistakes become increasingly rare, performance appears smoother, and individuals are not required to concentrate as much as before to perform at an acceptable level. After a limited period of training and experience—frequently less than 50 hr for most recreational activities such as skiing, tennis, and driving a car—an acceptable standard of performance is typically reached. As individuals adapt to a domain and their performance skills become automated, they may lose conscious control over execution of those skills and it may become difficult to intentionally modify these skills. In this case, further experience will not markedly improve learning. Consequently, the correlation between amount of experience and performance will be low in automated everyday activities.

In contrast, the performance levels of expert performers continue to improve as more experience is gained through deliberate practice. The key challenge for aspiring expert performers is to avoid the arrested development associated with automaticity and to acquire cognitive skills to support their continued learning and improvement. The expert performer actively counteracts the tendency toward automating performance by designing training activities, typically with the help of teachers that induce discrepancies between their actual and desired performance, as is shown by the top curve in Fig. 10.5. Raising performance standards causes experts to make mistakes. These failures force expert performers to keep refining their task representations so they continue regenerating the initial cognitive phase. Experts continue to acquire and refine cognitive mechanisms that mediate continued learning and improvement. These mechanisms are designed to increase the experts' control and ability to monitor the processes (Ericsson, 1998, 2001, 2002). However, the vast majority of expert performers tend to reach a satisfactory high level of performance and then allow their attained level of performance to become automated, as is illustrated by the middle curve in Fig. 10.5.

In the following section, I briefly summarize empirical evidence for cognitive control processes that support and mediate performance at the highest levels and for experts' use of acquired representations to support planning and reasoning. Then I discuss how future experts rely on these representations for continued learning through deliberate practice.

The Cognitive Mediation of Expert Performance

Expert performers are able to produce appropriate reactions in situations that they have never previously encountered and thus could not be explained by automatic retrieval of prepared responses. More generally, experts cannot rely on automated responses because they need to be able to make adjustments appropriate to specific performance situations. For example, experts routinely adjust to situational factors such as weather and new equipment. Experts also make fast, fluent adjustments in reaction to the behavior of competitors during matches in tournaments. These refined adjustments would not be possible if the expert performance was fully automated, nor would performers be able to encode and later recall detailed information about their competitive experiences.

The most compelling scientific evidence for preserved cognitive control of the critical aspects of expert performance comes from laboratory studies in which the experts' superior performance is reproduced with representative tasks that capture the essence of domain expertise (Ericsson & Smith, 1991). In his pioneering work on expertise, de Groot (1946/1978) instructed both good and world-class chess players to think aloud while selecting the best move in a set of unfamiliar chess positions. The task of selecting the best move appears very similar to the prototypical decision-making task in which players are asked to generate all of the alternative legal moves. The challenge posed to these players was to select the best among the 20 to 40 legal moves for most middle-game positions.

De Groot (1946/1978) found that the quality of the selected moves was closely associated with the performers' chess skill. Verbal report evidence revealed that all of the chess players first perceived and then interpreted the chess position to retrieve potential moves from memory. Promising moves were then evaluated by mentally planning consequences. During this evaluation, even the world-class players discovered better moves. Hence, the performance of experts is mediated by increasingly complex control processes. Although chess experts can rapidly retrieve appropriate actions for a new chess position (cf. Klein's, 1993, RPD model and Calderwood et al., 1988), their move selection can be further improved by planning, reasoning, and evaluation.

The same paradigm has been adapted to study other types of expertise in which experts have been presented with representative situations (such as simulated game situations) and asked to respond as rapidly and accurately as possible. In a recent review, Ericsson and Lehmann (1996) found that experts' think-aloud protocols revealed how their superior performance was mediated by deliberate preparation, planning, reasoning, and evaluation in a wide range of domains of expertise including medicine, computer programming, sports, and games.

Recent reviews (Ericsson, 1996; Ericsson & Kintsch, 1995) have shown that expert performers have acquired refined mental representations to maintain access to relevant information and to support flexible reasoning about an encountered task or situation. In most domains, better performers are able to rapidly encode and store relevant information for representative tasks in memory so they can efficiently manipulate the tasks in their minds. For example, with increased chess skill, chess players are able to plan deeper, mentally generate longer sequences of chess moves, and more fully evaluate the consequences of each potential move. Chess masters are even able to hold the image of the chess position in mind so accurately that they can play blindfold chess (i.e., play without perceptually available chessboards). Similar evidence for mental representations has been shown for motor-skill experts such as snooker players and musicians (Ericsson & Lehmann, 1996).

The performance of expert typists, who exhibit a faster rate and higher accuracy of typing movements, are also accounted for by mediating cognitive representations. The key to the expert typists' advantage lies in their looking beyond the word that they are currently typing (Salthouse, 1984). By looking further ahead, typing experts can prepare keystrokes in advance, moving relevant fingers toward their desired locations on the keyboard. This finding has been confirmed by analysis of high-speed films of expert typists and experimental studies in which expert typists have been prevented from looking ahead. Similarly, the rapid reactions of athletes such as hockey goalies, tennis players, and baseball batters have been found to reflect acquired skills involving the anticipation of future events. This evidence supports the hypothesis that expert athletes have a learned rather than an innate or biological speed advantage over their less accomplished peers (Abernethy, 1991). For example, when skilled tennis players are preparing to return a serve, they study the movements of the opponent leading up to contact between the ball and the racquet to identify the type of spin and the general direction of the ball. Given the ballistic and biomechanical nature of a serve, it is often possible for skilled players to make these judgments accurately. It is important to note that novice tennis players use an entirely different strategy and usually initiate their preparations to return the ball once it is apparent where the ball will bounce. Hence, the perceptual skill to anticipate future events is not part of the beginners' repertoire and thus could not possibly result from the automation of the beginners' cognitive processes. These anticipatory skills must therefore have been acquired later at more advanced level of skill. Expert athletes are not distinguished from their less skilled peers by any type of superior basic perceptual ability. Rather, expert athletes' superior performance has been shown to reflect specialized perceptual skills that cannot be explained by superiority on standard tests of basic visual ability (Williams, David, & Williams, 1999).

The deliberate acquisition of increasingly refined mental representations allows experts to attain greater control of relevant aspects of performance, including the ability to anticipate, plan, and reason about alternative courses of action. The development of these mental representations is directly linked to increases in the experts' ability to make decisions about superior actions within the time available. In the next section, I discuss how these mental representations are acquired and how they provide tools for experts to continue learning and to maintain their high level of performance.

The Development of Mental Representations

Because a theoretical framework of expert performance needs to offer a plausible account for how expert performers develop complex mental representations that are qualitatively different from those of amateurs, one must contrast how beginners and experts undergo the acquisition of skill in a given domain.

When introduced to a domain, beginners, whether they become future amateurs or expert performers, have to learn most of the basic rules and objectives to be able to participate successfully in the activities of the domain. Whereas amateurs rely on their initial representations to reach an acceptable level of performance, future experts use the initial representation as a starting point for developing more refined representations in response to instruction and deliberate practice.

Deliberate Practice Designed by Teachers in Instructional Settings.
Many types of sports, such as gymnastics, figure skating, and platform diving, are similar to the performance arts in that the acquisition of expert performance involves mastering increasingly complex and challenging sequences of motor actions. In all of these domains, guidance and instruction are crucial, and no performer reaches the elite levels without the help of coaches and teachers.

Music is a domain that has one of the longest traditions of successfully training expert performers and can therefore serve as a starting point to generalize findings to many different domains. The training of music students is not centered on mastery of a set of techniques and skills but rather focuses on helping students to perform complete music pieces in public. In the beginning of their training, music students begin by mastering simple pieces of music with a focus on accuracy of keystrokes. With increased mastery, music teachers select more challenging pieces for the students to play and have increased expectations for musical expression on the part of students.

When a student repeatedly practices a new, challenging piece, the difficulties that are experienced reveal weaknesses in the student's mental representations and technical skills. Depending on the type of problem, the teacher will then recommend a specific type of deliberate practice to improve that aspect of the student's performance. If the weaknesses reflect technical problems, the teacher might assign special exercises or études that have been constructed to improve mastery of a relevant finger combination at a specified speed. If the weakness refers to inability to express musical qualities, then the student might be assigned exercises of listening or of imitating recorded artists.

Over the years, many effective training methods have been devised to help musicians change their processes and representations. However, only the students themselves are able to address their own specific performance problem. Eventually, through problem solving, students can generate the specific modifications that, with extended practice, can be fully integrated with the complex representations that mediate their performance of a complete piece of music.

In other performance domains such as ballet, gymnastics, figure skating, and platform diving, there is a similar progression through increasingly difficult tasks in which the guidance of a teacher is critical for success. However, there are domains in which large improvements in performance are regularly attained without teachers and in which individuals increase the level of difficulty by seeking out more challenging situations on their own, for example, skiing more difficult slopes or playing with older or better tennis players.

Deliberate Practice That Leads to Increased Speed and Efficiency. In many domains of expertise such as skiing, tennis, and typing, the fundamental tasks remain very similar at every level of performance. Experts and novices differ markedly in their speed of initiating actions and their control of performance. The research described earlier shows that expert performers are more able than novices at anticipating performance demands and can thus better prepare actions in advance. How is it possible then to develop representations that mediate and extend anticipation?

Extensive research on typing provides the best insights into how performance speed can be increased through deliberate practice. The key finding is that individuals' typing speed is not completely fixed. It is possible for all typists to increase their typing speed by pushing themselves as long as they can sustain full concentration, which is typically only 15 to 30 min per day for untrained typists. While straining to type at a faster speed—typically around 10% to 20% faster than their normal speed—typists seem to strive to improve anticipation better, possibly by extending their gaze ahead further.

The faster tempo also serves to uncover keystroke combinations that are comparatively slow and poorly executed. These combinations are then trained in special exercises and incorporated in the typing of regular text to assure that any modifications can be integrated with the representations mediating regular typing. By successively eliminating weaknesses, typists can increase their average speed and practice at a rate that is still 10% to 20% faster then the new average typing speed. In domains in which speed and efficiency of performance presents the primary challenge, it is possible to attain high levels of performance with a minimum of instruction. Individuals can push their performance beyond its normal level—even if that performance can be maintained only for short time—and then identify and correct weaker components, enhancing anticipation to enhance speed and efficiency.

Deliberate Practice to Improve Selection of Actions in Tactical Situations. In many types of sports, individuals select or generate actions rapidly when they confront one or more opponents. We know that expert performers gain their advantage, at least in part, from being more able to plan and foresee consequences of their own actions and the opponents' responses to their actions in advance. However, is it possible to improve one's ability to generate superior plans? To improve the quality of planning rather than just the mere amount of planning, performers should conduct their deliberate practice by planning and selecting actions for representative situations in which the most appropriate actions are known. The performers can get feedback by comparing their selected actions with the correct actions. Based on an analysis of inferior choices, the performer can use problem-solving techniques to assess how the encountered situations should have been represented to allow identification and selection of the correct actions.

How is it even possible to know the best action in a given representative situation from everyday life? Chess players typically solve such problems by studying published games between the very best chess players. Chess masters play through the games one move at the time to see if their selected move matches the corresponding move originally selected by the masters. If the chess master's move differed from their own selection, it would imply that their planning and evaluation must have overlooked some aspect of the position. Through careful, extended analysis, the chess expert is generally able to discover the reasons for the chess master's move. Serious chess players spend as much as 4 hr every day engaged in this type of solitary study (Charness et al., 1996; Ericsson et al., 1993). By spending a longer time analyzing the consequences of moves for a chess position, players can improve the quality of their move selections. With further study, players refine their representations and can access or generate the same information faster. Chess masters can typically recognize an appropriate move immediately, whereas it will take a competent club player around 15 min to uncover a good move by successive planning and evaluation.

It is apparent that individuals' ability to select actions is not fixed, and when novices are permitted to extend the allowed time for analysis (such as during study and practice), novices are able to generate actions matching the qualities of more skilled players. Extended study can thus serve as a means to gradually elevate the quality of the selected actions—the level of performance.

Potential Generalizations to Professional Activities. There have been few empirical studies of deliberate practice in professional domains. However, Ericsson (1996, 2001, 2002) has described how the mechanisms for mastery and improving speed and accuracy discussed previously in this section can be generalized to any domain with objective criteria for assessing reliably superior performance.

Most professionals can assess their current performance and even know how their weaknesses could be overcome by various forms of training and changed procedures. Many of the remaining professionals know how they would be able to get feedback about their current performance and guidance for practice to eliminate their most important weaknesses. Those rare individuals who achieve at such a sufficiently high level that their weakness are not readily apparent to them can easily identify other individuals, nationally or internationally, that perform at a clearly higher level and then attempt to match the performance of such persons. Individuals can find records of outstanding individuals' achievements and then attempt to copy or reproduce these achievements. For example, Benjamin Franklin (2003) once described how he had learned to write in a clear and logical fashion by self-study. Franklin would read an argument in a well-written book and then try to reproduce it in writing from memory. By comparing his reproduction to the original, he was able to identify differences by iteration until his reproduced version matched the clearness and logic of the original. This general approach can be easily extended to learning how to reproduce the works of outstanding individuals and masters in one's respective domain of expertise such as painting, composing music, and even writing scientific articles.

Most professionals know how they could improve the quality of their achievements by preparing more in advance for a performance, such as thinking through a surgical procedure for a particular patient to anticipate and resolve potential problems. Similarly, most professionals know that seeking feedback from experts on how one could improve products, such as asking colleagues to comment on drafts and presentations, is an effective means to improve the final product. Professional workers also know that evaluation of and reflection on a completed performance is likely to allow one to identify weaknesses that can be eliminated and thus improve the quality of future performances and products. With the help of a mentoring master and the investment of additional time, professionals are able to focus on improving the most important aspects of their performance. With time, these professional workers may use these techniques to achieve superior levels of performance faster and more efficiently than before.

More generally, the goal of expert professionals should be to develop something innovative and general: an approach, a method, or an argument that can be gradually improved after feedback from colleagues and teachers (given that extensive opportunity for refinements are available). For example, when scientists informally and verbally describe their newest theories and ideas to colleagues, they often find that writing down their descriptions is a superior method of communication. Critical colleagues can then more easily point to problems with the written arguments so the author can respond to the criticisms by either more fully bolstering the argument or by making changes to the substance of the original argument. The revised document can then be circulated to other colleagues, and with further feedback, additional improvements to the structure of the argument can be made. The eventual result is a final work that can be integrated into contributions that extend the boundaries of knowledge in the given domain.

Most professionals' achievements are not constrained by fixed capacities that limit further increases in their performance. Instead, achievements are constrained by motivational issues related to the amount of time and effort that professionals are willing and able to invest in deliberate practice and continued improvement of their performance.

CONCLUDING REMARKS ON THE COMPLEX INTEGRATED MECHANISMS THAT MEDIATE SUPERIOR PERFORMANCE

The theoretical framework of expert performance makes the fundamental claim that improvements in superior reproducible performance of adult experts do not occur automatically or without discernible reason. Superior performance can be linked to changes in physiological adaptations or cognitive mechanisms mediating how the brain and nervous system control performance. However, it is difficult to attain stable changes that allow performance to be incrementally improved. The general rule (or law) of least effort predicts that the human body and brain have been designed to carry out activities at minimum cost to the metabolism. When physiological systems, including the nervous system, are significantly strained by ongoing activity, such systems initiate processes that lead to physiological adaptation and mediation of simpler cognitive processes to reduce the metabolic cost. This phenomenon is evident in most types of habitual everyday activities such as driving a car, typing, or strenuous physical work in which individuals tend to automate their behavior to minimize the effort required for execution of the desired performance.

Merely performing the same practice activities repeatedly on a regular daily schedule will not lead to further change once a physiological and cognitive adaptation to the current demand has been achieved. The central attribute of deliberate practice is that individuals seek out new challenges that demand concentration and effort as long as they want to keep improving their performance beyond its current level. As individuals' level of performance increases, the effort required to improve performance increases as well. Further improvement requires increased challenges and engagement in activities selected to improve current performance—deliberate practice.

Once one conceives of expert performance as being mediated by complex integrated systems of representations for the execution, monitoring, planning, and analyses of performance, it becomes clear that skill acquisition requires an orderly and deliberate approach. Deliberate practice focuses on improving a specific aspect of performance. By training under representative conditions, the performer will engage the complex cognitive mechanisms that mediate superior performance. The same mechanisms and representations also allow the expert to monitor, evaluate, and reason about their performance. During deliberate practice, it is possible for the performer to design training tasks that give feedback, and the mediating representations assist the performer in isolating specific problems and weaknesses. Through problem solving and deliberate practice, performers can make the required changes to their specific representations.

Improvements are always conditional on the performers' preexisting mechanisms and entail modifications of their own specific representations. The tight interrelation between representations that monitor and generate performance minimizes the risk of unwanted side effects from modifications. However, the complex integration of the mechanisms mediating expert performance makes it impossible to identify distinct processes of problem solving, decision making, and reasoning. In fact, the principal challenge of skill acquisition appears to be in developing representations that coordinate each of the following: selection of actions, monitoring, control of ongoing performance, and incremental improvement of performance.

If one is interested in improving decision making in particular domains, I suggest that the researcher first identify the crucial tasks that capture the essence of the relevant domain. By studying the development of reliably superior performance within that domain, the mechanisms and representations that are necessary for the generation of quality actions within the time available will be discovered.

Quality decision making in professional contexts in everyday life is an aspect of a functional system of mechanisms and representations. To abstract and to isolate the decision-making process away from the processes and representations that are responsible for the gradual improvement in performance is unlikely to make enduring contributions to theory or immediate contributions to applications. In my perhaps biased view, improvements in decision making will result from facilitating development of representative performance. This facilitation can be attained by identifying deliberate practice activities of successful performers. Researchers can then create opportunities and design environments in which individuals are given feedback about their representative performance and encouraged to engage in deliberate practice. These training environments would have considerable face validity and would even allow measurement of improvements in performance and thus provide explicit feedback on their ability to increase accuracy and speed of decision making.

I conclude by presenting a visual image of the essence of superior performance. I propose that we view expert performance metaphorically as one of the megalithic temples such as Stonehenge in England. These monuments with their massive stones arranged as arches were viewed as mysterious and seemingly required a magical explanation, such as the work of prehistoric giants or aliens from outer space. Even scientists believed for a long time that the Romans or some other highly developed civilization must have erected this monument, as the prehistoric farmers and hunters would seem to have been unable to do so.

With the advance of modern archeology, the assumption that Stonehenge was erected in a relatively short time was shown to be false—similar to the common myth about expert performance. Archeological evidence revealed that Stonehenge was the result of a series structures calibrated to correspond to astronomical events. These structures were successively refined over more than 1,000 years, with the final structure completed well before the reign of the Roman Empire (Hadingham, 1978). Detailed scientific analyses uncovered how the massive stones must have been transported long distances and how refined techniques had been developed to move the stones and arrange them into the form of arches. With these advances in human understanding, there is no remaining mystery about how Stonehenge could have been constructed by unexceptional human adults provided with appropriate knowledge and skills.

However, this account has revealed a new mystery. What factors motivated thousands of prehistoric individuals to sustain their efforts generation after generation to attain a major achievement such as Stonehenge? In similar vein, I argue that researchers' knowledge about how experts attain their performance during decades of intensive daily practice is rapidly increasing, but to gain all the benefits of that knowledge, we need to understand the motivational factors that drive individuals as they strive to achieve ever higher standards of performance in their domains.

As our scientific analysis of the highest levels of performance produces new insights into the complexity of skill that is attained after thousands of hours of deliberate efforts to improve, I believe that researchers will uncover deeper understanding of how effective learning and specialized practice methods can be used to target particular training goals. Instead of celebrating signs of innate talent and natural gifts, I recommend that people marvel at the discipline and the monumental effort that go into mastering a domain.

GENERAL CONCLUSIONS

1. Decision making is not a unique, fixed, basic ability. Rather, decision making in domains in which expertise is seen reflects an integrated aspect of a complex system that is able to reliably select and produce superior actions in representative situations.

2. Biological systems are capable of dramatic adaptation and change when physical and cognitive activity challenge metabolic balance. This homeostatic process includes human beings and is evidenced by their ability to change characteristics all the way down to the cellular level.

3. Reliably superior expert performance in domains is mediated by physiological adaptations and acquired complex mechanisms. These mechanisms and adaptations allow expert performers to plan and anticipate future actions and to control and monitor their actions.

4. Expert performance is gradually acquired over years after thousands hours of designed training called *deliberate practice*. Even the most talented individuals are able only to attain international levels of performance after a decade of regular training and practice.

5. Deliberate practice involves identification of specific aspects of the integrated performance that can be improved beyond their current level. These shortcomings in performance are typically brought to the attention of the student under the supervision of a master teacher or coach. Activities with feedback and opportunities for repetition and gradual refinement are also typically designed by the teacher against these shortcomings in performance.

6. Deliberate practice avoids the premature arresting of continuous development of performance by always reaching for a level of performance outside the performers' current ability. This type of practice requires full concentration and active problem solving, and thus, the amount of daily time required for deliberate practice requires a slow increase over years until the maximum of around 4 to 5 hr per day is achieved.

7. The complex integrated cognitive and physiological system necessary to produce expert performance requires deliberate, teacher-guided design and decades of sustained effortful construction. Hence, the real enigma surrounding expert performance and the requirement of deliberate practice may be to understand the motivational factors that support individuals' quest to reach these highest standards of performance.

ACKNOWLEDGMENTS

This research was supported by the FSCW/Conradi Endowment Fund of Florida State University Foundation. I thank Ray Amirault and Elizabeth Kirk for their most valuable comments on earlier drafts of this chapter.

REFERENCES

Abernethy, B. (1991). Visual search strategies and decision-making in sport. *International Journal of Sport Psychology, 22*, 189–210.
Beach, L. R. (1993). Image theory: Personal and organizational decisions. In G. A. Klein, J. Orasanu, R. Calderwood, & C. E. Zsambok (Eds.), *Decision making in action: Models and methods* (pp. 148–157). Norwood, NJ: Ablex.

Bernstein, N. A. (1996). Dexterity and its development. In M. L. Latash & M. T. Turvey (Eds.), *Dexterity and its development* (pp. 1–244). Mahwah, NJ: Lawrence Erlbaum Associates.

Bloom, B. S. (1985). Generalizations about talent development. In B. S. Bloom (Ed.), *Developing talent in young people* (pp. 507–549). New York: Ballantine.

Bühler, W. K. (1981). *Gauss: A biographical study.* New York: Springer.

Calderwood, R., Klein, G. A., & Crandall, B. W. (1988). Time pressure, skill and move quality in chess. *American Journal of Psychology, 101,* 481–493.

Camerer, C. F., & Johnson, E. J. (1991). The process-performance paradox in expert judgment: How can the experts know so much and predict so badly? In K. A. Ericsson & J. Smith (Eds.), *Toward a general theory of expertise: Prospects and limits* (pp. 195–217). Cambridge, England: Cambridge University Press.

Charness, N., Krampe, R. T., & Mayr, U. (1996). The role of practice and coaching in entrepreneurial skill domains: An international comparison of life-span chess skill acquisition. In K. A. Ericsson (Ed.), *The road to excellence: The acquisition of expert performance in the arts and sciences, sports, and games* (pp. 51–80). Mahwah, NJ: Lawrence Erlbaum Associates.

Chase, W. G., & Simon, H. A. (1973). The mind's eye in chess. In W. G. Chase (Ed.), *Visual information processing* (pp. 215–281). New York: Academic.

Cohen, G. (1996). *Memory in the real world* (2nd ed.). Hove, England: Psychology Press.

Dawes, R. M., Faust, D., & Meehl, P. E. (1989, March 31). Clinical versus actuarial judgment. *Science, 243,* 1668–1674.

de Groot, A. (1978). *Thought and choice in chess.* The Hague, Netherlands: Mouton. (Original work published 1946)

Ericsson, K. A. (1996). The acquisition of expert performance: An introduction to some of the issues. In K. A. Ericsson (Ed.), *The road to excellence: The acquisition of expert performance in the arts and sciences, sports, and games* (pp. 1–50). Mahwah, NJ: Lawrence Erlbaum Associates.

Ericsson, K. A. (1998). The scientific study of expert levels of performance: General implications for optimal learning and creativity. *High Ability Studies, 9,* 75–100.

Ericsson, K. A. (1999). Creative expertise as superior reproducible performance: Innovative and flexible aspects of expert performance. *Psychological Inquiry, 10,* 329–333.

Ericsson, K. A. (2001). The path to expert performance: Insights from the masters on how to improve performance by deliberate practice. In P. Thomas (Ed.), *Optimizing performance in golf* (pp. 1–57). Brisbane, Queensland, Australia: Australian Academic Press.

Ericsson, K. A. (2002). Attaining excellence through deliberate practice: Insights from the study of expert performance. In M. Ferrari (Ed.), *The pursuit of excellence in education* (pp. 21–55). Mahwah, NJ: Lawrence Erlbaum Associates.

Ericsson, K. A. (2003). The search for general abilities and basic capacities: Theoretical implications from the modifiability and complexity of mechanisms mediating expert performance. In R. J. Sternberg & E. L. Grigorenko (Eds.), *Perspectives on the psychology of abilities, competencies, and expertise* (pp. 93–125). Cambridge, England: Cambridge University Press.

Ericsson, K. A., Chase, W., & Faloon, S. (1980, June 6). Acquisition of a memory skill. *Science, 208,* 1181–1182.

Ericsson, K. A., & Kintsch, W. (1995). Long-term working memory. *Psychological Review, 102,* 211–245.

Ericsson, K. A., Krampe, R. T., & Heizmann, S. (1993). Can we create gifted people? In CIBA Foundation Symposium 178, *The origin and development of high ability* (pp. 222–231). Chichester, UK: Wiley.

Ericsson, K. A., Krampe, R. T., & Tesch-Römer, C. (1993). The role of deliberate practice in the acquisition of expert performance. *Psychological Review, 100,* 363–406.

Ericsson, K. A., & Lehmann, A. C. (1996). Expert and exceptional performance: Evidence on maximal adaptations on task constraints. *Annual Review of Psychology, 47,* 273–305.

Ericsson, K. A., & Lehmann, A. C. (1999). Expertise. In M. A. Runco & S. Pritzer (Eds.), *Encyclopedia of creativity* (Vol. 1, pp. 695–707). San Diego, CA: Academic Press.

Ericsson, K. A., & Simon, H. A. (1993). *Protocol analysis: Verbal reports as data* (Rev. ed.). Cambridge, MA: MIT Press.

Ericsson, K. A., & Smith, J. (1991). Prospects and limits in the empirical study of expertise: An introduction. In K. A. Ericsson & J. Smith (Eds.), *Toward a general theory of expertise: Prospects and limits* (pp. 1–38). Cambridge, England: Cambridge University Press.

Fitts, P., & Posner, M. I. (1967). *Human performance.* Belmont, CA: Brooks/Cole.

Franklin, B. (2003). *The autobiography of Benjamin Franklin.* New Haven, CT: Yale University Press.

Gibson, E. J., & Pick, A. D. (2000). *An ecological approach to perceptual learning and development.* Oxford, England: Oxford University Press.

Goldstein, W. M., & Hogarth, R. M. (1997). Judgment and decision research: Some historical context. In W. M. Goldstein & R. M. Hogarth (Eds.), *Research on judgment and decision making* (pp. 3–65). Cambridge, England: Cambridge University Press.

Gould, S. J. (1996). *Full house: The spread of excellence from Plato to Darwin.* New York: Harmony Books.

Hadingham, E. (1978). The secrets of Stonehenge. In *The world's last mysteries* (pp. 83–91). New York: The Reader's Digest Association.

Helsen, W. F., Starkes, J. L., & Hodges, N. J. (1998). Team sports and the theory of deliberate practice. *Journal of Sport and Exercise Psychology, 20,* 12–34.

Hodges, N. J., & Starkes, J. L. (1996). Wrestling with the nature of expertise: A sport specific test of Ericsson, Krampe and Tesch-Römer's (1993) theory of "Deliberate Practice." *International Journal of Sport Psychology, 27,* 400–424.

Howe, M. J. A., Davidson, J. W., & Sloboda, J. A. (1998). Innate talents: Reality or myth? *Behavioral and Brain Sciences, 21,* 399–442.

Kintsch, W. (1998). *Comprehension: A paradigm for cognition.* Cambridge, England: Cambridge University Press.

Klein, G. A. (1993). A recognition-primed decision (RPD) model of rapid decision making. In G. A. Klein, J. Orasanu, R. Calderwood, & C. E. Zsambok (Eds.), *Decision making in action: Models and methods* (pp. 138–147). Norwood, NJ: Ablex.

Klein, G. A., Orasanu, J., Calderwood, R., & Zsambok, C. E. (Eds.). (1993). *Decision making in action: Models and methods.* Norwood, NJ: Ablex.

Knapp, C. (1947). Achievement scales in six physical education activities for secondary school boys. *Research Quarterly, 18,* 187–197.

Krampe, R. T., & Ericsson, K. A. (1996). Maintaining excellence: Deliberate practice and elite performance in young and older pianists. *Journal of Experimental Psychology: General, 125,* 331–359.

Lehmann, A. C. (1997). Acquisition of expertise in music: Efficiency of deliberate practice as a moderating variable in accounting for sub-expert performance. In I. Deliege & J. A. Sloboda (Eds.), *Perception and cognition of music* (pp. 165–191). Mahwah, NJ: Lawrence Erlbaum Associates.

Lehmann, A. C., & Ericsson, K. A. (1996). Music performance without preparation: Structure and acquisition of expert sight-reading. *Psychomusicology, 15,* 1–29.

Lehmann, A. C., & Ericsson K. A. (1998). The historical development of domains of expertise: Performance standards and innovations in music. In A. Steptoe (Ed.), *Genius and the mind* (pp. 67–94). Oxford, England: Oxford University Press.

Lipshitz, R. (1993). Converging themes in the study of decision making in realistic settings. In G. A. Klein, J. Orasanu, R. Calderwood, & C. E. Zsambok (Eds.), *Decision making in action: Models and methods* (pp. 103–137). Norwood, NJ: Ablex.

McWhirter, N., & McWhirter, R. (1975). *Guinness book of world records.* New York: Sterling.

Montgomery, H. (1993). The search for a dominance structure in decision making: Examining the evidence. In G. A. Klein, J. Orasanu, R. Calderwood, & C. E. Zsambok (Eds.), *Decision making in action: Models and methods* (pp. 182–187). Norwood, NJ: Ablex.

Montgomery, H., & Svensson, O. (1976). On decision rules and information processing strategies for choices among multiattribute alternatives. *Scandinavian Journal of Psychology, 17,* 283–291.

Neisser, U. (1976). *Cognition and reality: Principles and implications of cognitive psychology.* San Francisco: Freeman.

Neisser, U. (1982). *Memory observed: Remembering in natural contexts.* San Francisco: Freeman.

Payne, J. W. (1976). Task complexity and contingent processing in decision making: An informational search and protocol analysis. *Organizational Behavior and Human Performance, 16,* 366–387.

Pennington, N., & Hastie, R. (1993). A theory of explanation-based decision making. In G. A. Klein, J. Orasanu, R. Calderwood, & C. E. Zsambok (Eds.), *Decision making in action: Models and methods* (pp. 188–201). Norwood, NJ: Ablex.

Robergs, R. A., & Roberts, S. O. (1997). *Exercise physiology: Exercise, performance, and clinical applications.* St. Louis, MO: Mosby-Year Book.

Salthouse, T. A. (1984). Effects of age and skill in typing. *Journal of Experimental Psychology: General, 113,* 345–371.

Schneider, W. (1993). Acquiring expertise: Determinants of exceptional performance. In K. A. Heller, J. Mönks, & H. Passow (Eds.), *International handbook of research and development of giftedness and talent* (pp. 311–324). Oxford, England: Pergamon.

Shanteau, J. (1988). Psychological characteristics and strategies of expert decision makers. *Acta Psychologica, 68,* 203–215.

Simon, H. A. (1955). A behavioral model of rational choice. *Quarterly Journal of Economics, 69,* 99–118.

Simon, H. A., & Chase, W. G. (1973). Skill in chess. *American Scientist, 61,* 394–403.

Sloboda, J. A. (1996). The acquisition of musical performance expertise: Deconstructing the 'talent' account of individual differences in musical expressivity. In K. A. Ericsson (Ed.), *The road to excellence: The acquisition of expert performance in the arts and sciences, sports, and games* (pp. 107–126). Mahwah, NJ: Lawrence Erlbaum Associates.

Starkes, J. L., Deakin, J., Allard, F., Hodges, N. J., & Hayes, A. (1996). Deliberate practice in sports: What is it anyway? In K. A. Ericsson (Ed.), *The road to excellence: The acquisition of expert performance in the arts and sciences, sports, and games* (pp. 81–106). Mahwah, NJ: Lawrence Erlbaum Associates.

Williams, A. M., Davids, K., & Williams, J. G. (1999). *Visual perception and action in sport.* New York: Routledge mot EF & N Spon.

Young, M. C. (Ed.). (1997). *The Guiness book of world records.* New York: Bantam Books.

SOCIAL DECISION MAKING

11

Intuitive Team Decision Making

Dee Ann Kline

Mason General Hospital, Washington

Two major forces driving current organizational decision-making strategies are a necessity to make faster, better decisions and the use of teams in making those decisions. Although both are driven by demands from the organization's internal and external environments, they are not necessarily compatible strategies. Making decisions in a team enhances the likelihood that these decisions will not only incorporate multiple perspectives but that new levels of understanding will develop (Ellis & Fisher, 1994). In reality though, the higher quality team decision is often made at the expense of speed, as team decision-making cycles are generally shown to be longer than those of individual decision makers (Ellis & Fisher, 1994). As a result, the organizational development literature has suggested that as organizations foster the development of teams, they simultaneously develop new methods of team decision making, including the use of intuition (Peters & Waterman, 1982; Senge, 1990).

Expert teams are described from the cognitive standpoint of shared mental models (Cannon-Bowers, Salas, & Converse, 1993) and the social aspect of cohesiveness (Cartwright, 1968). Group cohesiveness is a widely studied phenomenon, which is developed through the conditions promoting interdependence of group task work, teamwork, and pride in and identity of the group (Cartwright, 1968; Festinger, 1950; Zaccaro & Lowe, 1988). Cannon-Bowers et al. (1993) proposed that shared mental models also form from the expectations of shared teamwork and team tasks and are based on team members' collective knowledge, experiences, and skills. Both cohesiveness and shared mental models describe a level of development in which groups become interdependent in accomplishing their task.

Newell and Simon (1972) examined the relation between intuition and shared mental models. While studying differences between expert and novice chess players, they found a strong correlation between the quality of mental models and the level of proficiency. This expertise was dependent on content-specific knowledge about the task; however, an equally important aspect was the expert's ability to access and use the knowledge in ways that were more efficient than that of the novice. Simon (1983) described this efficiency in accessing and using the knowledge in mental models as intuition. If shared mental models can be considered a group-level representation of mental models, then the research associating intuition with highly developed mental models can be extrapolated to explain shared mental models as the construct by which teams make intuitive decisions.

The data for this research on intuitive team decision making was derived from interviews with five cohesive teams involving a total of 22 individuals. The teams came from health care, a consulting firm, a privately held corporation, and two governmental agencies. All were asked to recall and then asked questions about a decision that "just seemed to happen" or "a difficult decision where there was initially a struggle to get a perspective and then the answer just seemed to be there." The data was analyzed utilizing the constant comparative method of grounded theory (Glaser & Strauss, 1967).

Intuitive decision making is at first glance difficult to distinguish from a cohesive team's routine decision-making processes. Teams are able to describe an intuitive decision they have made, yet the terminology and some of the activities resemble what I consider to be analytical, nonintuitive decision making. However, close analysis reveals some distinct conditions and activities of intuitive decisions.

When a team is faced with a decision event, they intuitively compare the event to the knowledge in the shared mental models. If the event can be understood by this comparison, the team implicitly knows the solution and is able to rapidly reach intuitive consensus. The decision is described as something that just happened, an agreement to just act, or something that the team just knew. The team reaches consensus with limited discussion of the decision event. If they do suggest different courses of action, they do not stop and evaluate each one.

Discussions and evaluations do play a role in intuitive team decision making; however, they largely take place after the decision is made. The intuitive decision is followed by a period in which the team both validates the decision and concurrently plans the course of implementation.

The model of intuitive decision making in cohesive workplace teams (Fig. 11.1) is shown as having distinctive steps, yet intuitive team decision

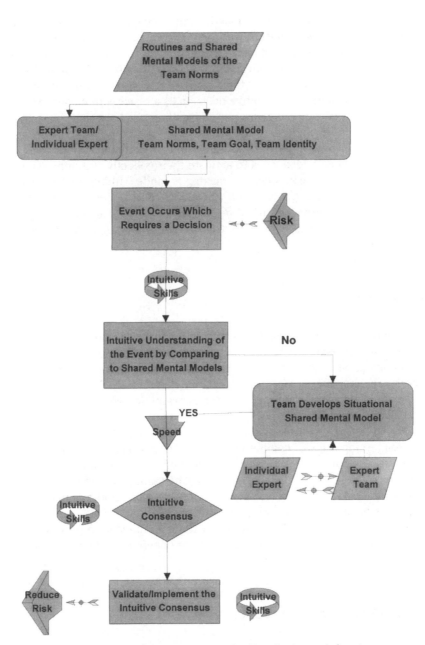

FIG. 11.1. Model of intuitive decision making in cohesive workplace teams.

making is holistic in nature. A holistic experience is characterized by the awareness of an overall pattern without being able to explain the individual components of the pattern. It has no sequential steps, is experienced rather than conceptualized, and known rather than rationalized (Gardner, 1985).

Team members experienced their intuitive decision in a holistic fashion. They did not describe the intuitive decision as a process that started at "A" and ended in "B" but rather as something that just happened, something they just knew, or an action they just took.

Yet each team was able to describe significant common experiences and knowledge that served as cues to activate their collective intuition. The actual act of intuitive team decision making is made without specific verbal reference to the team's common knowledge and experience. However, when team members were asked to describe what was happening as they made the decision, they were able to elucidate thoughts, feelings, and factual knowledge.

ROUTINES

The routines that are associated with intuitive team decision making largely center on the ways in which teams communicate with one another. This communication is often informal, as when team members tell one another about projects, discuss work problems, or ask for advice. More formal communication occurs during meetings when teams follow a planned agenda and share information that is specific to an agenda item.

The literature (Corner, Kinicki, & Keats, 1996; Kim, 1997) has described the reinforcing roles of communication and shared mental models.

Teams use different processes to communicate with one another and build or reinforce shared mental models. One team described keying off of one another. One person thinks of something and another builds on it. Members of another team use each other as sounding boards for new ideas, situations, or problems. No matter how the communication process is described, the end result is the same: a greater common understanding of the team goals, identity, and norms.

SHARED MENTAL MODEL: TEAM NORMS

Cohesive workplace teams have a shared mental model of team norms that are most concisely described as behaviors that team members expect one another to exhibit. Team members described norms such as honesty, trust, respect for others, accountability, empathy, belonging, and open communication. Coming to consensus was consistently mentioned as a decision-making norm.

When team members have a shared mental model of their norms, they foster an environment that enables them to work interdependently to achieve their goals. For one team member, a norm means something as simple as team members being able to count on one another. In other teams, the shared mental model of the team norms may be instrumental in making an intuitive decision. One team from a governmental agency intuitively selected a new team member because they felt the applicant would best fit with the team. They intuitively recognized that the candidate's values were very similar to those of the current team members.

SHARED MENTAL MODEL: TEAM GOALS

Cohesive workplace teams have a shared mental model of goals that are closely aligned with the vision or philosophy of their team. The shared mental model of team goals includes common understandings about the objectives that the team is trying to achieve. It is also comprised of common understandings about the tasks and experiences that have helped them achieve or prevented them from achieving these goals. Unless the vision or purpose changes, the shared mental model of the goals does not change. However, the shared mental model of how to achieve the goals will change as the team shares new experiences and tasks.

The shared mental model of the goal is the most predominant shared mental model described in relation to intuitive team decision making In explaining why their professional oversight team did not adopt a survey that was being adopted in other parts of the country, a team member of a governmental agency said, "Our gut feeling was that it is not going to work. They would rather have us come in for an hour and get this over with rather than sitting there spending an hour or 2 hours filling out paperwork and then sending us a stack of paper because they don't know what we want or what we are asking for."

SHARED MENTAL MODEL: TEAM IDENTITY

The identity is either aligned with and therefore a close resemblance to the identity of the system in which they reside or a special niche that is perceived to be superior to their referent groups. The special niche is often commonly understood as the exceptional manner in which the team meets their goals, and it is therefore not always easy to differentiate the identity from the goals.

Whatever the source, decisions will be made that intuitively reinforce that identity. A team member from a privately held corporation depicted this concept when she talked about the selection of a team logo. "I don't think any of us said out loud 'I want to be proud of this because it represents us,' but certainly that's what we were going for and I wanted it to be dignified and yet bright and eye-catching. Those weren't things we said out loud, but I think just intuitively we all knew."

EXPERT TEAM/INDIVIDUAL EXPERT

Supported by and supporting shared mental models is the concept of the expert team. The expert team is formed when the knowledge and experience of the team has reached a level at which the whole team is considered an expert. They know the expectations and roles of the team and one another. They operate fairly autonomously, yet with an understanding of their role within the larger system/organization. They have committed themselves to working as a team to achieve their goals. They are able to reach consensus on the majority if not all of their decisions.

Embedded in the expert team is the expert individual. This team member has extensive knowledge or experience about a certain aspect of the work of the team. The expert is expected to share that information to educate the other team members. One team member from a governmental agency provided an example of how the individual expert operates within the expert team. "So much of what we do is governed by laws and civil service laws and that system's rules. Something for me that might come by itself, something that I might have experienced that you haven't, or something you have experienced that I haven't. I talk with you and share the knowledge that we have instead of reinventing the wheel each time."

Drawing on the expert–novice studies of intuition (Simon, 1983) as a model for intuitive team decision making, the expert team has developed strong shared mental models that cause them to intuitively understand an event in relation to prior events they have experienced together. Collectively understanding the event allows each team member to make an intuitive decision that is the same as his or her teammates.

INTUITIVE SKILLS

The use of intuition during team decision making is represented by a variety of intuitive skills that are based on Cappon's (1993) anatomy of intuitive skills and Dreyfus and Dreyfus' (as cited in Benner & Tanner, 1987) six key aspects

of intuitive judgment. Because intuition is largely outside of conscious awareness, the team members act without realizing they are doing so.

However, in the team setting, they are described from a collective standpoint. One team member related an example of this collectivity. "The thing
that impressed me was that again we, the four of us, independently all saw
the same thing." Four new intuitive skills—just knew, just happened, just
acted, and creation—were defined directly from this research. Although intuitive skills permeate all steps of team intuitive decision making, some are
more closely aligned with a certain part of the intuitive decision. Figure 11.2
shows the relation of the intuitive skill to a particular step.

INTUITIVE UNDERSTANDING

Using one or more intuitive skills, the team compares the decision event to
their shared mental models. If the event can be understood by this comparison, the team implicitly knows the solution and implicitly makes the decision. They do not engage in common team decision-making tasks such as
clarifying the goal, evaluating the alternatives, or collecting more information. They immediately and collectively come to consensus about a course
of action.

When the newest member of a consulting firm raised the question,
"What business are we in?," the team members all intuitively realized there
was a dissociative match between the work they were doing and the shared
mental model of the vision and goal of the firm. According to one member,
"I knew this is what we needed to do and what we wanted to do and I knew
intuitively if we didn't do it, we'd never build our business."

By intuitively putting themselves in the place of the facilities they sur-
veyed, an oversight team from a governmental agency understood a new survey in relation to the shared mental model of their goal of establishing realistic regulations for their clients. When members of a team from health care
discovered a serious clinical error made by a nurse, they intuitively realized
the salience of the incident, describing it as a "red flag" in relation to their
shared mental model of how to provide exceptional clinical care and safety.
They also intuitively understood the patterns of the relations between the error, their experiences with this person, and their knowledge of established
clinical routines.

RISK

Although only one team that was interviewed for the research specifically
mentioned risk, there was an element of risk in the intuitive decision that
four teams described. In analyzing the risk aspect of intuitive decision mak-

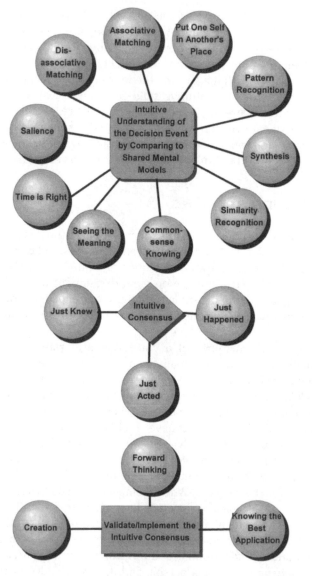

FIG. 11.2. Intuitive skills.

ing, Agor (1986) found that uncertainty is a condition of intuitive decision making. The greater the uncertainty, the higher the risk that the goal will actually be obtained. It is possible that the riskiness of the decision is somehow sublimated when cohesive teams come to intuitive consensus. The risk of the decision is then brought to conscious awareness after the decision has been made.

SITUATIONAL SHARED MENTAL MODELS

Shared mental models also provide the basis for the construction of another more dynamic type of mental model: the situational shared mental model. When an event cannot be completely understood by intuitively comparing it to the shared mental models, the cohesive team still has an opportunity to make an intuitive decision. They will attempt to construct a situational shared mental model of the piece they do not understand. If they are successful, they can rapidly reach intuitive consensus.

During the construction of a situational shared mental model, the team members may suggest alternative courses of action. They do not stop and evaluate any one alternative but instead share information to clarify the decision event. When the right course of action is suggested, the team instantly and collectively agrees to it. A team member from a privately held corporation succinctly described this process. "I think that we had talked ourselves through it and knew pretty much what we wanted by the time we got to it conceptually."

The dynamic relationship between the expert team and the individual expert plays an important role when building situational shared mental models. The information shared by the member who has the most knowledge or expertise about the event will sometimes intuitively be perceived as more salient than information shared by team members with less expertise.

INTUITIVE CONSENSUS

Intuitive team decision making is characterized by an ability to speed up decision making and quickly reach a decision. The team from the consulting firm epitomized the speed of intuitive team decision making when within 3 minutes maximum they said, "Of course, we are not in that business anymore." Yet although the decision was fast, it was not made without a shared mental model. As one team member related, "God knows we had been thinking about it and talking about pieces of it before we met. But the decision to just drop it was in those 3 minutes for me."

Intuitive consensus is not reached by negotiating disparate views; these disparate views are already incorporated into the shared mental models. Rather, intuitive consensus occurs when the team uses one or more intuitive skills to understand the event by comparing it to the shared mental models. The team instantaneously and collectively reaches an agreement and implicitly and explicitly agrees to support it.

Intuitive consensus is commonly described as just knowing, just happening, or agreeing to just act. Participants from two different teams described it as almost being a nondecision but with no question that they would be proceeding with it.

VALIDATING AND IMPLEMENTING THE INTUITIVE DECISION

According to Agor (1986), when people make an intuitive decision, they often try to disguise the fact that it was made intuitively. They do this by collecting information after the decision is made and then using that information to justify it to others. A team follows this pattern, only they also validate the decision to one another. Validation brings the commonly shared knowledge into conscious recognition, thereby confirming that the decision is consistent with the shared mental models from which the decision was made.

At this point, the team also designs the implementation of the decision. Validating and implementing are closely interwoven because as the team plans the implementation, they simultaneously reinforce the decision they made. The intuitive skills that teams use at this stage include forward thinking, that is, envisioning what will happen next, finding creative ways to think about or implement the decision, and knowing how to best apply the decision.

During the validation period, risk is recognized and cognitively reduced. Although the riskiness of the decision is sublimated at the time the decision event is assessed, it is discussed as the team begins to validate the decision.

SUMMARY

From an organizational standpoint, the value of this model will be in "teaching" teams to make intuitive decisions. This will involve guiding teams in the development of certain skills as well as teaching them to recognize the significant cues of an intuitive decision. Because intuitive decisions are made outside of conscious awareness, teams cannot be taught to make intuitive decisions. They can, however, learn to develop the strong shared mental models that enable intuitive decision making.

Because communication is necessary for the development of strong shared mental models, teams will need to learn basic communication skills such as active listening, fashioning an argument, and representing personal viewpoints. Once these skills are in place, teams can be taught to use dialogue as a means of questioning assumptions and developing a deeper understanding of one another's viewpoints, roles, and experiences. As they begin to develop these understandings, they will begin to establish norms and come to agreement about the values they expect one another to exemplify. They will devise and commit to communication routines that support and enhance their existing communication channels.

Teams must have a strong goal based on the vision and purpose of their work as well as their relation to and role in the larger system. They will develop evaluation tools to help them assess how well they are achieving these goals and skills that will enable them to continually assess the basic assumptions of the goal.

Teams should be taught the more traditional methods of team decision making. As they become comfortable with these patterns, they will begin to learn how they might also be able to make intuitive decisions. They will learn the following:

1. How and when to share unique or commonly held information and develop trust in one another's knowledge and expertise.
2. To recognize the antecedent conditions of an intuitive decision: a high level of uncertainty, multiplicity of solutions, little previous precedent, and short time frame.
3. To understand the role of risk in a decision.
4. To recognize when the team has made an intuitive decision and raise important issues about the decision while simultaneously planning the implementation.

REFERENCES

Agor, W. H. (1986). *The logic of intuitive decision making: A research-based approach for top management.* Westport, CT: Quorum.

Benner, P., & Tanner, C. (1987, January). How expert nurses use intuition. *American Journal of Nursing, 87*(1), 23–31.

Cannon-Bowers, J. A., Salas, E., & Converse, S. (1993). Shared mental models in expert team decision making. In N. J. Castellan, Jr. (Ed.), *Individual and group decision making: Current issues* (pp. 221–246). Hillsdale, NJ: Lawrence Erlbaum Associates.

Cappon, D. (1993, May/June). The anatomy of intuition. *Psychology Today, 26*(3), 41–92.

Cartwright, D. (1968). The nature of group cohesiveness. In D. Cartwright & A. Zander (Eds.), *Group dynamics: Research and theory* (pp. 91–109). New York: Harper & Row.

Corner, P. D., Kinicki, A. J., & Keats, B. W. (1996). Integrating organizational and individual information processing perspectives on choice. In J. R. Meindl, C. Stubbart, & J. F. Porac (Eds.), *Cognition within and between organizations* (pp. 145–172). Thousand Oaks, CA: Sage.

Ellis, D. G., & Fisher, B. A. (1994). *Small group decision making: Communication and the group process.* New York: McGraw-Hill.

Festinger, L. (1950). Informal social communication. *Psychological Review, 57*, 271–282.

Gardner, H. (1985). *The mind's new science: A history of the cognitive revolution.* New York: Basic Books.

Glaser, B., & Strauss, A. L. (1967). *The discovery of grounded theory.* Chicago: Aldine.

Kim, P. H. (1997). When what you know can hurt you: A study of experiential effects on group discussion and performance. *Organizational Behavior and Human Decision Processes, 69*, 165–177.

Newell, A., & Simon, H. A. (1972). *Human problem solving.* Englewood Cliffs, NJ: Prentice Hall.

Peters, T. J., & Waterman, R. H., Jr. (1982). *In search of excellence: Lessons from America's best run companies.* New York: Warner Books.

Senge, P. M. (1990). *The fifth discipline: The art and practice of the learning organization.* New York: Doubleday/Currency.

Simon, H. A. (1983). *Reason in human affairs.* Stanford, CA: Stanford University Press.

Zaccaro, S. J., & Lowe, C. A. (1988). Cohesiveness and performance on an additive task: Evidence for multidimensionality. *Journal of Social Psychology, 128,* 547–558.

CHAPTER 10

12

Bridge Resource Management Training: Enhancing Shared Mental Models and Task Performance?

Wibecke Brun
Jarle Eid
Bjørn Helge Johnsen
Jon C. Laberg
Belinda Ekornås
Therese Kobbeltvedt
University of Bergen, Norway
and The Royal Norwegian Naval Academy

The main goal of this study was to develop a course for crew members that would facilitate crew interaction and shared mental models (SMM) among the crew members, with the ultimate goal of enhancing team performance. Recent studies on crew resource management training of cockpit crew members have concluded that such training is beneficial (Salas, Fowlkes, Stout, Milanovich, & Prince, 1999; Salas, Prince, et al., 1999), and there is no reason to suppose that it would not be beneficial also in the naval context we studied.

The main goal of the training was to enhance qualities that characterize well-performing teams (Salas, Cannon-Bowers, & Bowers, 1995; Stout, Cannon-Bowers, & Milanovich, 1999; Tannenbaum, Beard, & Salas, 1992; Urban, Bowers, Monday, & Morgan, 1995). Communication and decision making in operative environments, the importance of well-defined roles in teamwork, and team-interaction processes in tactical decision making were among the topics of the course. The importance of knowing the task and the specific roles of the team members, having the skills to meet the task requirements, and taking responsibilities was demonstrated through specially designed problem-solving tasks and so forth. The course followed a learning-by-doing philosophy. An important goal was to enhance the SMM of the bridge teams through practical team training.

The first focus of this pilot study concerned the degree of SMM within each team in accordance with the four models Cannon-Bowers, Salas, and Converse (1993) suggested: equipment model, task model, team-interaction model, and team model. We were interested in whether there was a higher degree of SMM within some domains than other domains. It seems reasonable to assume that our participants would have a more common understanding in areas regarding standard operating procedures (SOPs) and technical matters than in areas involving personal skills and characteristics. The next objective was to study the effect of a bridge resource management (BRM) program on the degree of SMM in bridge teams and the effect of the program on team performance in a sailing task. We expected to find that the BRM course would facilitate SMM and improve team performance. The last objective was to study a possible relation between degree of SMM in the teams and their performance in a simulated sailing task. We expected to find that teams with a higher degree of SMM would perform better than teams with a lower degree of shared understanding.

METHOD

Participants

Twenty-two male and 2 female cadets from The Royal Norwegian Naval Academy served as participants. They participated in the experiment as part of their ordinary cadet training and received no payment for participating. The participants were already operating as four teams (each with six team members) before the experiment started.

Procedure and Experimental Design

The experiment was composed of three parts. The first part was a planning session in which the participants were asked to plan a specific sailing route. The second part of the experiment was a BRM training program. The third part was a practical sailing task in a navigation simulator (to sail the route they had planned in the planning phase).

All participants first participated in the planning session. After this session, both groups (experimental and control) filled in a questionnaire designed to measure SMM pretest (SMM1). Afterward, the experimental group was exposed to the BRM course, and finally, they conducted the sailing task in the navigation simulator. The order was reversed for the control group (sailing task first, then BRM training). The experimental group filled in the posttest questionnaire (SMM2) after having been exposed to the BRM course and the simulator sailing session. The control group filled in their SMM2 questionnaire after the sailing task, followed by the BRM training.

Equipment

A full-scale navigation simulator produced by STN ATLAS Electronic GmbH was used in the study. The simulator includes all kinds of modern nautical equipment such as track control, satellite-based communication, and GMDSS communication. It has a powerful, real-time visual system DISI 5 M. It provides a realistic impression of the outside world, including traffic movements in ports and sea areas.

Planning Session

The teams were asked to plan a specific sailing route (from Brekstad to Stokkesund in Norway). The commanding officer was free to delegate the subtasks within his team. During the planning phase, the cadets had sailing charts available as well as weather and tide reports. The planning phase lasted for about 2½ days.

BRM Program

The weeklong BRM program used in the study was developed at the Royal Norwegian Naval Academy. The course is a mandatory part of the cadets' naval education. The training program combines practical tasks, case studies, lectures, and simulator training.

The main purpose of the course is to integrate theoretical and practical training of cadets, thereby trying to enhance the SMM of the bridge teams. The course is theoretically based on general psychological literature (i.e., theories of perception, information processing, problem solving, stress, leadership, communication, etc.) as well as on findings from the naturalistic decision making (NDM) domain. The aim is to enhance qualities that, according to the NDM literature, characterize teams that perform well. The program deals with individual, dyadic, group, and team processes. Throughout the course, human perception and the way it may be influenced and biased by expectations and stress is highlighted. The first part, "Coping in critical situations," treats mental preparedness, situation awareness, and individual coping strategies. Information about mental processes and typical behavior in stressful situations are central topics.

The second part of the course focuses on dyadic and group processes. Communication and decision making in operative environments are the main foci. The importance of precise communication and well-defined roles in decision-making processes is demonstrated for the cadets.

The third session is assigned to teamwork. Focus is on team-interaction processes in tactical decision making. The concept of SMM is introduced, and the importance of knowing the task and each team member's roles, responsibilities, and skills to meet the task requirements is demonstrated.

To assess the cadets' subjective experience of the BRM program, all participants filled in an evaluation questionnaire. The cadets rated the course on a number of factors, such as relevance for the other navigation education they have had, their preference for the different components, and how well the different sessions were integrated, as well as their overall satisfaction with the course rated on 6-point scales ranging from 0 (*not satisfied*) to 5 (*very satisfied*).

SMM Questionnaire

An SMM questionnaire was developed to measure four different categories of mental models:

1. Equipment model: functioning of equipment, operating procedures, equipment limitations, and possible failures.
2. Task model: task procedures, likely contingencies, likely scenarios, task strategies, and environmental constraints.
3. The team-interaction model: perceptions of roles/responsibilities, information sources, interaction patterns, and role interdependencies.
4. Team model: team members' understanding of each other with respect to knowledge, skills, abilities, preferences, and emotional reactions.

For the shared mental equipment model, the participants rated the following types of equipment regarding suitability when sailing outside the lead: electronic chart, paper chart, loran C, log, echo sounder, navtex, and global positioning system (GPS). A 5-point rating scale was used, and the seven types of equipment were rated from (*very satisfying*) to (*very unsatisfying*). The degree of shared mental equipment model was scored by counting the percentage of identical scoring of team members in each team for the seven equipment types.

The degree to which the participants shared a mental task model was measured by asking the cadets to check the two most important tasks to execute in a critical situation among seven alternative tasks (give signal, turn the ship, stop the ship, check charts, turn on lights, use VHF Channel 16, other actions). The question read as follows: "You are sailing in a narrow passage at 25 knots and get fuzzy radar signals 0.10 NM in front of the ship. You are not observing anything visual in the nearby waters. What are the first two tasks you would execute?" The degree of shared mental task model was scored by counting percentage of participants choosing—or not choosing—each of the given task procedures and by estimating shared agreement (in percent) of choosing or not choosing to perform a task.

The shared mental team-interaction model was measured by asking the participants to name the information sources needed on the bridge for each team member. The question read as follows: "Which information sources does your commander/second commander/operational officer/navigation officer/helmsman/lookout need?" The cadets made a choice among the same nine sources of information for all the different roles in the team. The available information sources were communications, paper chart, loran C, echo sounder, binocular, electronic chart, GPS, log, and navtex. The percentage of identical scorings indicated shared mental team-interaction model.

To get information about the extent to which the participants had a shared mental team model, the cadets answered questions regarding anticipated emotional reactions and characteristics of their teammates. For example, "Consider that the helmsman makes a mistake and this leads to a critical situation. How would you and the members in your team react in this situation?" Each team member was rated regarding three different emotional reactions (calmness, frustration, support) on 5-point scales ranging from 1 (*not at all*) to 5 (*very strong reaction*). Another question was related to stress; it read, "To what extent would you think you and the other members of the team would feel stressed if you ran aground/experienced a shipwreck?" The same 5-point scale was used to rate anticipated stress reaction. The percentage of identical scoring of anticipated emotional reactions of the team members indicated the shared mental team model.

Sailing Task in Simulator

The cadets acted as the bridge crew on KNM Skudd, an MTB of the Hauk class, using the simulator described previously. They were to sail from "Brekstad harbour" to "Stokkesund" where the commander of the naval forces of southern Norway would embark. The teams should have been able to sail this distance within an hour, and they were informed that the quality of their task performance would be evaluated.

Officers at the Navigation unit, in cooperation with researchers in military psychology at the naval academy, developed a set of criteria for rating performance in the simulated sailing task. Two experienced naval training officers observed and judged the performance of the bridge teams in the simulator as they were performing the interactive navigation task.[1] The expert judges subjectively rated the cadets' and teams' performance on a

---

[1]Unfortunately, these experts also participated as instructors at the BRM training program and were aware of the experimental design.

scale ranging from 0 (*no correct performance*) to 10 (*100% correct perform-ance*) for nine different domains. The domains were preparations, distribu-tion of workload, navigation, consequential awareness, SOPs, teamwork, ship handling, communication, and decision making.

RESULTS

The degree of SMM in each team was measured in accordance with the four kinds of mental models in Cannon-Bowers et al. (1993). The frequency of identical ratings (on a 5-point rating scale) of the suitability of different kinds of equipment indicated the extent to which there existed a shared team understanding in the equipment domain. Table 12.1 shows the mean percentages of identical scoring for experimental and control teams regard-ing the suitability of different kinds of equipment.

The pretests showed a medium degree of shared understanding within the teams (from 38% to 52% across the teams). The results with respect to the SMM2 were quite similar to that in the pretest (ranging from 48% to 50% identical scores on average). The ratings mean that up to half of the partici-pants shared the same attitudes compared to the a priori probability that any given rating was given just by chance (i.e., 20%). Contrary to our expec-tation, we did not find that participating in the BRM course facilitated a shared understanding of equipment needs. On average, there was only a 2.8% change in the direction of higher agreement in the posttest.

To find an indication of the amount of shared task understanding within a team, we calculated the number of team members who ticked each of

TABLE 12.1
Overview of Results on Shared Mental Models (SMM): Change
From Pretest to Posttest Administration of SMM in Experimental
and Control Groups, Total Mean of SMM, and Total Change
From Pretest to Posttest Administration of the SMM Questionnaire

| SMM | Experimental Group | | | Control Group | | | M SMM | M Change |
	Pretest	Posttest	Change	Pretest	Posttest	Change		
Equipment (a priori 20%)	45.6	48.8	3.2	46.4	48.7	2.4	47.4	2.8
Task (a priori 50%)	76.1	78.6	2.4	72.9	70.3	−2.6	74.5	−0.1
Team interaction (a priori 50%)	84.0	85.1	1.1	84.1	88.8	4.7	85.5	2.9
Team (a priori 20%)	45.6	53.0	7.4	47.4	53.8	6.4	49.9	6.9

Note. All values are percentages.

seven tasks to be implemented first in case of a critical situation. The critical situation refers to "sailing in a narrow passage at 25 knots while suddenly getting fuzzy radar signals 0.10 NM in front of the ship." The a priori probability that a task should be ticked just by chance was 28.5% (six team members ticking two tasks each over seven tasks). We found a 37% agreement on the choices (averaged over the control and experimental groups and over pretest and posttest administration of the questionnaire). In addition, there were tasks that no participants chose (i.e., perfect agreement about lack of suitability).

The four teams exhibited quite similar scores on task priorities. The most popular task was to turn on the lights (9 participants ticked this option in the pretest and 12 in the posttest questionnaire), followed by checking the charts (9 participants in both test administrations), and turning the ship (9 in the pretest and 8 in the posttest administration).

Calculating the total percentage of agreement on the choices (that a task should be implemented or not) gives an indication of the extent to which the team members shared a common understanding of the need to implement given tasks in the scenario. On average, there was a 74.5% agreement on the choices. This number should be compared to an a priori probability of 50%. There were no systematic changes in overall agreement between the pretest and the posttest administration of the questionnaire. The most frequently chosen tasks were turn on light, check chart, and turn or stop the boat, irrespective of whether the participants gave their answers before or after the BRM course and regardless of whether the teams had participated in the simulator task. It seems reasonable to regard these task priorities as reflections of SOPs. These are procedures in which the cadets are more or less drilled before the experiment; therefore, we should not have expected the experimental manipulation to have too much effect.

Did the participants agree on what information sources different crew members need? The shared team understanding regarding the other crew members' technical needs—team-interaction mental model—is shown in Table 12.1. The frequency of identical scorings over eight different information sources for six different crew roles is presented. Average team scores varied from 83% to 85% agreement on the SMM1 questionnaire and from 83.5% to 91% agreement on SMM2. There was, on average, only a 2.9% increase in agreement in the posttest compared to the pretest.

Almost all participants agreed that the commander and the second in command would need nearly all the available equipment (with a possible exception of electronic charts). The helmsman and the lookout, on the other hand, would need very little equipment. The participants were in total agreement about the fact that the lookout would need binoculars, and that the helmsman would need communication. The operational officer would need communication and paper charts, whereas the navigation offi-

TABLE 12.2
Team Model: Mean Percentage Identical Ratings[a] of Anticipated Reactions
Within Each Team Averaged Over Six Different Team Roles

Anticipated Reactions	Control Group				Experimental Group			
	Team 1		Team 2		Team 3		Team 4	
	Pretest	Posttest	Pretest	Posttest	Pretest	Posttest	Pretest	Posttest
Stress	42	53	50	67	39	42	44	50
Calmness	56	61	42	47	39	58	42	42
Social support	50	44	47	53	56	69	39	58
Frustration	50	58	42	47	56	61	50	44
M	49.5	54.0	45.3	53.5	47.5	57.5	43.8	48.5

Note. $N = 24$ (6 in each team).
[a]Ratings were on a scale ranging from 1 (not at all) to 5 (high degree).

cer would need paper charts, GPS, and the log. To conclude, there was a very high agreement among the team members on what were considered relevant information sources for each team member, with very little change from pretest to posttest administration of SMM. No systematic differences between the control and experimental groups were found.

Anticipated emotional reactions of the other team members in response to a given critical scenario were studied to find an indication of shared team understanding of personal characteristics and emotional reactions of the crew. Table 12.2 shows the degree of shared understanding regarding different emotional reactions expected from other team members.

Inspection of the table reveals a weak tendency toward more identical scorings of what the participants anticipated to be their team members' emotional reactions to a critical event in the posttest administering of the SMM questionnaire compared to the pretest. This held true for all groups.

The judgments made by the expert judges concerning performance of the bridge teams are shown in Table 12.3.

According to our experts, all participants performed well on the preparation phase of the task. Also, navigation, ship handling, and following standard procedures received relatively high ratings.

The results on a questionnaire used to assess the cadets' subjective perception of the BRM course showed that generally the cadets were very pleased with the course as a whole (M rating 4.1 on a scale ranging from 0 to 5). They were particularly positive to using simulator training to highlight the processes, which were introduced in the lectures. Thirteen out of 21 cadets (3 missing) mentioned this as one of the reasons for their satisfaction with the course.

TABLE 12.3
Team Performance: Mean Expert Rating
of Team Performance on the Simulated Sailing Task[a]

	Control Group		Experimental Group	
Criteria	Team 1	Team 2	Team 3	Team 4
Preparations	7.0	7.0	7.0	7.0
Distribution of workload	2.0	5.0	6.0	6.5
Navigation	3.0	7.0	6.0	7.0
Consequential awareness	2.0	5.0	4.0	6.0
Procedures	4.0	6.0	6.0	6.0
Teamwork	2.0	5.0	6.0	6.0
Ship handling	5.0	5.0	6.0	7.0
Communication	2.0	4.0	6.0	6.0
Decision making	2.0	5.0	4.0	6.0
M	3.2	5.4	5.7	6.3

Note. $N = 24$ (6 in each team).
[a]Scores based on agreement between two expert raters on a scale from 0 (*0% correct performance*) to 10 (*100% correct performance*).

DISCUSSION

With respect to the degree of SMM in the teams concerning the four domains equipment, task, team interaction, and team, the results showed only small—and nonsystematic—differences among teams in the degree of shared team understanding in these four domains. Generally, the degree of agreement was high in all domains. We expected a higher agreement in the technical domains (regarding equipment and task models) than in the personal domains (team and team-interaction models), but no clear tendencies in this direction were found.

The answers to our questions regarding the task model (prioritize action in a critical situation) showed that there did not exist a common understanding of what to do in such a critical situation. It seems reasonable to think that the mental model of task priorities should be less a matter of personal preferences and attitudes and more a matter of what are defined as SOPs. Taking this into consideration, it is surprising that the answers regarding task priorities in a critical situation were not more uniform than found in this study.

There were no systematic differences between SMM pretest and posttest scores in the experimental group, compared to the control group. The team model questions showed the clearest tendency toward a change from pretest to posttesting, but there were no systematic differences between the control and experimental groups.

Did teams with a higher degree of SMM perform better in the simulated sailing task? One of the teams (Team 1) received lower performance ratings than the rest of the teams. Due to low sample size of teams, it is not possible to conclude from our data that this was due to a lower degree of SMM in this team than the others. It seems reasonable to conclude that if the BRM course does enhance performance, it is not because the course systematically increases the SMM of the teams. This is an important result to test further in new studies with a larger sample of bridge teams.

The results showed no systematic increase of ratings on the SMM questionnaire from pretest to posttest in the experimental group across the domains. In most cases, we found an increased SMM score in the posttest administration of the questionnaire. However, these results are not exclusive for teams in the experimental group. In fact, all groups had a higher degree of shared understanding of the emotional reactions of their team members in the posttest compared to the pretest. This means that we cannot attribute an increase in SMM to the BRM training program.

One can always question whether a simulated environment such as the one used in our study can provide a realistic experience and learning environment for the participants. Earlier studies have shown, however, that the effect of simulator perception is very similar to that of real life (see, e.g., Rudle, Payne, & Jones, 1997), and experiments in interactive environments such as simulators have considerable ecological validity.

Experimental teams in earlier research have often been put together for the sake of the study, not being part of existing work teams. The cadets in this study were part of teams that had been together for 2 years before participating in this study. This may explain why we generally found SMM among the participants but no systematic increase in SMM due to participating in a specific BRM training program.

REFERENCES

Cannon-Bowers, J. A., Salas, E., & Converse, S. A. (1993). Shared mental models in expert team decision making. In N. J. Castellan, Jr. (Ed.), *Individual and group decision making: Current issues* (pp. 221–246). Hillsdale, NJ: Lawrence Erlbaum Associates.

Rudle, R. A., Payne, S. J., & Jones D. M. (1997). Navigational buildings in "desk-top" virtual environments: Experimental investigations using extended navigational experience. *Journal of Experimental Psychology Applied, 3*, 143–159.

Salas, E., Cannon-Bowers, E. S., & Bowers, J. A. (1995). Military team research—Ten years of progress. *Military Psychology, 7*, 55–75.

Salas, E., Fowlkes, J. E., Stout, R. J., Milanovich, D. M., & Prince, C. (1999). Does CRM training improve teamwork skills in the cockpit? Two evaluation studies. *Human Factors, 41*, 326–343.

Salas, E., Prince, C., Cannon-Bowers, E. S., Stout, R. J., Oser, R. L., & Cannon-Bowers, J. A. (1999). A methodology for enhancing crew resource management training. *Human Factors, 41*, 161–172.

Stout, R. J., Cannon-Bowers, E. S., & Milanovich, D. M. (1999). Planning, shared mental models, and co-ordinated performance: An empirical link is established. *Human Factors, 41*, 61–71.

Tannenbaum, S. I., Beard, R. L., & Salas, E. (1992). Team building and its influence on team effectiveness: An examination of conceptual and empirical developments. In K. Kelley (Ed.), *Issues, theory, and research in industrial/organizational psychology* (pp. 117–153). Amsterdam: North-Holland.

Urban, J. M., Bowers, C. A., Monday, S. D., & Morgan, B. B., Jr. (1995). Workload, team structure, and communication in team performance. *Military Psychology, 7*, 123–139.

13

Research on Decision Making and New Technology: Methodological Issues

Björn Johansson
Rego Granlund
Yvonne Waern
Linköping University

Naturalistic decision-making (NDM) research tries to explain how humans (individuals) make decisions in real-world settings and how this decision making is affected by various factors such as stress, complex information, and technical systems. Findings are usually based on performance measurements. Recently, other paradigms such as distributed cognition, activity theory, and communication studies have entered the field of NDM. Such studies include a wider context, such as collaboration, organization, and artifacts. Little effort has been put into merging the results from these various fields. In this chapter, we propose a more holistic view on research concerning the implementation of new technology in team decision making. We use our research on the new Swedish effort to create a joint mobile command and control concept (JMC2) as an example of how such a view can be applied and how methods have to be adapted to the more holistic approach. We also present results from a preliminary study to illuminate some methodological points.

THE ROLF ENVIRONMENT

The ROLF (joint mobile command and control concept) 2010 project is planned to be the next generation of command environment for the Swedish armed forces and is also meant for commanding in peace-crisis situations. The vision includes a high-technical environment where teams of com-

FIG. 13.1. The ROLF joint mobile command and control concept.

manders work around shared representations (see Fig. 13.1). The environment also provides information from a vast number of sensors and communication equipment constantly feeding the commanding staff with a "view" of the world outside (Sundin & Friman, 1998). When information is needed, the staff is supposed to obtain this directly by computer agents or by subscriptions to various databases. The staff is gathered around a Visioscope™, which is a table-resembling screen that projects a three-dimensional (3-D) view of a map. This setup comes from a camp-fire metaphor.

At this time, the vision includes adding 3-D features to the Visioscope, but this has not yet been implemented. Something that we find necessary to ask is how the work around Visioscope should be organized. This technology is supposed to provide easily accessible information in visual form, but how should it be accessed?

There are basically two ways of dealing with the updating of an artifact like the Visioscope, and this was also the main focus of our research for this example. One possibility is to provide the input from the outside directly (through sensors and various filtering on the input of these sensors) on the Visioscope in some kind of symbolic form. An alternative possibility is to give the information to the staff, meaning they would have to insert it into the Visioscope themselves.

THEORETICAL BACKGROUND

In one of his recent books on NDM, Klein (1998) listed eight different "sources of power" lying behind people's ability to make good decisions. They are the following: intuition, mental simulation, the power to see the invisible, the power of stories, the power of metaphors and analogues, the power to read minds, the power of the team mind, and the power of rational analysis.

All the sources mentioned by Klein (1998), except one, concern an isolated individual who solves the problem independent of other people or other resources than an expert mind. When the resource of other people is mentioned, it is in terms of the team mind that in all its essentials reflects the workings of an individual mind. The role of the artifact is also neglected.

In contrast to this individual and mental focus, several theories and approaches stress that no individual works in isolation, that other people as well as artifacts serve in an intricate combination in routine work as well as in problematic situations. In particular, Hutchins (1995) pointed out that performance emerges from the functioning of a system as a whole, consisting of people and their artifacts. It is not possible to sort out a single individual and attribute a good or bad performance to this person, nor is it possible to study the effect of an artifact in isolation. We decided to study the effect of two different ways of performing a task with the same artifact.

What is then learned when one approaches work as a system? Approaches under the heading of "workplace studies" may be subsumed under the system approach (Engeström & Middleton, 1996; Garbis & Waern, 1999; Heath & Luff, 2000; Theureau & Filippi, 1993). Such studies show that people at a workplace organize their work by various more or less subtle (and unconscious) ways. This means that studies of a formal organization do not at all tell the whole picture of the work process. These studies have been mainly descriptive.

There is one remaining weakness of descriptive research: It cannot predict anything, and we as researchers cannot test ideas about the future. Imagine, for instance, that one wants to test the effects of a new technology. Note that by effects we mean the way in which people change their social organization of work including the new artifact. Then researchers cannot only study current work. We know already from the beginning that work practice will change. How can we then tell if ideas about the new technology and its role in the organization are "good" or "bad"? Our approach was experimental to enable predictions.

METHOD

We used the C³Fire microworld for this study. C³Fire is a command, control, and communication experimental simulation environment (Granlund, 2002). The primary purpose of the system is to allow researchers to experiment with different strategies for studies and training of team activities. The domain, which is forest fire fighting, is of subsidiary interest and has been chosen simply to demonstrate the principles.

The system generates a task environment in which team members cooperate to extinguish a forest fire. The simulation includes the forest fire, houses, different kinds of vegetation, computer-simulated agents such as reconnaissance persons, and fire-fighting units. The people who take part in the simulation are part of a fire-fighting organization and take the roles of staff members or fire-fighting unit chiefs (see Fig. 13.2).

Both the fire and the fire-fighting organization are complex dynamic systems, which change both autonomously and as a consequence of actions made on them (Svenmarck & Brehmer, 1991). Decision making is distributed over a number of persons and can be viewed as team decision making in which the members have different roles, tasks, and items of information available in their decision process (Orasanu & Salas, 1993). As in most hierarchical organizations, the decision makers work on different time scales.

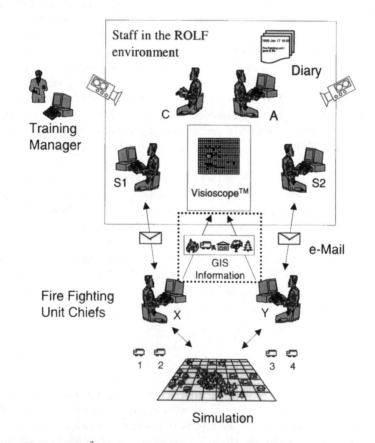

FIG. 13.2. The C³Fire microworld. The commander (C) is sitting next to an assistant. Two staff officers (S1 and S2) communicate with ground chiefs (X and Y) who control the fire brigades.

The fire-fighting unit chiefs are responsible for the low-level operations such as the fire fighting. The staff is responsible for the coordination of the fire-fighting units and strategic thinking.

The two ground chiefs are supposed to receive orders from the staff, meaning they also have to report their observations to the staff. This is done differently depending on the current independent variable. Either they report via e-mail or by updating a map they have on their screen (see Fig. 13.2). The map corresponds to the map on the Visioscope, and they have access to all occurring symbols in the microworld.

The staff consists of four persons, two who communicate with the outer world (the ground chiefs) and two who serve as commanding officer and assistant (see Fig. 13.2). These four persons decide together where the ground chiefs are to move their fire-fighting units.

The variable used as an indication of success in this example was the number of squares (the simulation consists of a total of 400 squares) and houses (a total of five houses) that were consumed by the forest fire in the trials.

Participants

A total of five experiments in each condition were planned, with 6 participants performing three trials in each condition. This sums up to a total of 60 participants, all of them professional military officers of at least the grade of lieutenant. The analyzed data came from 4 conducted experiments out of 10 planned.

Design

The study was conducted in line with a 2 (direct vs. manual updating) × 3 (trials) design with repeated measures of the latter factor.

There were two experimental conditions involving updating of the Visioscope map: either directly from the ground chief (who controls the fire brigades), which creates an illusion of automatic updating (the ground chief is the sensor), or manual updating based on e-mail communication. All the trials were exactly the same with the exception that the map used was rotated and mirrored between the trials. The participants did not know this. The participants were randomly assigned to the conditions.

Procedure

All teams were informed about the nature of their task and the roles within the team. They were able to choose which roles to take for themselves. The following roles were available: two ground chiefs (who control the fire brigades), two communications officers, one commander, and one assisting commander (who is responsible for the documentation of their actions; see Fig. 13.2). After having selected the roles, the team went through a 15-min training session.

After the training session, participants got 5 min to discuss within the team before the first trial started. Each trial lasted 30 min and was followed by a 5-min evaluation session in which the trial was replayed for the participants 10 times the normal speed, giving them an opportunity to reflect on their actions and to discuss strategies. The participants were encouraged to express how they were thinking during the scenario when it was being replayed.

Data

The C³Fire microworld gives an excellent support for data retrieval. Figure 13.3 shows the data that can be extracted from a scenario.

Beyond this source of data, everything was recorded from two different angles using several microphones and digital video, providing a high-quality recording for further analysis (see Fig. 13.2).

PRELIMINARY RESULTS

The following results are preliminary and should not be seen as the results of the study as a whole.

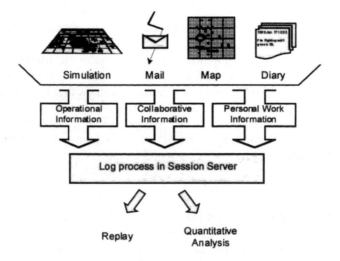

FIG. 13.3. The C³Fire microworld log process.

Quantitative

The most obvious finding in the preliminary data was the abnormal learning curve for the trials with direct updating (map) of the Visioscope (see Figs. 13.4 and 13.5). During manual update (text) by the commander, the results indicate a traditional learning effect, meaning fewer burned-down squares in each trial. In the direct update from the ground chiefs, however, we saw a strong learning effect between Trials 1 and 2, whereas the performance decreased in Trial 3. The results for the third trial were worse than in the first trial for both of the tested groups using direct updating (map). In Figs. 13.4 and 13.5, the text represents the manual update, and the map represents the direct-update condition.

Qualitative

To go further into the data, we used the first videotapes for interaction analyses. This was inspired by the studies of Heath and Luff (2000). In the map condition, the teams tended to gather very much around the Visioscope, and the discussion within the staff was orientated toward this, and hypotheses about what was going on were proposed. The text condition

Saved area

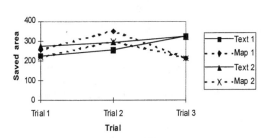

FIG. 13.4. Number of not-burned squares at the end of the trials.

Saved houses

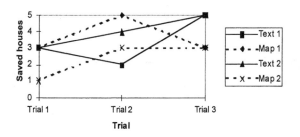

FIG. 13.5. Number of not-burned houses at the end of the trials.

creates a somewhat different approach from the team, which seems to have been more orientated toward the messages from the ground chiefs. A simple interaction analysis of referring gestures toward the different screens in the ROLF environment was conducted. Three different screens were used, two ordinary PC screens used by the staff officers and the Visioscope, the large table-like screen. The data were collected from the last trial from each team. Table 13.2 shows the number of times someone in the staff pointed toward a screen with the intention of showing, explaining, or suggesting something to the other members of the staff (see also Table 13.1).

The data indicate that the use of the Visioscope as a shared representation was more common in the direct-update condition. There are also differ-

TABLE 13.1
Pointing, Direct-Updating (Map) Condition, Last Trial

Map Team 1	Visioscope™ Screen	Staff Officers' Screens
Commander	68	0
Assistant	8	0
Staff Officer 1	15	0
Staff Officer 2	19	0
Map Team 2	Visioscope Screen	Staff Officers' Screens
Commander	36	2
Assistant	21	0
Staff Officer 1	12	0
Staff Officer 2	1	1

Note. The table describes how many times the individuals in the staff have used the maps shown on the screens as a reference in their work. The Visioscope™ is the large shared screen in the room (see Figs. 13.6 and 13.7). The table does not discriminate between the two screens used by the staff officers.

TABLE 13.2
Pointing, Manual Updating (Text) Condition, Last Trial

Text Team 1	Visioscope™ Screen	Staff Officers' Screens
Commander	28	2
Assistant	12	0
Staff Officer 1	24	0
Staff Officer 2	0	0
Text Team 2	Visioscope Screen	Staff Officers' Screens
Commander	2	2
Assistant	3	0
Staff Officer 1	0	0
Staff Officer 2	0	0

ences in the communicative practices during the two conditions. Whereas the text condition created a very verbal communicative environment, the direct updating created a more silent environment (see Figs. 13.6 and 13.7). The styles were also very different. In the text condition, the participants communicated in very clear patterns. Staff Officer 1 or Staff Officer 2 reported what the ground chiefs had done and the commander confirmed that he understood.

This was only interrupted when the commander gave a new order or something needed to be clarified. In direct update, the communication was focused around events. The staff was silent until something happened and then everyone gathered around the Visioscope to discuss the event and generate proposals about possible actions.

FIG. 13.6. Typical work situation during the text condition.

FIG. 13.7. Typical work situation during direct-update (map) condition.

One explanation of the different results in the two conditions is that the text condition is more similar to the conditions under which the staff is used to working. Another important aspect of this view is that the direct-update condition is a simulation of a vision, not a real work practice. There are no empirical findings that support that direct updating really is more efficient; this is merely an assumption from the ROLF vision.

Another possible explanation is that the direct updating simply does not provide the staff with a "feeling" of what is really happening in the field. Heath and Hindmarsh (2000) pointed out that the meaning of an object is assembled by the participants at particular moments in the developing course of interaction; its significance is bound to the moment of action in and through which it occurs. If we study the way in which information reaches the Visioscope (see Fig. 13.8), we find something interesting. Be-

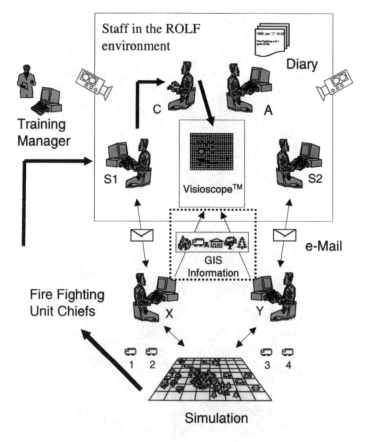

FIG. 13.8. The arrows indicate the way information moves to the Visioscope™ from the "field" during the text condition.

cause the meaning of the object has to be announced before it reaches the Visioscope (through one of the staff officers), it might be interpreted in a different way than in the direct updating.

As seen in Fig. 13.8, the information is first interpreted by one of the staff officers and then verbally communicated to the commander and his assistant before it is finally added to the map. This means that there has to be a shared understanding of the meaning of the information if it is to be placed in correct form at the correct place on the map. Therefore, it is important to study the way in which information flows through the system to understand why and how the meaning of a shared representation is created.

The participants in this study had the opportunity to discuss their work process between the trials. We are planning to study these conversations to a greater extent to find which aspects of the situation the participants focus on and see what kind of reflections are made. It is also possible that a longer series of trials would show a change in the learning effects, which might eliminate the differences between the conditions. This was the case in a study reported by Artman and Granlund (1998). However, the effect of the organization in qualitative terms will probably lead to different work procedures, which get more fine-tuned with learning.

DISCUSSION

Although this study is not yet finished, we want to point to some important considerations for imagining and testing future systems for supporting NDM. The results give us an idea about the problems that face the researcher. The quantitative data were in this case mostly confusing, maybe telling us something about learning effects and performance. The qualitative data, on the other hand, described how the working procedure was being shaped but did not tell us very much about the effect of the different conditions. However, together they gave us an opportunity to analyze the team and study the work of their artifacts.

The groups differed in terms of work procedure within each of the conditions. This fact makes it difficult to compare the groups based on these statistics. Instead, a careful qualitative analysis should be performed. Other researchers have also pointed out that differences between individuals often disturb the effect of the independent variable (M. Omodei, personal communication, 2000). The differences in working procedures observed so far do not necessarily have to affect the outcome of a scenario, although they seem to have some kind of impact on learning.

We suggest that there is a possibility to test new ideas by a systematic approach built on the following main pillars:

1. Study simulations. Simulation of a complex environment is a research tool that is proposed to give researchers some handle on such situations without getting lost in the blue sea of natural field studies (cf. Brehmer & Dörner, 1993). What is needed is for the simulations to simulate "relevant" aspects of the situation targeted. The meaning of relevant can be determined by prior field studies (cf. Artman, 1998).

2. Study professional people. Although much simulation research has been performed with nonprofessional people, Johansson, Persson, Granlund, and Mattson (2003; see also Hoc, 1995, and Rogalski, 1999) have found that there are systematic differences between professional and nonprofessional people in their approach to a simulated environment. In particular, the differences concern the parts of the simulation that are close to their professional expertise.

3. Consider organizational factors when studying the effect of technology. It is too simpleminded to study the contrasts between a situation with and without a certain technology. Because the technology will interact with the organization (see, for instance, Artman, 1998), researchers have to consider the technology together with the fixed (as well as socially constructed) organization.

4. Study process as well as outcome. In a complex environment, there are too many factors affecting the outcome to enable conclusions to be drawn from performance, whatever the number of experiments or case studies. Various factors as well as processes that counteract these affect the performance. Researchers need to consider these processes and this is best done by a qualitative approach.

ACKNOWLEDGMENTS

This work was supported by the Board for Working Life Studies and the Graduate School of Human–Machine Interaction, Linköping, Sweden.

REFERENCES

Artman, H. (1998). Team decision making and situation awareness in military command and control. In Y. Wærn (Ed.), *Co-operative process management—Cognition and information technology* (pp. 55–68). London: Taylor & Francis.

Artman, H., & Granlund, R. (1998). Team situation awareness using graphical or textual databases in dynamic decision making. In T. Green, L. Bannon, C. Warren, & J. Buckley (Eds.), *Cognition and cooperation. Proceedings of 9th Conference of Cognitive Ergonomics* (pp. 149–155). Limerick, Ireland: University of Limerick.

Brehmer, B., & Dörner, D. (1993). Experiments with computer-simulated microworlds: Escaping both the narrow straits of the laboratory and the deep blue sea of the field study. *Computers in Human Behaviour, 9,* 171–184.

Engeström, Y., & Middleton, D. (Eds.). (1996). *Cognition and communication at work.* Cambridge, England: Cambridge University Press.

Garbis, C., & Wærn, Y. (1999). Team coordination and communication in a rescue command staff—The role of public representations. *Le Travail Humain, 62,* 273–291.

Granlund, R. (2002). *Monitoring distributed teamwork training* (Dissertation No. 746). Linköping Studies in Science and Technology, Linköping University, Linköping, Sweden.

Heath, C., & Hindmarsh, J. (2000). Configuring action in objects: From mutual space to media space. *Mind, Culture and Activity, 7,* 81–104.

Heath, C., & Luff, P. K. (2000). *Technology in action.* Cambridge, England: Cambridge University Press.

Hoc, J.-M. (1995). Ecological validity of dynamic micro worlds: Lessons drawn from experiments on NEWFIRE. In L. Norros (Ed.), *Proceedings of 5th European conference on Cognitive Science Approaches to Process Control, ASCAP, 1995* (pp. 126–137). Espoo, Finland: Technical Research Centre of Finland.

Hutchins, E. (1995). *Cognition in the wild.* Cambridge, MA: MIT Press.

Johansson, B., Persson, M., Granlund, R., & Mattson, P. (2003). C3 fire in command and control research. *Cognition, Technology & Work, 5*(3), 191–196.

Klein, G. (1998). *Sources of power. How people make decisions.* Cambridge, MA: MIT Press.

Orasanu, J., & Salas, E. (1993). Team decision making in complex environments. In G. A. Klein, J. Orasanu, R. Calderwood, & C. E. Zsambok (Eds.), *Decision making in action: Models and methods* (pp. 327–346). Norwood, NJ: Ablex.

Rogalski, J. (1999). Decision making and dynamic risk management. *Cognition, Technology & Work, 1,* 247–256.

Sundin, C., & Friman, H. (Eds.). (1998). *ROLF 2010. A mobile joint command and control concept.* Stockholm, Sweden: The Swedish National Defence College.

Svenmarck, P., & Brehmer, B. (1991). D3Fire: An experimental paradigm for the studies of distributed decision making. In B. Brehmer (Ed.), *Distributed decision making. Proceedings of the third MOHAWC workshop. 1991* (Vol. 7, pp. 47–77). Roskilde, Denmark: Risø National Laboratory.

Theureau, J., & Filippi, G. (1993, August). *Metropolitan traffic control activities and design of a support system for the coordination of actions in future control rooms.* Paper presented at the 4th European Meeting on Cognitive Science Approaches to Process Control: Designing for Simplicity, Copenhagen, Denmark.

14

Cognitive Control Processes Discriminate Between Better Versus Poorer Performance by Fire Ground Commanders

Jim McLennan
La Trobe University, Melbourne

Olga Pavlou
Swinburne University of Technology, Melbourne

Mary M. Omodei
La Trobe University, Melbourne

One of the defining features of a naturalistic decision-making (NDM) perspective is rejection of prescriptive accounts of decision making, notably subjective expected utility (SEU) theory, as the model of rational decision making against which human, real-world decision behavior must be compared (e.g., Beach & Lipshitz, 1993). NDM accounts have, in turn, been subject to criticism on various grounds by behavioral decision researchers (e.g., Doherty, 1993; Howell, 1997). Perhaps the most cogent attack on NDM to date has been that of Yates (1998). Yates proposed that NDM accounts of decision making were merely descriptive and unlike SEU theory did not permit the quality of a decision to be evaluated prior to the results of the decision becoming apparent. In other words, NDM accounts are inadequate in that they do not differentiate between the processes involved in a good decision and the processes involved in a poor decision. In this chapter, we attempt to respond to Yates' criticism by (a) offering some suggestions about processes likely to be associated with good versus poor decision making by experts in their particular decision task domain then (b) reporting a study of decision making by fire officers who were in command of fire crews at large-scale, simulated fire ground exercises. The primary cognitive process data for the study were generated by means of head-mounted, video-cued recall debriefing interviews with each fire ground commander (FGC) at the conclusion of his exercise.

An important aspect of NDM methodology is an assumption of the need for researchers to study domain-expert decision makers. Concentrating on expert decision makers suggests two important issues in relation to good versus poor decisions. The first is that experts, by definition, will make mostly good (or at least adequate) decisions and few poor decisions. This is not to suggest that expert decision makers never make mistakes. However, their errors are most likely to be of the kind categorized as "slips" by Norman (1998, pp. 105–110): capture errors, description errors, data-driven errors, and so forth. Such errors are the result of fundamental characteristics of the human information-processing system. The second issue is that field investigations suggest that experts generally know when they are performing poorly: Their very expertise enables them to be aware of this (McLennan, Omodei, & Wearing, 2001). Accordingly, we suggest that the most important characteristic of those processes that underlie poor decision making by an expert is concurrent awareness of his or her poor performance.

The important role of concurrent awareness of performance on a complex dynamic task has been discussed by Brehmer (1992) and by Doerner and Schaub (1994). These researchers proposed that a decision maker operating in a complex, dynamic task environment should be thought of as an engineer endeavoring to control adaptively a complex dynamic system, a key element of which is the decision maker's own self-regulatory processes. Thus, self-monitoring and self-regulation processes are a source of feedback about the effectiveness of the decision maker's adaptive task system control endeavors. These ideas were developed further by Omodei and Wearing (1995). Omodei and Wearing (1995) proposed that in a real-world decision situation, a decision maker takes actions that are intended to establish control over the task environment by progressively incorporating and adapting to feedback from the environment. A decision situation becomes problematic when the decision maker perceives some actual or threatened loss of control over the task environment. Effective decision making in a dynamic situation involves motivational, cognitive, and affective factors. To maintain control over the developing task situation, a decision maker must exercise self-regulatory, or self-management, skills. Effectiveness in a decision task environment in which the decision maker has some personal stake in the outcome resides in an ability to control cognitive and attentional resources and also emotional states and motivated cognitive effort so the decision maker can continue to maintain efforts to carry out effective actions. Failure in self-regulation will allow intrusions of irrelevant or disruptive emotions and cognitions that may degrade the effectiveness of ongoing decision actions. Omodei and Wearing's (1995) model received some empirical support in a study of decision making by telephone crisis counselors reported by Bobevski and McLennan (1998).

The preceding discussion suggests that one ought to be able to predict a poor decision outcome by a domain expert when his or her decision processes manifest two qualities. The first is a relatively high proportion of cognitive activities associated with self-monitoring and self-regulation of negative self-evaluations of performance. The second is a relatively high proportion of cognitive activities associated with attempting to develop an adequate conceptualization of the most important features of the task environment (as distinct from cognitive activities associated with forming intentions and generating actions).

In attempting to test our hypothesis, we had to confront two questions. The first was how to obtain adequate data. The second was how to establish an independent standard of quality of an expert's decision making. Our answer to the first question was to use head-mounted, video camera, cued-recall debriefing interviews to generate verbal protocols of cognitive activities underlying observed decision-making behaviors. This procedure has been used previously to investigate decision making by athletes (Omodei & McLennan, 1994; Omodei, McLennan, & Whitford, 1998), football umpires (McLennan & Omodei, 1996), and FGCs (McLennan et al., 2001). The procedure is described in detail in Omodei, Wearing, and McLennan (1997). Our answer to the second question was to take advantage of a naturally occurring situation in which FGCs' command and control decision-making performances were to be assessed by an expert panel. This took place during a sequence of large-scale fire ground simulation exercises conducted by the Melbourne Metropolitan Fire and Emergency Services Board (MFESB) as part of preparation for a competitive examination to promote fire officers to a higher rank.

METHOD

Participants

There were 12 candidates for promotion from Station Officer to Senior Station Officer. They had between 10 and 20 years of experience in the MFESB and between 4 and 12 years of experience as Station Officers.

Equipment

A Sony DXC–LS1P CCD color "lipstick camera" was mounted in a protective fiberglass shell fitted over a standard Bullard Firedome safety helmet. The camera was connected via a cable to the camera control unit and a 12 V power cell, both secured in a small "bumbag." A microphone was located under the rim of the helmet and both video and audio were recorded by means of a Sony CCD–TR1E video Hi8 Handycam also carried in the bumbag.

Procedure

Prior to each candidate being assessed on his incident command skills by means of a practical examination, candidates spent 5 days undertaking a range of role-playing simulation exercises in which each in turn took the role of incident commander (FGC) in charge of four 3-person fire crews. During each exercise, the candidate in the FGC role (a) listened to the initial radio turnout message, (b) heard the "wordback" message from the first-on-scene appliance, and then (c) assumed control of the incident establishing an on-site control center. In each exercise, scripted role players provided reports and carried out the FGC's instructions. After the incident had developed for about 20 min, an instructor assumed the role of a senior officer who had arrived on scene to take charge, and the candidate was required to brief the (notional) superior officer on the situation. This handover briefing concluded the exercise. Two experienced observers each independently rated each FGC's command and control decision-making skill on a 10-point scale ranging from 1 (*unacceptable*) to 10 (*superior*). Immediately following an exercise, there was a postincident discussion involving the FGC, instructors, and other participants during which the most salient events of the exercise were noted. The FGC then moved to a small room nearby for a video-cued recall debriefing and sat in front of a TV monitor. The FGC and debriefer each wore a small tiepin microphone. Both microphones were connected to a video/audio mixing unit. An 8-mm video player/recorder to replay the helmet camera footage also fed into the mixing unit. Outputs from the 8-mm video unit and both tiepin microphones were copied onto ½-in. videotape on a VHS recorder.

Prior to a video-cued replay debriefing, each FGC was instructed with the following:

> We are going to watch a replay of footage of the exercise taken from the helmet camera. As you watch, I want you to take yourself back to being in the role of the FGC. I want you to recall as much as you can of what was going on in your mind when you were managing the incident. I want you to speak these recollections out loud—just begin talking and I will pause the tape so you have plenty of time to recall as much as you can. Your recollections will be recorded onto a VHS copy of the original footage of the incident as you saw it and all the radio and voice communications, plus your recollections of the things that were going on in your mind that "drove" your actions, decisions, and communications. Later, you can

then replay this tape with your instructors and fellow candidates to get their feedback and suggestions.

The 8-mm (helmet camera) tape was forwarded to the beginning of the exercise, and the image was paused. The FGC was then instructed, "Now, as you watch this picture of the start of the exercise take yourself back—what do you recall thinking just as the exercise was about to begin?" This began the recall process. When the FGC finished verbalizing his initial recollections, the tape was started, and the cued recall session proceeded. The debriefer encouraged the FGC to recall as much as possible, occasionally using nondirective probes and when necessary reminding an FGC to recall rather than engage in after-the-fact self-criticism.

Each of the 20 video-cued recall interviews was transcribed and the FGCs' cognitive activities during the course of the incident were categorized using a cognitive process tracing scheme (CPTCS). The cognitive activity categories incorporated in the CPTCS are listed in Table 14.1. In brief, there are 11 cognitive activities associated with situation assessment and understanding, 11 cognitive activities associated with intention formation and action generation, 4 cognitive activities associated with self-monitoring, and 4 cognitive activities associated with self-regulation. The CPTCS was initially conceptualized on the basis of several theoretical formulations of aspects of dynamic decision making, notably those of Brehmer (1992), Doerner and Schaub (1994), Kuhle (1985), and Omodei and Wearing (1995). Subsequent empirical refinement of the CPTCS was carried out in the course of several research studies using head-mounted, video-cued recall. An initial version of the CPTCS was first described at the 1996 Judgement and Decision Making Society Annual Meeting (McLennan, Omodei, & Wearing, 1996).

RESULTS

Reliability of Command and Control Decision-Making Performance Ratings

The 12 fire officers provided a total of 20 cued-recall interviews. Each officer wore the head-mounted camera during one, two, or three exercises scheduled over 3 successive days. When an officer wore the camera was largely a matter of random chance depending on the simulation exercise schedule, equipment readiness, and researcher availability. The independent ratings by the two observers of the candidates' 20 command and control decision-making performances were found to be quite reliable overall: $r = .81$, $p < .01$. However, the two observers disagreed about the performance of one FGC:

TABLE 14.1
Percentage Frequencies of Use of Cognitive Activity Categories Recalled by the Six Superior Fire Ground Commanders (FGCs) Compared With Those of the Six Inferior FGCs

Cognitive Activity Category	FGC Performance (%)	
	Inferior[a]	Superior[b]
Situation assessment and understanding (SAU)[c]		
SAU1. Identify relevant cues/features	5	1
SAU2. Identify need for specific information	19	25
SAU3. Attend to specific information	10	7
SAU4. Recognize demand/threat/need	21	17
SAU5. Anticipate a demand/threat/need	9	16
SAU6. Inventory available resources	6	6
SAU7. Anticipate future resources	6	7
SAU8. Evaluate effects of an action	2	2
SAU9. Evaluate conformity of an action to SOPs	5	0
SAU10. Anticipate a situation development	0	0
SAU11. Formulate/revise a situation conceptualization	17	18
Intention formation and action generation (IN)[d]		
IN1. Remembering fact/rule/procedure	19	4
IN2. Prioritizing actions	18	11
IN3. Evaluating demands/resources balance	4	6
IN4. Planning actions	17	17
IN5. Generating a specific action	16	20
IN6. Presimulating a possible scenario	0	0
IN7. Simulating a possible development	1	0
IN8. Evaluating and choosing among alternative actions	1	3
IN9. Analyzing and problem solving	23	30
IN10. Distributing/delegating decision making	2	9
IN11. Nonproductive cognitions—Wishing, wondering	0	0
Self-monitoring (SM) and self-regulation (SR)[e]		
SM1. Maintaining self-view consistency	3	0
SM2. Noting negative affect	8	15
SM3. Noting positive affect	6	8
SM4. Noting level of control/mastery versus overload	29	31
SR1. Control attentional focus	27	14
SR2. Engage in self-talk	8	0
SR3. Restrict information inflow	20	31
SR4. Engage in physical activity	0	0

Note. Percentages were calculated within each of three major classes of cognitive activity categories. SOPs = standard operating procedures.

[a]$N = 256$. [b]$N = 229$. [c]$n = 81$ (32%) inferior performances; $n = 87$ (38%) superior performances. [d]$n = 96$ (37%) inferior performances; $n = 116$ (51%) superior performances. [e]$n = 79$ (31%) inferior performances; $n = 26$ (11%) superior performances.

Whereas one observer rated his performance as being among the superior six, the other observer rated his performance as being among the inferior six. Because of the disagreement, this performance was discarded from further consideration. Apart from this case, the two observers' ratings were in agreement about the composition of the superior six performances and the six inferior FGC performances.

Behavioral Markers of Superior Command and Control Decision-Making Performance

The technique of identifying behavioral markers associated with superior performance during the course of multiperson decision tasks was developed originally within the context of aviation psychology (Helmreich, Butler, Taggart, & Wilhelm, 1994) and extended to hospital operating rooms (Gaba, Fish, & Howard, 1994). Videotapes of the six superior FGCs (taken from an external observer's perspective using a handheld camera) were examined and compared with those of the six inferior FGCs. Behaviors exhibited by at least four of the six superior FGCs and exhibited by no more than one of the six inferior FGCs were identified by two independent observers. Table 14.2 lists these, organized using the scheme proposed by Gaba, Howard, and Fish. The overall impression created by the pattern of superior FGCs' observable behaviors is one of being in control: of proactively managing the incident rather than merely reacting to unfolding events.

FGCs' Command and Control Decision-Making Cognitive Activities

The helmet-mounted video footage of each FGC's field of view (plus communications) during each simulated incident proved a powerful cue for officers to recall and verbalize in considerable detail the bases of their fire ground command decision making. Candidates also verbalized uncertainties, self-questioning, and self-doubt during the course of an exercise. The recorded cued-recall interviews were transcribed, and the verbal protocols were categorized by two independent raters. The interrater agreement was high: Cohen's $\kappa = .94$. Overall, the 20 head-mounted, video-cued recall interviews generated a total of 1,020 verbalized recalled cognitive activities: 344 (33%) concerned with situation assessment and understanding, 435 (43%) concerned with intention formation and action generation, and 241 (24%) concerned with self-monitoring and self-regulation. Of the 435 cognitive activities concerned with intention formation and action generation, only 10 (2%) involved choosing among alternative courses of action: clear evidence of the inappropriateness of behavioral or managerial models of decision making—generation of alternatives and choice of the optimal course of ac-

TABLE 14.2
Behavioral Markers of Superior Fire Ground Command Performance[a]

Anticipation and planning
  Used "dead time" to study site plans and diagrams
  Prepared for "worst case" scenario early, took precautions, and called for additional resources
  Warned crews of likely developments and tasks
Communication
  Used site maps and diagrams to explain intentions to subordinates
  Clear, controlled speech to subordinates
  Maintained eye contact when speaking/listening to subordinates
  When using a radio, paused after subordinate acknowledged call before proceeding to give instructions
Leadership and assertiveness
  Spoke clearly, firmly, and decisively
  Greeted key (role) players (e.g., building supervisor) warmly but decisively
Management of workload
  Used control center white board to record incoming information and to write reminder notes to self
  When interrupted by an incoming radio message, asked sender to "wait" until the current task was completed
  Requested new arrivals at control center to wait outside until he was ready to speak with them
Reevaluation of situations
  At first indication of deterioration in the situation, raised the "level" of the incident so as to call out additional resources
Use of available information
  Used multiple sources of information, actively sought information from subordinates, experts, and site plans and diagrams

[a]Following Gaba, Fish, and Howard's (1994) categories.

tion—for understanding time-pressured decision situations such as emergency incident management.

Table 14.1 compares the pattern of cognitive activities recalled by the six superior FGCs with those of the six inferior FGCs. Overall, the six superior FGCs reported a different pattern of cognitive activities: $\chi^2(5, N = 485) = 27.4$, $p < .001$. The significant difference was accounted for by the inferior FGCs' relatively more frequent reporting of self-monitoring and self-regulation cognitive activities. When self-monitoring and self-regulation cognitive activities were excluded from consideration, the six superior FGCs' relative proportions of situation assessment and understanding and intention formation and action generation cognitive activities were 0.43 and 0.57, respectively. For the six inferior FGCs, the corresponding proportions were 0.46 and 0.54, respectively, suggesting that the six inferior FGCs engaged in relatively more situation assessment and understanding cognitive activities. However, the difference was not statistically significant.

Self-Monitoring and Self-Regulation Cognitive Activity

Inspection of Table 14.1 indicates that overall, the six superior FGCs reported a relatively greater proportion of "noting negative affect" activity and "restrict information inflow" activity and a relatively smaller proportion of "control attentional focus" activity and "self-talk" activity. This pattern of relative activity is consistent with superior FGCs experiencing more effective control of events. Inspection of the content of the cued-recall verbalization protocols of the six inferior FGCs and comparison with those of the six superior FGCs provided additional support for this interpretation. Two representative recollections from the inferior FGCs follow:

> I haven't taken in what the other [officer] has actually told me. He's told me we've got a large fire on the container ship. Now the Captain walks in and I tell him that it appears we've got a fire. So I can see that things are being told to me and I'm not taking it in because maybe I'm too nervous or I can't visualize it.

> At this stage I've sort of lost it too because I think I should have gone back and spoken to the Station Master and got everyone evacuated through the emergency evacuation system and started the smoke ventilation straight away. I wasn't thinking clearly. I'm focusing on things in general and I'm not clearly identifying tasks and carrying them out. Then confusion reigns because in the short time span I've let things build up and I haven't been able to prioritize things. I've just let it get away a bit.

There were no such recollections by the six superior FGCs but rather a clear picture of feeling in control emerged:

> So at this stage I thought "Right, that's the next thing I have to do is I have to give him some manpower for a start so he can start operations." I wanted to establish early on that he was going to be in charge over there so that's why I said to him "You're the Operations Officer." So I could just send him resources and he would delegate the tasks because he had the big picture and he could see what was going on.

Situation Assessment and Understanding
Cognitive Activity

Inspection of Table 14.1 indicated that the six superior FGCs reported relatively more "anticipate a demand/threat/need" cognitive activity. Inspection of the cued-recall verbalization protocols confirmed the importance given by the superior FGCs to anticipation of future developments rather than simply responding to unfolding events. "Well I'm going to need some resources. We're going to need to get the people off so we're going to need bodies to help get people off the ship and also to combat the fire. So we're going to need crews here quickly so I sent a wordback in fairly brisk time."

Intention Formation and Action Generation Cognitive Activity

Inspection of Table 14.1 suggests that the superior FGCs reported relatively more "analyzing and problem-solving" activity. "Like I said before, being lighter than air, the ammonia's going to rise so I want to see the engineer to see if we can pressurize the area to stop the fumes from coming up in there." In contrast, the inferior FGCs devoted relatively more attention to remembering correct procedures. "I kept reminding myself, 'I have to appoint a safety officer.'"

Summary of the Most Salient Characteristics of the Superior FGCs

Overall, the superior FGCs took charge of the incident: They behaved like leaders and generally felt in control. They made maximum use of available information to inform decisions. They kept track of events often by making notes on a white board and followed up assigned tasks to check progress. They anticipated developments rather than simply reacting to unfolding events. They accepted the necessity to revise plans and generally reacted to negative developments without undue concern or irritation.

Summary of the Most Salient Characteristics of the Inferior FGCs

Obviously, the defining characteristics of the inferior FGCs could most simply be described as the absence of those qualities that defined superior performance as an FGC. However, inspection of the content of the cued-recall interviews suggested that poor performance seemed to be related to one of two decision "themes." First, it was being overwhelmed from the start of the exercise by the demands of the unfolding emergency: being unable to form a plan and always struggling to manage the cognitive work load associated with the stream of incoming information. Second, it was focusing on one very salient feature of the emergency and failing to take into account other emerging developments that rapidly nullified the FGC's action plan.

DISCUSSION

Overall, the data generated by the head-mounted, video-cued recall interviews provided considerable support for our initial hypothesis. It seemed
that in these emergency incident simulations, effective fire ground command decision making was closely associated with effective task-related
control processes, notably those involved in anticipating future developments rather than simply reacting to emerging events. As predicted, experts—judged subsequently by an external panel to have been less effective
FGCs—quickly recognized that they were not performing well. They experienced negative affects and intrusive self-critical thoughts. They made considerable, although unsuccessful, efforts to develop an adequate situation
understanding to take effective actions. We conclude that in response to
Yates' (1998) criticism of NDM theories as being merely descriptive, it may
be possible, in a given task domain, to specify processes associated with a
good as opposed to a poor decision. A good decision by a domain expert involves (a) effective control of task-related mental activities, (b) adequate
understanding of the key aspects of the emerging situation, and (c) awareness of a felt sense of being in control. Conversely, poor decision making by
a domain expert is characterized by a felt sense of not being in control and
awareness of intrusive negative affects and self-critical thoughts.

What perhaps remains problematic is the degree to which the findings
can be generalized: from fire ground simulation exercises to field operations and from emergency incident management settings to other task environments. A previous study (McLennan et al., 2001) of operational fire
ground command decision making provides some support: Greater subjective difficulty in fire ground command was associated with higher levels of
uncertainty inhering in the emergency situation. Ultimately, further conceptual analysis and field research hold the key.

Questions might be raised about the validity of the verbal protocols generated by the cued-recall procedure. Are these genuine reports of underlying psychological processes that generated the observed behaviors, or are
they postexercise reconstructions at best tenuously related to the original
decision making? This is a complex issue discussed in some detail in
Omodei et al. (1997). We suggest that there is no reason to doubt that the
cued-recall procedure taps reasonably well rule- and knowledge-based decision processes (Rasmussen, 1983), but the procedure is unlikely to tap skill-
based processes, which are by definition routinized and unavailable to
introspection. Because fire ground command involves mostly high-level
cognitive skills such as reasoning, judgment, and decision making (rather
than lower level psychomotor skills), video-cued recollections constitute
useful data.

In passing, all 12 candidates were successful at their practical examinations following the conclusion of the research study suggesting that they all had "the right stuff" regardless of variable performance during their simulation exercises.

By way of concluding, the research suggests a promising line of theory and research that goes beyond current somewhat static NDM theoretical concepts such as pattern recognition. It seems to us that what is needed is theoretical and empirical work that takes greater account of the real-time experiences of decision makers seeking to control adaptively dynamic task environments. Adaptive self-control models of decision making of the kind suggested by Omodei and Wearing (1995) and others may well provide useful lines of inquiry.

ACKNOWLEDGMENTS

This research was supported by an ARC–SPIRT Grant. We are very appreciative of the assistance given by personnel from the Melbourne Metropolitan Fire and Emergency Services Board (MFESB), especially Commander Philip Kline. The opinions expressed in this chapter are ours and do not necessarily represent those of the MFESB.

REFERENCES

Beach, L. R., & Lipshitz, R. (1993). Why classical decision theory is an inappropriate standard for evaluating and aiding most human decision making. In G. A. Klein, J. Orasanu, R. Calderwood, & C. E. Zsambok (Eds.), *Decision making in action* (pp. 21–35). Norwood, NJ: Ablex.

Bobevski, I., & McLennan, J. (1998). The telephone counseling interview as a complex, dynamic, decision process: A self-regulation model of counselor effectiveness. *Journal of Personality, 132*, 47–60.

Brehmer, B. (1992). Dynamic decision making: Human control of complex systems. *Acta Psychologica, 81*, 211–241.

Doerner, D., & Schaub, H. (1994). Errors in planning and decision-making and the nature of human information processing. *Applied Psychology: An International Review, 43*, 433–453.

Doherty, M. E. (1993). A laboratory scientist's view of naturalistic decision making. In G. A. Klein, J. Orasanu, R. Calderwood, & C. E. Zsambok (Eds.), *Decision making in action* (pp. 362–388). Norwood, NJ: Ablex.

Gaba, D. M., Fish, K. J., & Howard, S. K. (1994). *Crisis management in anesthesiology.* New York: Churchill-Livingston.

Helmreich, R. L., Butler, R. A., Taggart, W. R., & Wilhelm, J. A. (1994). *The NASA/University of Texas/FAA Line/LOS Checklist: A behavioral marker-based checklist for CRM skills assessment* (University of Texas Aerospace Crew Research Project Tech. Rep. No. 94–02).

Howell, W. C. (1997). Progress, prospects, and problems in NDM: A global view. In C. E. Zsambok & G. A. Klein (Eds.), *Naturalistic decision making* (pp. 37–46). Mahwah, NJ: Lawrence Erlbaum Associates.

Kuhle, J. (1985). Volitional mediators of cognition-behaviour consistency: Self-regulatory processes and action versus state orientation. In J. Kuhle & J. Beckman (Eds.), *Action control: From cognition to behaviour* (pp. 295–307). New York: Springer-Verlag.

McLennan, J., & Omodei, M. M. (1996). The role of prepriming in recognition primed decision making. *Perceptual and Motor Skills, 82*, 1059–1069.

McLennan, J., Omodei, M. M., & Wearing, A. J. (1996, November). *A new methodology for naturalistic decision making research: Investigating fire officers' decisions*. Paper presented at the Judgment and Decision Making Society Annual Meeting, Chicago, IL.

McLennan, J., Omodei, M. M., & Wearing, A. J. (2001). Cognitive processes of first-on-scene fire officers in command at emergency incidents as an analogue of small-unit command in peace support operations. In P. Essens, A Vogelaar, E. Tanercan, & D. Winslow (Eds.), *The human in command: Peace support operations* (pp. 312–332). Amsterdam: Mets & Schilt.

Norman, D. A. (1998). *The design of everyday things*. Cambridge, MA: MIT Press.

Omodei, M. M., & McLennan, J. (1994). Studying complex decision making in natural settings: Using a head-mounted video camera to study competitive orienteering. *Perceptual and Motor Skills, 79*, 1411–1425.

Omodei, M. M., McLennan, J., & Whitford, P. (1998). Using a head-mounted video camera and two-stage replay to enhance orienteering performance. *International Journal of Sport Psychology, 29*, 115–131.

Omodei, M. M., & Wearing, A. J. (1995). Decision making in complex, dynamic settings: A theoretical model incorporating motivation, intention, affect, and cognitive performance. *Sprache und Kognition, 14*, 75–90.

Omodei, M. M., Wearing, A. J., & McLennan, J. (1997). Head-mounted video recording: A methodology for studying naturalistic decision making. In R. Flin, M. Strub, E. Salas, & L. Martin (Eds.), *Decision making under stress: Emerging themes and applications* (pp. 137–146). Aldershot, England: Ashgate.

Rasmussen, J. (1983). Skills, rules, and knowledge: Signals, signs, and symbols, and other distinctions in human performance models. *IEEE Transactions on Systems, Man and Cybernetics, 15*, 234–243.

Yates, J. F. (1998, May). *Observations on naturalistic decision making: The phenomenon and the "framework."* Paper presented at the Fourth Conference on Naturalistic Decision Making, Warrenton, VA.

15

Adjusting New Initiatives to the Social Environment: Organizational Decision Making as Learning, Commitment Creating, and Behavior Regulation

Carl Martin Allwood
Lund University

Lisbeth Hedelin
Göteborg University

It is often maintained that the most important function of decision making is to regulate behavior. Decisions, however, do not always lead to behavior, and sometimes the effects of a decision being taken can be something else than behavior regulation, for example, an increase in the understanding of the social environment. In this chapter, we argue that when applied to social contexts such as the business world, the concept of decision making should be broadened so as to include the functions of learning and of establishing commitment. In addition, we assume that the functions of decision making will also include the functions of the decision-making process itself.

In an organization, decision making occurs at two levels: the formal level and the informal. When decisions are made at the formal level, they are taken in the name of the organization, the company, the board, or whatever. In contrast, decisions taken at the informal level are more private. These two levels tend to be associated with the collective group level and the individual level, respectively. This is not a necessary association, however, because individuals (e.g., the chief executive officer [CEO]) may well take decisions at the formal level, and groups (e.g., a division) take informal decisions (e.g., to get together for a party on the weekend). In this contribution, *decision making* is defined as formal decision making. Note that even with formal decisions, the decision process as a whole need not be regulated in a formal way, although parts of it may be regulated.

In addition to the three functions of decisions alluded to previously—those of regulation, learning, and creating commitment—formal decisions can have the function of providing power or control, of giving one part of an organization (a certain degree of) power or control over other parts. Thus, the requirement that some types of decisions be taken formally ensures that certain types of actions can only (properly or legally) be executed after the agreement of an appointed set of persons formally responsible has been reached, such as the company board. In addition, the fact that a certain action has been formally decided on normally obviously increases its chances of being implemented.

Briefly put, we define a *decision process* as the sum of the activities performed to prepare for the product that the process aims at, that is, the formal decision to be taken. We also regard decision making primarily as a form of planning. Obviously, this is not a necessary approach. Nutt (1984), for example, defined the decision process as "the set of activities that begins with the identification of an issue and ends with an action" (p. 415). An effect of Nutt's definition is that it excludes decisions that our definition includes, that is, decisions that are not implemented. However, we claim that decisions that are not implemented may still be of importance in an organization. Nutt's definition nevertheless has a positive consequence in that it helps to focus attention on how implementation is prepared for in the decision process. This can be done, for example, by building up commitment for a particular decision alternative during the decision process. How this is done is an important research question in itself.

THE LARGER CONTEXT OF DECISION PROCESSES

Decision making in organizations occurs within a larger context of activities. At different levels in the organization, there are strivings to achieve either institutional or more privately oriented goals. The group of top-level managers we studied (Hedelin & Allwood, 2001) can be described as faced with the challenge of seeing to it that successful outcomes are achieved for the tasks they were responsible for, the idea being that a successful manager "gets things done." Thus, when decisions are part of the route to completion of a task, it is usually important that any stalling of the decision process be avoided. To this end, it is important that the manager is a skilled social negotiator. Managers can therefore be seen as entrepreneurs who need to skillfully navigate their current set of tasks to a successful completion.

Mintzberg (1975) described the work situation of managers and showed them to often spend very little time with each of the issues that they deal with: "Brevity, fragmentation and verbal communication characterize his work" (p. 54). Managers need to allocate their time between different tasks opportunistically to accomplish as many as possible of their most important tasks.

The job of managers of driving on to completion those tasks of which decision making is a part includes the construction or identification of one or more highly attractive decision alternatives. This involves identifying the relevant properties of the alternative(s) and assigning appropriate values to these properties. A successful alternative is one that works well, and working well usually (or often) demands that it be well accepted by those who come in contact with it and feel concerned about it, both within and outside the organization.

SELLING A DECISION AS LEARNING AND CREATING OF COMMITMENT

The process of constructing the alternative(s) involves communicating with different categories of actors. An important part of the properties of an alternative is how various categories of actors react to the alternative when it is presented to them. These actors may also contribute new perspectives or other information about the alternative that the decision maker was not aware of. Through communicating with different categories of actors, the manager can learn more about the alternative and about the possibilities for improving it and of getting it approved. This can help the manager to develop the alternative further.

Through learning from the actors' reactions and comments and adjusting an alternative in response to these reactions, the manager also increases the chances of achieving the commitment of the actors with whom he has communicated. One reason for this may be that it is easier for these actors to identify with an alternative and feel responsible for it if their own views have contributed to its very construction. At the same time, if the manager encounters too much resistance in the course of the communication process, he or she may realize that it will be difficult to get the alternative formally accepted (formally decided on) or that even if formally decided on, it would be difficult to implement. Communicating with others in the organization about the alternative has the function, therefore, of both developing and testing it.

These considerations show that learning and the creation of commitment to an alternative are intertwined in the decision process. In our study (Hedelin & Allwood, 2001), we identified important categories of actors managers communicated with about an alternative. These persons were found both inside and outside the organization. Examples of such actors inside the organization were the CEO and other senior executives, board members, coworkers, union leaders, and staff at lower levels in the organization. Examples of categories of actors external to the organization were customers, suppliers, members of parliament, and representatives of different state authorities. The activity of communicating about an alternative with different categories of actors to learn from them and to use what has been learned to develop the alternative, and in the process accomplish that these actors become committed to the alternative, is termed *selling* the alternative.[1] To sum up, we suggest that efforts to sell a particular alternative can be regarded as an important part of many decision processes in organizational settings.

RESULTS OF A STUDY (HEDELIN & ALLWOOD, 2001) ON THE DECISION PROCESSES OF TOP-LEVEL MANAGERS

In the study of ours referred to previously (Hedelin & Allwood, 2001), we analyzed the reports of top-level managers on decisions selected in advance by the informants themselves. Our findings were based on three interviews with each of 41 experienced decision makers, all in leading positions in large organizations (both private and state organizations; the number of employees in the organizations ranged from 251 to 10,108 in 1994).

Prior to the first interview, the informant was asked to select two forthcoming decisions, data concerning one of which we analyzed. Informants were instructed that the decisions should be strategic in character (i.e., nonroutine, usually with long-term consequences) but that they could differ considerably in content. The decisions should also be ones for which it was likely that a formal decision would be taken within 3 months of the point in time when the interview was conducted. The fact that the time span between the first and the third interview with each participant ranged between 28 and 369 days suggests the unpredictability (and thus openness) of the context in which the decision processes studied were embedded.

The first and the third interview were conducted face to face, each taking about 2 hr. The second interview was a phone interview lasting between about 30 and 60 min. In the first interview, informants were asked to estimate a point in time representing approximately the middle of the decision process. When this time arrived, we phoned them up and set the time for the second interview, reminding them that it was to occur when the decision process had reached halfway.

---

[1]This term (*selling*) represents our attempt to translate into English the Swedish term *förankring* that was often used by our informants. (We avoided the term *anchoring* because it could easily be confused with the anchoring and adjustment effect dealt with in the biases and heuristics tradition of Tversky & Kahneman, 1974.) Please note that Hedelin and Allwood (2001) used the term *selling-in* for the same concept.

The 41 decisions concerned matters of investments (12), personnel issues (10), organizational change (8), marketing policy (4), mergers with another company (3), economic matters (3), and information technology (IT)
policy (1). Most of the interview questions concerned the informants' experiences with use of IT within the framework of the decision process. The informants also reported on current plans regarding the decision process
and what had been done in connection with it. The last interview included a
request that the informants describe what they considered to have been
most difficult during the decision process.

Analysis of the interviews showed that the majority of the informants
had focused on one alternative only in the decision process. Nearly half of
the decision processes (19) were reported to involve only one decision alternative. The following quote provides an example: "We have to show that
the project is profitable [a project in which two long-existent waterpower
stations were to be replaced by a new one]" (Hedelin & Allwood, 2001, p.
267). Three informants reported on a decision process in which one alternative had been replaced by another, two informants reported pushing for
one alternative among many, whereas other parties were pushing for the
other alternatives.

Five of the informants reported that the decision process had involved a
choice between two alternatives, such as buying a company or not buying
it, recruiting people externally or internally, or attempting or not attempting to expand the market. Finally, 12 of the informants stated either explicitly or implicitly that they considered several alternatives during the decision process. Most of them reduced the number of alternatives, however,
as the process developed.

The outcomes of the decision processes as far as we followed them (369
days) are shown in Table 15.1.

For 2 of the 33 cases in which the decision was taken formally, the content of the decision changed into something quite different than it had been
at the start. In the 1 case in which only a part of the decision was taken and
that concerned the development of a policy for IT-data communication, one
dysfunctional alternative was ruled out during the decision process and
two alternatives remained. The stalled decision referred to in Table 15.1

TABLE 15.1
Outcome of the Decision Processes After 369 Days

Outcome	No. Cases
Decision formally taken	33
Only part of the decision taken	1
Stalled decision process	1
No decision	6

concerned the sponsoring of a sports event. The manager was unable to contact the office responsible for the event, and this stopped the decision process momentarily. An example of a delayed decision ("No decision" in Table 15.1) concerned the creation of a new model for the price setting of goods. This decision was postponed due to unforeseen consequences linked to structural changes on the market.

The extent to which the informants reported searching for information on an alternative in ways not involving selling the decision (e.g., through searching in a database) was compared to how often they reported having made efforts to sell an alternative. For each of these, the extent to which informants reported having planned for the activity in question was also analyzed. It was found that planning and activities associated with a search for information in ways other than selling the decision were reported by a greater number of informants (40 and 36 informants, respectively) than planning and activities directly associated with the selling activity (28 and 20 informants, respectively). Although fewer informants reported on selling activities than on other information-seeking activities, the fact that about half of them did report on selling activities indicates that this aspect of the decision process in an organizational setting is worthy of greater attention than it has received thus far in the decision-making literature and within the naturalistic decision-making tradition.

A further result of our study suggests that selling activities are even more important. In response to the question, "What was the greatest difficulty in the decision process?" (posed near the end of Interview 3), the largest category of answers given were those concerning difficulties connected with the selling process. Note in this connection that most of the interview questions dealt with the use of IT during the decision process and that informants may thus have been strongly geared toward reporting on their IT-related difficulties, which makes the high incidence of reports of selling activities the more remarkable. Table 15.2 shows the categories employed for coding the difficulties mentioned and the number of informants mentioning them.

Some of the cases in the selling the decision category had to do with difficulties in selling the decision alternative to the appropriate persons, such as upward in the organization. In other cases, informants felt that they had not sold the proposal in the right way. The category of interactions with an opposing party had to do with preconditions for selling decisions in which negotiations with one or more opposing parties were necessary.

To exemplify the category finding information, one informant stated that the worst difficulty had been to correctly interpret information about sales, market development, and competition. Other informants mentioned difficulties in finding information due to poor routines for storing or retrieving information, unsuitable information banks, or the presence of "lots of soft data."

TABLE 15.2
Difficulties During the Decision Process Mentioned by the Informants

Difficulty	No. of Informants	Explanation of Difficulties
Selling the decision	14	Mostly difficulties in achieving consensus in support of a decision alternative the informant was seeking acceptance for
Selling—Peripheral aspects	2	Difficulties peripherally related to selling
Interactions with an opposing party	2	
Finding information	5	
Predicting the development of events	2	
Management of time	2	Adhering to the time plan and correctly predicting the time that would be needed
Choosing the best alternative	9	Mainly a restatement of the decision problem
Miscellaneous	5	

The difficulty coded as predicting the development of events can be seen as involving difficulties in finding information. One informant said, for example, that the greatest difficulty was in assessing the "lifetimes" of equipment in technical and economic terms.

The difficulties referred to in the miscellaneous category concerned such matters as "realizing that we were actually on the wrong track," "getting the other party to react as I wish," "working with an imprecise goal," and "deciding whether the change should occur at once or gradually" (Hedelin & Allwood, 2001, p. 275). The last instance in this category concerned problems in agreeing about whether the organization could live up to the quality demands of another organization.

DISCUSSION

In conclusion, the results of the study by Hedelin and Allwood (2001) indicate that to give an adequate description of managers' decision processes, managers' communications regarding decision alternative(s) with other persons, both within the organization and outside of it, should be attended to. Naturalistic studies of communication processes in connection with formal (and informal) decision making in organizations clearly appear to be promising areas for future research.

Although communicative aspects of decision making have often been neglected in previous research on decision making, other research has pointed to the importance of such aspects. For example, Heath and Gonzalez (1995) asked a large group of university students about important decisions and found that as many as 91% of their informants indicated that they sought the advice of others while making these decisions.

In another context, Sonnentag (2000), when investigating excellent performers in the domains of software development and engineering, found that excellent performers compared to moderate performers "more often regarded cooperation as a useful strategy" and more often participated "in work-related communication and cooperation processes than did moderate performers" (p. 483). Two interesting questions for future research are if more expert (or more excellent) decision makers in organizations also communicate more as part of identifying the decision alternative(s) in the decision process and if they do so in a different way compared with less expert (or less excellent) decision makers.

Communications regarding decision alternatives fulfill at least two functions. First, they allow a manager to learn more about the alternatives under consideration. Second, through negotiating with others about the properties of the decision alternative, the commitment of these other persons to the version of the alternative that evolved during the communicative process often increases as well. Both of these functions are part of what we term here the selling of a decision alternative.

Seeking and interpreting information are important parts of the decision-making process. Selling activities can be seen as an important form of information seeking. Hedelin and Allwood's (2001) results indicate that selling activities took a large part of the informants' time and attention during the decision process. However, the decision makers' selling activities are important not only because of the information they unearth. They are also important because of their other functions, in particular those of increasing commitment to the decision alternative(s) constructed by the manager and of testing how the actors in question react to these alternative(s). Both of these functions serve to improve the chances that the favored alternative, as finally constructed, will be formally decided on by the executive authority in the organization. It can also serve to support the behavior-regulating effects of the decision alternative decided on.

Our notion of selling a decision should not be confused with the concept of selling previously used by Tannenbaum and Schmidt (1958). Selling a decision, as we use the concept, concerns the coordination of wills, intentions, and desires within and outside the organization. It can be seen as helping to adjust the managers' (or the organization's) decision-making initiatives to the specific properties of the organization and to its external environment. In contrast, the notion contained in Tannenbaum and Schmidt's selling concept was developed in a context concerned with describing different forms of leadership. Selling concerns a decision already taken by the manager and has to do with the type of manager leadership where he or she "rather than simply announcing [the decision] takes the additional step of persuading his subordinates to accept it" (Tannenbaum & Schmidt, 1958, p. 97). Likewise, the concept of alignment as discussed by Kotter (1990; see also Kotter, 1995) appears to concern getting people to be inspired, loyal, and committed to a vision or to a decision. We see the selling efforts of the managers with whom we had contact as being carried out by champions, so to speak, who endeavored to complete their task successfully. This usually includes the completion of the processes that lead to the formal decision they are responsible for. However, note that in some cases their task might be more successfully completed in ways that do not involve a formal decision.

The degree to which selling activity is engaged in may be greater in Sweden and in Scandinavia generally than in the United States, for example, where managers have more of a reputation for their action-oriented attitude. We suggest, however, that the selling aspect of the decision process is of general importance in strategic decision making in organizations. The most important argument for this is that the selling process concerns more than simply the implementation of a decision. Involving learning, as the selling process does, it can also contribute to the creation of high-quality decision alternatives so the decisions that are implemented are more likely to have favorable consequences for the organization. Communication with experts and others helps to achieve this, irrespective of country and culture.

ACKNOWLEDGMENT

This research was supported by a grant from The Bank of Sweden Tercentenary Foundation.

REFERENCES

Heath, C., & Gonzalez, R. (1995). Interaction with others increases decision confidence but not decision quality: Evidence against information collection views of interactive decision making. *Organizational Behavior and Human Decision Processes, 61*, 305–326.

Hedelin, L., & Allwood, C. M. (2001). Managers' strategic decision processes in large organizations. In C. M. Allwood & M. Selart (Eds.), *Decision making: Social and creative dimensions* (pp. 259–280). Dordrecht, The Netherlands: Kluwer Academic.

Kotter, J. P. (1990, May–June). What leaders really do? *Harvard Business Review, 68*, 103–111.

Kotter, J. P. (1995, March–April). Leading change: Why transformation efforts fail. *Harvard Business Review, 73*, 59–67.

Mintzberg, H. (1975, July–August). The manager's job: Folklore and fact. *Harvard Business Review, 53*, 49–61.

Nutt, P. C. (1984). Types of organizational decision processes. *Administrative Science Quarterly, 29,* 414–450.

Sonnentag, S. (2000). Excellent performance: The role of communication and cooperation processes. *Applied Psychology: An International Review, 49,* 483–497.

Tannenbaum, R., & Schmidt, W. H. (1958, March–April). How to choose a leadership pattern. *Harvard Business Review, 36,* 95–102.

Tversky, A., & Kahneman, D. (1974). Judgment under uncertainty: Heuristics and biases. *Science, 185,* 1124–1130.

16

Observing Situational Awareness: When Differences in Opinion Appear

Peter Berggren
Swedish Defence Research Agency

The purpose of this work was to investigate and describe how the concepts situational awareness (SA), pilot performance (PP), and pilot mental workload (PMWL) relate to each other in the aviation context. The concepts were studied using the methods developed by the European project Visual Interaction and Human Effectiveness in the Cockpit (VINTHEC) and the Swedish Defence Research Agency (FOI). The research platform used was the simulator at The Swedish Commercial Flight School (TFHS) in Ljungbyhed, Sweden. Since 1985 TFHS has used simulators in their teaching of new pilots. The methodology, proved in military settings, was adjusted to civilian flights in cooperation with FOI researchers and one of the most experienced instructors at TFHS.

SA, or situation awareness, has been defined by Adams (1995) as "Knowing what's going on so you can figure out what to do!" (p. 9/2). Endsley (1995b) described three levels of SA: perception (Level 1) of the physical elements of the situation, comprehension (Level 2) of the significance of objects and events within the situation, and projection (Level 3) of future events based on the current situation. There exist a variety of techniques to measure SA. According to Endsley (1995a) SA can be measured through performance techniques, physiological techniques, questionnaires, and through subjective techniques. Among the subjective techniques, self-rating is easy to use and inexpensive. The problem with self-rating during a test trial is that the operator may not be aware of what is really happening in the environment; he or she has only a semiconfirmed picture of what is happening. When an observer rates the operator's SA, he or she might have more information about the world. Yet, the observer would have limited knowledge about the operator's concept of the world. The only thing that might indicate the operator's SA is his or her actions and verbalizations. In isolation, the observer's rating of an operator's SA is limited. Endsley (1995b) wrote about how SA is related to performance: "Good SA can therefore be viewed as a factor of good performance but cannot necessarily guarantee it" (p. 40). This says nothing about whether there exists a difference between SA as rated by an observer and SA as self-rated by the operator.

Mental workload is sensitive to changes in information load. Those changes therefore predict later changes in performance and SA (Svensson, Angelborg-Thanderz, & Wilson, 1999; Svensson, Angelborg-Thanderz, Sjöberg, & Olsson, 1997). Mental workload is a measure of the capacity of the human information processing resources (Wickens, 1992). When those resources are overused, performance declines. Workload is defined "as the effort invested by the human operator into task performance" (Hart & Wickens, 1990, p. 258), a definition that was adopted for use within the pilot domain; hence the concept PMWL. PMWL can be measured through different techniques. Those can be categorized into subjective, performance, and psychophysiological measures (VINTHEC, 1997). The Bedford Rating Scale (BFRS) gives a static subjective measure of PMWL (Roscoe, 1987; Roscoe & Ellis, 1990). The method is a modified version of the Cooper–Harper Aircraft Handling Characteristic Scale (Cooper & Harper, 1969). The BFRS is a unidimensional decision tree scale in which the participant has to choose which level is representing his workload effort for the moment. Berggren (2000) argued that when using a static measure at several different occasions, this gives a pseudodynamic measure of the course of events. This pseudodynamic measure gives an overall picture of the course of events regarding the rated measure.

Human performance can be measured in a multitude of ways with respect to the quantity and quality of the action (Sanders & McCormick, 1992). This can be done by subjective responses (Svensson, 1997; Svensson et al., 1997; Wickens, 1992). The performance can either be rated by the operator or by an independent observer. This is a method that is easy to use and inexpensive. PP is the rated performance of the pilots.

SPECIFIC AIMS OF THE STUDY

As stated before, the purpose of this work was to investigate and describe how the concepts SA, PP, and PMWL relate to each other in the aviation context. In other words

1. Explore how the concepts SA, PP, and PMWL rated by both pilots and the observer relate to each other.
2. Determine whether rated performance can be predicted from the ratings of the concepts SA, PP, and PMWL.
3. Explore how the ratings differ between the observer and the pilots concerning SA and performance.

METHOD

Participants

Eighteen male pilots, considered to be experts, participated in the study. Their mean age was 50.8 years (SD = 6.18). Their mean experience of flying was 8,300 hr varying between 3,000 to 20,800 hours. They were all instructors at the TFHS in Ljungbyhed, Sweden. Their mean experience of the aircraft represented by the simulator varied between 375 and 3,000 hr (M = 1,500 hr). Ten of the pilots used glasses. One half of the pilots had a military background and one half of the pilots had a civilian background. The test leader was one of the most experienced instructors at TFHS. He had been a military pilot for 35 years and a civilian pilot for 25 years. For 30 years he had worked as an instructor, and he had a total of 6,500 flight hours. He had participated in the design of the simulator that was used in this study. During the whole study, he was in charge as instructor, supervisor, observer, and rater.

Apparatus

A simulator of the type Piper Navajo PA31 (Essco, Inc., Barberton, Ohio) was used in the study. It consists of a cockpit where all instruments and controls are replicas with identical functionality to the originals. The displays and meters are connected to the program running the simulation; all feedback from them is realistic. The feedback, apart from the meters and displays, is audible. There is no visual feedback of the world outside of the airplane in terms of front screen displays. The simulation is considered to resemble instrument flying.

Two scenarios were used: A and B. Each scenario consisted of seven stages with different levels of difficulty. The two scenarios were balanced in terms of difficulty, that is, they were reversed. The scenarios were, apart from being data collection opportunities, also used as the pilots' Periodic Flight Training (PFT), the test that every pilot has to take every 6 months to keep his or her certificate.

Design

The study was a within group construction with two occasions for data collection. The dependent variables that are in focus are PP, PMWL, and SA. The independent variables are type of scenario (A and B), the raters (pilots and observer), and stage.

Operationalization of Variables

PP. The pilots rated their performance on four different occasions during scenario A and on five different occasions during scenario B. PP was graded from 1 to 7, where a low number indicated *bad performance* and a high number indicated *exceptionally good performance*. The observer rated the pilots on seven different occasions during each scenario using the same scale as the pilots used.

PMWL. The pilots rated their mental workload after each of the seven different stages during each of the scenarios. A Swedish adaptation of the BFRS was used.

SA. Both pilots and observer rated the pilots' SA after each of the seven different stages of each of the scenarios. A modified version of the Cooper–Harper Scale, adjusted for use with the SA concept, was employed.

Procedure

All pilots flew each scenario. The order of the scenarios was held constant among the pilots. Scenario A always came before Scenario B. Each session started with the pilot answering a questionnaire on background variables such as personal status, motivation, and mood. After that, the observer who sat in the copilot's seat started the simulation. At the end of each stage throughout the scenario, the observer took notes on PP and SA. After each stage of the scenario, the pilot rated his performance, mental workload, and SA. When the scenario was finished (after about 40 to 50 min) the pilot answered another questionnaire concerning motivation, difficulty, and mood. Then again, when it was time for the next scenario, the same procedure was repeated.

RESULTS AND DISCUSSION

The pilots were all motivated to participate in the study. They rated the scenarios as realistic. All participants succeeded in their PFT; their knowledge and skill was considered good. The remaining issues in the question-naires from before and after the scenarios is left out of this analysis.

How Do the Concepts SA, PP, and PMWL as Rated by Pilots and Observer Relate to Each Other?

Correlations between ratings of different concepts are presented in Table 16.1. SA rated by the observer is OSA, performance rated by the observer is OP, PMWL is rated by the pilots, SA rated by the pilots is PSA, and PP is rated by the pilots. These abbreviations will be used throughout.

The correlation matrix shows that the ratings given by the observer and the pilots did not correspond to a greater extent. Performance says something about the quality of a skill. How performance correlates with SA and mental workload therefore says something about flying skill. The correlation between OSA and PP is strong ($r = .75$, $p < .001$) as it was between PSA and PP ($r = .77$, $p < .001$). About 55% of the performance variance was explained by the PSA in the pilots' ratings. That the correlation between performance and SA is not perfect suggests that SA was not the same as performance in this study. The correlation between PMWL and PP was strong ($r = -.65$, $p < .001$) as was the case with the correlation between PMWL and PSA ($r = -.71$, $p < .001$). It is interesting that PSA and PMWL correlated strongly, a finding suggesting that the more mental resources are used, the harder it is to keep high SA. SA seems to have been affected by how heavy the mental workload was. That is, the more the situation demanded from the operator (i.e., variation of difficulty in the sessions), the more he had to focus to keep up the performance. No sign of the hysteresis phenomenon (Cumming & Croft, 1973) was found when standardizing the scenarios individually. This implies that there was no difference between the two scenarios regarding their difficulty.

TABLE 16.1
Correlations Among the Ratings of the Concepts

Concepts	OSA	OP	PMWL	PSA	PP
OSA					
OP	.754				
PMWL	−.224	−.275			
PSA	.362	.409	−.711		
PP	.305	.307	−.651	.765	

Note. SA = situational awareness. OSA = SA rated by the observer. OP = performance rated by the observer. PMWL = pilot mental workload rated by the pilots. PSA = SA rated by the pilots. PP = performance rated by the pilots. All values significant using Bonferroni ($p < .01$).

Can Rated Performance Be Predicted From the Ratings of the Concepts SA, PP, and PMWL?

A multiple regression analysis, with OP as criterion and the OSA, PSA, PP, and PMWL as predictors, was performed. The multiple correlation (R) was .81 $(p < .001)$, which explains about 65% of the criterion variance. The only significant β weight was the OSA predictor $(\beta = .76, p < .001)$.

Another multiple regression analysis using the pilots' rating of PP as criterion and the OSA, PSA, OP, and PMWL as predictors, was performed. The multiple correlation (R) was .78 $(p < .001)$ which explains about 61% of the criterion variance. There were two significant β weights, PMWL $(\beta = -.20, p < .01)$ and PSA $(\beta = .61, p < .001)$.

These two regression analyses show that there was a slight difference between what the observer was looking at when rating performance and what the pilots were looking at when they rate performance.

How Do the Ratings of OSA and OP Differ From Those of PSA and PP Rated by the Pilots?

The correlation between the ratings of PP made by pilots and OP for both sessions was weak $(r = .31, p < .001)$; the common variance was about 10%. These findings show that there was a difference between OP and ratings of PP by the pilots.

The variability of the ratings of SA was larger for the pilots than for the observer $(SD = 1.285$ for observer and 1.697 for pilots respectively). The pilots rated their SA slightly lower than the observer did, $t(474) = 3.53, p < .001$ $(M = 8.10$ for observer and 7.61 for pilots). The correlation between OSA and PSA was weak $(r = .36, p < .001)$. The opinions of pilots and the observer differed with respect to how high the pilots' SA was. The significant difference between observer and participants may have occurred because the observer rated SA from what he saw, whereas the pilot rated from both how he performed and how he felt about the situation. The observer could not possibly see if the participants were aware of every aspect of the situation, but he could see how they performed (cf. Endsley, 1995a). It is also possible that the observer interpreted the participants' SA from the SA he thought he would have had if he had acted as the participants did in a given situation. Both of these hypotheses lead to the conclusion that you cannot know about someone else's SA on all levels of SA (cf. Endsley, 1995b). Thus, to rate the pilot's level of SA, the observer was probably using his previous experience of how he handled a similar situation when making an assumption about the pilot's SA. On the other hand, no implicit components were described this way.

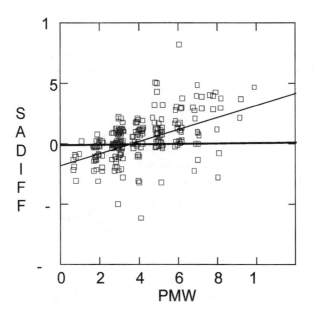

FIG. 16.1. The relation between SADIFF and PMWL. *Note.* SADIFF = difference between pilot rating of SA and observer rating of SA. PMWL = pilot mental workload rated by the pilots.

When comparing the PSA with the OP the OP increased with the PSAs (*r* = .41, *p* < .001). This accounts for 18% of the variance in the criterion performance. This relation was important because it suggested that the PSAs were valid.

There was a difference in ratings between the observer and the pilots when rating SA. A new variable that shows the difference in opinion (OSA–PSA) was created: SADIFF. Figure 16.1 shows the relation between this variable and PMWL.

The difference in opinion of SA increased as a function of workload. The correlation between SADIFF and PMWL was .54 (*p* < .001). When the mental workload increased, the pilots said that their SA declined. Yet, the observer still saw how well the pilots performed and therefore rated their SA as good.

A similar difference was also found when examining performance. A new variable (OP–PP) was created: PFDIFF.

PFDIFF increased as a function of workload. The correlation between PFDIFF and PMWL was .43 (*p* < .001). When the mental workload increased, the pilots said that their performance declined. The observer rated the pilots' performance higher than the pilots did.

GENERAL DISCUSSION

The results can be summarized in two points: (a) there were differences between the pilot and the observer when comparing ratings on performance and SA; and (b) this difference was most apparent when mental effort was rated high on the pilot, that is, when mental effort was rated high, PFDIFF and SADIFF increased between observer and pilot. It is therefore suggested that the pilot and the observer use different criteria when they rate performance and SA, that is, they base the rating on different aspects.

The observer can possibly conclude from the pilot's behavior what the pilot can perceive and comprehend, but it might be harder to infer anything about the things that are going on in the pilot's head, that is, planning of future events. Thus, the observer does not know all aspects that the pilot includes in his rating of performance and SA. However, the pilot himself may not be aware of aspects of his behavior, that is, his knowledge and skills may be implicit. Thus, the pilots may base their rating more on meta-cognitive aspects (existing in consciousness) and the observer may base the rating more on observable behavior. This might explain why there exists a difference in opinion between pilots and observers. Hence, different persons involved in rating SA and performance might base their rating on different aspects. These differences in opinion between pilots and observers might be related to levels of SA (cf. Endsley, 1995b), concerning the projection of future states (Level 3) and perception (Level 1). This discrepancy implies that depending on who is asked to make the rating, different answers about what is happening will be given.

REFERENCES

Adam, E. C. (1995). Tactical cockpits—flat panel imperatives. *Situation awareness: Limitations and enhancement in the aviation environment* (AGARD–CP–575; pp. 9/1–9/7). Brussels, Belgium: NATO–AGARD.

Berggren, P. (2000). *Situational awareness, mental workload, and pilot performance—relationships and conceptual aspects* (Scientific Report FOA-R—00-01438-706—SE). Linköping, Sweden: Defence Research Establishment.

Cooper, G. E., & Harper, R., Jr. (1969). *The use of pilot rating in the evaluation of aircraft handling qualities* (Report No. NASA TN-D-5153). Moffett Field, CA: NASA.

Cumming, R. W., & Croft, P. G. (1973). Human information processing under varying task demand. In L.-O. Persson & L. Sjöberg (1978). The influence of emotions on information processing. *Göteborg Psychological Reports, 8*(7).

Endsley, M. R. (1995a). Measurement of situation awareness in dynamic systems. *Human Factors, 37*(1), 65–84.

Endsley, M. R. (1995b). Toward a theory of situation awareness in dynamic systems. *Human Factors, 37*(1), 32–64.

Hart, S. G., & Wickens, C. D. (1990). Workload assessment and prediction. In H. R. Booher (Ed.), *Manprint: An approach to systems integration* (pp. 257–296). New York: Van Nostrand Reinhold.

Roscoe, A. H. (1987). In-flight assessment of workload using plot ratings and heart rate. *The practical assessment of pilot workload*. AGARDograph No. 282.

Roscoe, A. H., & Ellis, G. A. (1990). *A subjective rating scale for assessing pilot workload in flight: A decade of practical use* (Technical Report 90019). Royal Aerospace Establishment.

Sanders, M. S., & McCormick, E. J. (1992). *Human factors in engineering and design* (7th ed.). Singapore: McGraw-Hill.

Svensson, E. (1997). Pilot mental workload and situational awareness—Psychological models of the pilot. In R. Flin, E. Salas, M. Strub, & L. Martin (Eds.), *Decision making under stress: Emerging themes and applications* (pp. 261–270). Aldershot: Ashgate Publishing.

Svensson, E., Angelborg-Thanderz, M., & Wilson, G. F. (1999). *Models of pilot performance and mission evaluation—Psychological and psychophysiological aspects* (Technical Report AFRL-HE-WP-TR-1999-0215). Wright-Patterson AFB, OH: USAF/RL.

Svensson, E., Angelborg-Thanderz, M., Sjöberg, L., & Olsson, S. (1997). Information complexity—mental workload and performance in combat aircraft. *Ergonomics, 40*(3), 362–380.

VINTHEC. (1997). *Review of eye point-of-gaze equipment and data analysis* (Technical Report VINTHEC-WP2-TR-02). Amsterdam, NL: NLR.

Wickens, C. D. (1992). *Engineering psychology and human performance* (2nd ed.). New York: HarperCollins.

Cultural Differences in Cognition: Barriers in Multinational Collaborations

Helen Altman Klein
Wright State University

Multinational military, peacekeeping, and humanitarian collaborations require that allies make decisions in complex, high-stakes situations under time pressure and uncertainty and with organizational constraints. These situations require problem identification, risk assessment, planning and replanning, and situational awareness. Practitioners have increasingly turned to naturalistic decision-making (NDM) models. NDM is a theoretical perspective for describing how people make decisions and how they use their experience to perform effectively in complex situations (Orasanu & Connolly, 1993).

Although NDM researchers study the dynamics of decisions, there is a critical piece missing. Researchers have assumed that decision making is the same across national groups. Researchers do not know, however, if research outcomes based on data collected mainly with Americans and Western Europeans will hold for other national groups. We even lack data on the commonality of cognition and decision making for Americans and Western Europeans. There is a growing body of research suggesting profound differences in psychosocial functioning but also in cognition and managing uncertainty across national groups.

Multinational collaborations can bring together professionals with very different views of the world. Worldviews influence how people react to uncertainty, balance power, and make decisions. Worldviews shape anticipations and coordinate actions. When multinational allies differ in their view, it can lead to conflict and be counterproductive during complex, ill-structured, high-stakes decision making. National differences challenge the unexamined use of NDM models during multinational collaborations.

In this chapter, I make the case that national culture differences are important as allies use NDM findings to understand multinational collaborations during military, peacekeeping, and humanitarian operations. I describe a framework for understanding culture. I then present seven cultural dimensions in terms of potential impact on multinational collaboration. Finally, I review broader implications of culture for NDM.

A FRAMEWORK FOR UNDERSTANDING NATIONAL CULTURES

Cultures differ because each has evolved in a distinct physical and social context. Culture, as I conceptualize it in this chapter, has three defining characteristics. First, cultures are functional blueprints for a group's behavior, social patterns, and cognition in the same way that DNA is the individual's blueprint for physiology and anatomy. Culture provides guidelines ranging from rules about verbal interactions to acceptable social behavior and from the expression of emotion to the cognitive tools for assessing risk and deciding between alternatives. Cultures retain characteristics that have enhanced the ability to survival and to raise a next generation.

Second, cultures are dynamic systems that emerge from a particular ecological context. People who share an ecological context often share features of culture (Berry, 1986; Segall, Dasen, Berry, & Poortinga, 1990). Context includes the physical environment and also the social and political environment. Cultures evolve as context changes over time and experience. When food sources are altered by changing climate, successful cultures alter their subsistence patterns to ensure continued survival. With the technological changes of the industrial revolution, cultures shifted toward new roles for urban dwellers. They adopted new patterns of land use, concepts of time, and reasoning styles.

Finally, cultures are composed of integrated components. Each culture includes a range of harmonious elements that operate together for the goals of survival, interaction, and propagation. Traditional anthropology has emphasized the physical features, language, and customs of culture. Patterns of subsistence, language elements, and marriage customs all fit together. These physical and behavioral features are useful but not sufficient for understanding complex cognition and decision making. Psychosocial and cognitive characteristics are necessary for the integrated whole. Agrarian cultures tend to share behavioral, psychosocial, and cognitive characteristics because these characteristics work together for the success of farming. National groups maintain characteristics that serve the particular needs of their environment.

Experiences in a culture, beginning from birth, frame the individual's view of the world. Because culture is an integrated whole, it is difficult to see and to change. Triandis (1994) asserted that we see the world less as "how it is" and more as "how we are." Culture is like a lens through which we see the world (Klein, Pongonis, & Klein, 2000). We interact most effectively with people when we can see the world as they do. This allows communication, shared situational awareness, and effective coordinated action. Problems arise when we assume that others interpret and react as we do.

Different world views present barriers in multinational military, peacekeeping, and humanitarian operations. These operations require complex coordination and decision making. Personnel respond to the uncertainties of emerging natural threats, such as the earthquake in Turkey, and of political instability, such as the shifting power structure in Somalia. Time pressure and stakes can be high. Complexity is compounded when decision making is distributed (e.g., Hutchins, 1995; Woods & Patterson, 2001). Different worldviews make it hard for allies to maintain communication, anticipate decisions, and coordinate actions in the face of uncertainty and unpredictability. It is time to look to cultural differences for help in increasing effectiveness during multinational operations.

WHAT CULTURAL DIFFERENCES AFFECT MULTINATIONAL COLLABORATIONS?

Cultural differences, beyond the well-studied behavioral and linguistic ones, can interfere with decision making in complex naturalistic settings. I describe seven potentially important cognitive and psychosocial differences here. Although the universality of cognition is often assumed, recent research has questioned this. Dialectical reasoning (Peng & Nisbett, 1999) and hypothetical thinking (Markus & Kitayama, 1991; Tetlock, 1998) have been suggested as potentially importance cultural differences for complex decision making. Markus and Kitayama (1991) provided a case for the importance of independence versus interdependence. Hofstede (1980) found power distance and uncertainty avoidance to be values that differ across national groups. Finally, Kluckholn and Strodtbeck (1961) suggested that time orientation and activity orientation are influential contributions to decision-making differences between cultures.

Dialectical Reasoning

Many planning activities are concerned with exploring possibilities and understanding choices. Dialectical reasoning researchers have found cultural differences in handling contradiction (Peng & Nisbett, 1999). The difference is linked to distinct epistemologies. Some national groups like to differenti-

ate—polarizing contradictory perspectives to select the single best. They
sharpen differences to clarify objectives, goals, and methods. Other national groups resolve seeming contradictions by seeking compromise—retaining elements of both perspectives. Their reasoning style is to seek a
middle ground and deny dichotomous descriptions. These different styles
provide different paths to resolving conflicts.

Multinational partners face complex situations in which there are competing goals. This may relate to the timing of operations, the prioritizing of tasks,
or the selection of strategies. Decisions are difficult when they include inherent contradictions and incompatible goals. Resolving conflicts between opposing positions is not easy when some team members seek the best position and optimize alternatives, whereas other members seek an intermediate
goal. This difference can plague multinational decision making. Ignoring the
difference can leave some participants feeling disenfranchised. When operation personnel complain about the lack of clear directives, it is often because
the coalition partners lacks a common reasoning style to resolve differences.

Hypothetical Thinking

Hypothetical or counterfactual thinking uses mental representations of past
or future events to consider alternate outcomes (Markus & Kitayama, 1991;
Tetlock, 1998). Hypothetical thinking is not universal (Markus & Kitayama,
1991). Independent cultural groups including most Westerners frequently
use hypothetical thinking to make plans and to examine the implications of
these plans. They separate reasoning from context and consider options in
an abstract, hypothesis-driven manner (Markus & Kitayama, 1991). In other
groups, particularly those from interdependent cultures, reasoning is contextually grounded in personal experience. They work to improve future
performance in context.

Hypothetical thinking differences present barriers in multinational collaborations. To improve collaborative efforts, past actions, goals, and ac-
complishments need to be reviewed. Some groups use mental simulation to
play out alternate strategies and imagine how they might have resulted in
different outcomes. They identify ways to enhance future performance by
changing a troublesome aspect of plans. The essence of mental simulation
is hypothetical thinking. Other national groups, however, find this to be a
useless activity because it is not grounded in reality. This difference can inhibit learning exercises and pose barriers to effective collaboration.

Independence Versus Interdependence

This dimension is related to Hofstede's (1980) earlier notion of the collective versus individual dimension. It is constructed on the conceptual base
of self (Markus & Kitayama, 1991). An independent view stresses the unique

and autonomous character of the individual in which people view themselves as complex aggregates of attributes, emotions, motivations, and aptitudes that are distinct from those of others. Interdependent people see themselves not as separate entities but as connected to and undifferentiated from others. Feelings, thoughts, and motivations are related to others and are meaningful only when considered in relationships. In the same way that people do not think of their lungs and their kidneys as having independent goals and motivations, in interdependent cultures, individuals are not thought of as having special rights or directions outside those of the group. The group with which an individual feels interdependent varies by culture.

It is easy for members of independent national cultures to make decisions as individuals. They view themselves as self-contained and personally responsible. They may even ignore the ideas of other relevant players. In contrast, members of interdependent national groups may seek a consensus with appropriate segments of their group. Even if the individual has the knowledge to make the best decision, a sense of interdependency might delay or prevent action. These patterns can generate very different and sometimes incompatible decision-making patterns.

Power Distance

Power distance is the extent to which members of a group expect the uneven distribution of power (Hofstede, 1980). This includes the acceptance of unequal distribution of power by cultural institutions (Dorfman & Howell, 1988). The differences in interpersonal power and influence between superior and subordinate team members reflect differences in this dimension. Hofstede (1980) described the variability in power distance over national groups. Low power distance was associated with egalitarian working patterns and team interchanges. Leaders with low power distance typically listen to the ideas of others on the team, even those of lower rank. In contrast, those with high power distance maintain rank in decision making.

In multinational collaborations, allies may have limited experience working together and they may need to work in different locations. The structure and the lines of command for decision making and for implementation may cross national boundaries. Several leaders may need to coordinate actions to avoid counterproductivity. This works best when everyone adheres to and respects the same command structure regarding responsibility for decisions. If participants differ in power distance, they may struggle to define working relationships rather than to accomplish goals. Further, because operations are often so complex, no one person can have all the needed knowledge. Sometimes the lower ranking technical staff have the expertise to make the best decision. When collaboration partners have dis-

crepancies in power distance, it can interfere with the utilization of expertise and can delay or compromise action.

Uncertainty Avoidance

Uncertainty avoidance is the extent to which members of a culture experience uncertainty as stressful and take actions to avoid it (Hofstede, 1980;
Dorfman & Howell, 1988). Some cultures have a high need for predictability
and perceive uncertain or ambiguous situations as threatening. Other cultures are more comfortable with uncertainty. This cultural difference is manifested in decision-making styles. Uncertainty avoidance influences how ready
a national group is to adapt in the face of new and unexpected developments.
Differing judgments of urgency can result in uncoordinated actions. It is hard
for professionals who value flexibility, spontaneity, and last-minute decisions
to work with those who need firm, committed plans of action.

Making decisions and carrying out plans in the face of uncertainty, high
stakes, and time pressure are central issues for NDM (Lipshitz, 1997). Multinational missions include uncertainties. The disease risks from flooding in
Mozambique and the potential of biological warfare agents in Iraq both carried an unknowable risk and made planning difficult. With all this uncertainty, it was still necessary to act. Different groups have a different point at
which they are comfortable acting, and these differences in uncertainty
avoidance complicate multinational operations. Differences in tolerance for
uncertainty create tension and fear.

Time Orientation: Past, Present, or Future

Kluckholn and Strodtbeck (1961) considered time orientation as a pervasive
way in which people view the world. A focus on the past leads a culture to
pay respect to traditions and customs. Past issues and relationships are a
living part of the present. Thus, historical animosities and land holdings are
important today and into the future. Past patterns are used as models for
planning and tradition plays a large role in decision making. In contrast,
present-oriented cultures are less interested in traditions and more in immediate goals. They are not burdened by accumulated weight from the past
nor by long-term projected needs. They have immediate concerns in planning and weigh the initial rather than the long-term impact of actions. Fu-
ture-oriented cultures focus on long-term goals even at the cost of short-
term gains. They are less concerned with limits imposed by the past.

Time orientation is a driving force when priorities and goals need to be
set. It is important for decisions concerning resources and actions. Some
groups prefer a less optimal solution achievable in the near future, whereas
others would prefer solutions that would reach fruition years later. Opera-

tion forces create resentment when they fail to acknowledge or respect traditional practices and hereditary leaders. Do we spend the time drinking coffee with the elders or talk to the mayor? Do roads go through historic towns and cemeteries or weave around such sites? During multinational operations, these conflicts can emerge. Differences in time orientation create difficulties and misunderstandings in multinational interactions.

Activity Orientation

Kluckholn and Strodtbeck (1961) proposed that activity orientation, "doing" or "being," affects the way people approach life, work, and relationships. National groups characterized as doing view work-related activities as a central focus and accomplishment as their defining goal. Western thinkers are generally "do-ers." They look for task demands and how best to accomplish them. They give less attention to the interpersonal dynamics of the situation. In being cultures, on the other hand, relationships are the central focus. Being is characterized by working for the moment, stress release, and the nurturance of relationships.

Differences in decision making and planning styles linked to different activity orientations present obstacles to effective multinational functioning. Barriers can arise when do-ers who opt for pragmatic decision making work with "be-ers" whose decision making is based on emotions and tied to relationships. Do-ers, furthermore, believe that they can hasten change when plans are outlined, target dates set, and frequent reports made (Adler, 1991). Be-ers, in contrast, believe in allowing change to happen at its own pace without rushing things to attain short-term results. Speeding up change is considered unwise. These differences interfere with ongoing operations when they are not recognized and managed.

WHAT ARE THE BROADER IMPLICATIONS OF CULTURE FOR NDM?

NDM is concerned with the challenges facing professionals as they work in complex, ill-defined, time-restricted, and high-stakes operations. Shared psychosocial and cognitive processes are the basis of effective communication, coordination, and action. In multinational collaborations, however, the participants may differ in their worldview. I have explored seven national cultural differences that can introduce problems. These differences contribute to the risk of coordination breakdown because they interfere with decision making from initial strategic planning to ongoing replanning to tactics in the field. These differences demand careful attention and additional research. NDM researchers need to incorporate national culture differences into models, research, training, and other applications.

It is time for NDM researchers to consider how national culture differences create vulnerability during complex multinational operations. The occurrence of unexpected events are the source of such vulnerability. This happens, for example, when a computer failure disrupts communication or when the delayed arrival of troops forces replanning. Vulnerability occurs with changes in the middle of an operation. Changes in staff, for example, require that the team redefine roles and relationships. Distributed teams are vulnerable when they must act in the face of uncertainty. Groups may interpret events differently because they lack a shared worldview. These common occurrences in multinational operations are vulnerable because of the dimensions I have discussed here.

Several methods are available for studying the vulnerable points that occur during collaborative operations. Archival records provide a picture of communication, coordination, and conflicts during past multinational missions. Such missions are common as the North Atlantic Treaty Organization, the United Nations, and other coalition forces replace national forces in many international arenas. Records can be selected and analyzed for the presence of specific national groups and evidence of culture-driven actions, conflicts, and outcomes. The data can provide a picture of the impact of national cultures on complex operations. A limitation of archival data stems from the invisibility of national cultural differences beyond those of language and custom. Unless specific attention is given to cognition and to psychosocial dimensions, culturally relevant information may not have been recorded.

Critical Incident Technique (Flanagan, 1954) with key informants can provide in-depth probes for incidents with multinational professionals at culturally vulnerable points. These points might include unexpected delays or staff changes. Informants from different national groups who took part in the incident can be included so different views of the same incident can be captured. This would provide the varied perspectives of a single incident and the commonality across different incidents. Researchers could learn how cues are interpreted and how different groups understand option generation, assessment, and selection. Finally, researchers need to learn how some multinational teams have succeeded, whereas others have failed. The limitation of this Critical Incident Technique is that cultural differences are confounded with individual differences reducing the ability to generalize.

Critical incidents can provide the basis for laboratory simulations presenting standard challenges to representatives of a range of national groups. Researchers now have the technology to simulate distributed task scenarios with variations in the national culture dimensions. Such simulations, if grounded in actual field experiences, would extend our understanding of the role of national cultural differences for NDM. The use of standard simulations would reduce the confounding factors of the other methods. These include the individual personalities involved, unique aspects of the mission, and contextual biases. The reduction of confounding is done at the cost of reduced fidelity.

Because multinational collaborations are here to stay, a long-term goal will be to train for more effective interactions. Practitioners need to go beyond national differences in holy days and terms of address. A knowledge of national culture differences related to psychosocial and cognitive functioning is a step toward effective collaboration during complex operations. Researchers know that training in specific rules and procedures will not be effective in complex environments. Rather, it will be necessary to train those involved to be able to take the perspective of the others. Such training must extend beyond the demands of a particular mission or those of a specific simulation. It requires that individuals come to see the world as if through the cultural lens of others. In this way, trainers could increase common vision in the face of divergent views. Although some individuals can do this, our missions will be more effective if more can.

In this chapter, I have emphasized the importance of national cultural differences among allies during multinational operations. Multinational operations also involve citizens at the site of the action and nongovernmental organization (NGO) personnel. When the cognitive and psychosocial characteristics of those served by peacekeeping and humanitarian missions are ignored, unforeseen consequences are more likely. American forces were careful to accommodate the customs of their hosts in the Gulf War but were not always attuned to cognitive differences in judgment style. Military personnel also need to understand the NGOs that are a common part of humanitarian missions.

In a similar way, national culture differences are useful for understanding and predicting the actions of adversaries. Accurate forecasting of future actions must be built on an understanding of thinking style, power relationships, uncertainty avoidance, and the like. Good cultural knowledge can prevent new incidents and ease escalating conflicts. Cultural differences in cognition are particularly important in asymmetrical warfare in which little specific knowledge is available about the enemy (Klein & Klein, 2000). National differences in cognition and psychosocial functioning can contribute to international stability.

Multinational collaboration extends far beyond the military. Multinational corporations are seeking ways to establish an effective workforce that includes different worldviews. Commercial aviation and medicine both often require coordination among operators of different nationalities (Helmreich & Merritt, 1998). Advances in science ranging from gene identification to space exploration are accomplished by teams from around the world. In this context, it is critical to bridge the gap between national differences and the challenges of complex planning and decision making. Whenever collaborative decision making is undertaken in a multinational context, culture and the worldviews they spawn are potential barriers. It is time to lower the barriers and harness the strength provided by multinational collaboration in complex domains.

ACKNOWLEDGMENT

I acknowledge the contributions made by David D. Woods to my understanding of laboratory simulations for use in complex situations.

REFERENCES

Adler, N. (1991). *International dimensions of organizational behavior* (2nd ed.). Boston: Kent.
Berry, J. (1986). The comparative study of cognitive abilities: A summary. In S. E. Newstead, S. H. Irvine, & P. L. Dann (Eds.), *Human assessment: Cognition and motivation* (pp. 57–74). Dordrecht, Netherlands: Martinus Nijhoff.
Dorfman, P., & Howell, J. (1988). Dimensions of national culture and effective leadership patterns: Hofstede revisited. *Advances in International Comparative Management, 3*, 127–150.
Helmreich, R. L., & Merritt, A. C. (1998). *Culture at work in aviation and medicine.* Aldershot, England: Ashgate Publishing.
Hofstede, G. (1980). *Culture's consequences: International differences in work-related values.* Newbury Park, CA: Sage.
Hutchins, E. (1995). *Cognition in the wild.* Cambridge, MA: MIT Press.
Klein, H. A., & Klein, G. (2000, May). *Cultural lens: Seeing through the eyes of the adversary.* 9th Annual Conference on Computer Generated Forces and Behavioral Representation, Orlando, FL.
Klein, H. A., Pongonis, A., & Klein, G. (2000, June). *Cultural barriers to multinational C2 decision making.* Paper presented at the meeting of the 2000 Command and Control research and technology symposium proceedings, Monterey, CA.
Kluckhohn, F., & Strodtbeck, F. (1961). *Variations in value orientations.* Evanston, IL: Row, Peterson.
Lipshitz, R. (1997). On-line coping with uncertainty: Beyond the reduce, quantify and plug heuristic. In R. Flin, E. Salas, M. Strub, & L. Martin (Eds.), *Decision making under stress* (pp. 149–160). Aldershot, England: Ashgate Publishing.
Markus, H., & Kitayama, S. (1991). Culture and the self: Implications for cognition, emotion, and motivation. *Psychological Review, 98*, 224–253.
Orasanu, J., & Connolly, T. (1993). The reinvention of decision-making. In G. A. Klein, J. Orasanu, R. Calderwood, & C. E. Zsambok (Eds.), *Decision-making in action: Models and methods* (pp. 3–20). Norwood, NJ: Ablex.
Peng, K., & Nisbett, R. (1999). Culture, dialectics, and reasoning about contradiction. *American Psychologist, 54*, 741–754.
Segall, M., Dasen, P., Berry, J., & Poortinga, Y. (1990). *Human behavior on global perspective.* New York: Permagon.

Tetlock, P. (1998). Close-call counterfactuals and belief-system defenses: I was not almost wrong but I was almost right. *Journal of Personality and Social Psychology, 75,* 639–652.

Triandis, H. (1994). *Culture and social behavior.* New York: McGraw-Hill.

Woods, D. D., & Patterson, E. S. (2001). How unexpected events produce an escalation of cognitive and coordinative demands. In P. A. Hancock & P. Desmond (Eds.), *Stress, workload and fatigue* (pp. 290–302). Mahwah, NJ: Lawrence Erlbaum Associates.

18

The Normalization of Deviance: Signals of Danger, Situated Action, and Risk

Diane Vaughan
Boston College

Barry M. Turner, in *Man-Made Disasters* (1978), analyzed 85 different disasters and found a common pattern. They were typified by failures of foresight: long incubation periods with signals of potential danger that were either ignored or misinterpreted. What causes these failures of foresight? How can they be prevented? In this chapter, I highlight the main findings of my research on the U.S. National Aeronautics and Space Administration's (NASA) 1986 Space Shuttle *Challenger* tragedy, an example of a failure of foresight (Vaughan, 1996). The case reveals that history, the political environment, and complex organization structures and processes intersected to normalize signals of potential danger. Drawing general lessons from the case, I make suggestions about preventive strategies and risk reduction for other organizations that do risky work. To orient readers to the theoretical, methodological, and policy orientation of this project, I begin by introducing four principles that guided my research.

1. Situated action. My research took a situated-action approach that included macrolevels and microlevels of analysis. A fundamental sociological understanding is that interaction takes place in socially organized settings. Rather than isolating action from its circumstances, the task of scholars is to uncover the relation between the individual act and the social context. To study individual action and decisions requires research that goes beyond daily interaction with others to the socially organized setting and larger environment in which individual action is situated. Consequently, taking a situ-

ated-action approach requires examining the connection between three levels of analysis: environment, organization, and individual choice and action. The goal was to explore the relation between the three, as they affect the phenomenon of interest (Vaughan, 1998).

2. Culture as connecting link. The link between macrolevel structure and microlevel processes has long been established by sociological theorists (see, e.g., Giddens, 1984). Culture connects these three levels of analysis, serving as a mediating link in the macro–micro relation (DiMaggio, 1997; Vaughan, 1998). By *culture* I mean the tacit understandings, habits, assumptions, routines, and practices that constitute a repository of unarticulated source material from which more self-conscious thought and action emerge. A full grasp of the situated nature of any event, activity, or outcome calls for an examination of the role culture plays: how it is created and, reciprocally, how it affects everyday life.

3. Qualitative methods. Qualitative methods—ethnographic observations, interviews, document analysis, and video and photographic methods—are particularly appropriate for studying situated action because the focus on individual interaction and choice also reveals to the attentive observer the functioning of the organization and its environment: People explain and do their work in reference to these larger structures. The modern organization is a wonderful site for a cultural analysis: It is like the old anthropological studies of islands and isolated communities because it ties people to a particular territory marked by boundaries and sharing (to a greater or lesser extent) a common concern. However, in addition, the modern organization produces a paper trail that allows researchers to study its history, culture, structure, decision making, and interaction with other organizations in its environment, supplementing the ethnographic methods and interviewing.

4. The connection between cause and control. The final guiding principle is that the discovery of causes, whether researchers are investigating how things go wrong or how things go right, points out directions for policy. In the public realm, the typical response to a mistake or error leading to some harmful outcome is to blame it on individual failure: some version of operator error. It is also typical to attribute success to individual hard work, wise decisions, and excellent planning (Starbuck & Milliken, 1988). The perceived cause of a problem is the basis for the strategy for control. When individual action is identified as the cause of a problem, then the corrective policy is directed toward individuals: better training, higher levels of education, more frequent breaks to prevent fatigue, or firing, demotion, and punishment. However, a situated-action approach expands the target: Policy needs to be directed toward all three levels of analysis—environment, organization, and individual choice and action. Thus, when a situated-action approach guides research on risky work, accidents, failures, or high-reliability organizations, the result is a more complete and effective strategy for control that targets to

the greatest extent possible the relevant aspects of environment, organization, and individuals. Moreover, attention must be paid to the relation between culture, risk, and safety.

I turn now to an overview of the causes of the *Challenger* accident. The
published research from which this is drawn is a historical ethnography
based on original NASA documents and interviews used by the Presidential
Commission appointed to investigate the tragedy (currently stored at the
National Archives, Washington, DC), the Presidential Commission's (1986)
report (five volumes) and the 1986 report of the U.S. House Committee on
Science and Technology, personal interviews, analysis of original documents, and other sources. To present the full theory of the normalization of
deviance in these pages, I present the findings without the supporting data.
Interested readers should see the original (Vaughan, 1996). Relevant references selected from the original, supplemented by some recent works with
similar themes, are in the bibliography of this text.

THE *CHALLENGER* ACCIDENT:
RISKY TECHNOLOGY, CULTURE,
AND DEVIANCE AT NASA

The Presidential Commission investigating the *Challenger* disaster discovered the technical cause was an O-ring failure on the Solid Rocket Boosters.
However, the NASA organization was implicated. The accident was pre-
ceded by questionable management actions and decisions that smacked of
at the least, incompetence, and at the most, evil. First, the Commission
learned that in the years preceding the January 28, 1986, tragedy, NASA continued shuttle launches in spite of recurring damage on the O-rings. Second,
they learned that production pressures at the agency had made launching
the shuttle on schedule a priority over safety. Finally, they learned of a midnight hour teleconference on the eve of the *Challenger* launch in which contractor engineers located at Morton Thiokol, Utah, protested launching
Challenger in the unprecedented cold temperatures predicted for launch
time the next morning. Nonetheless, NASA managers at the Marshall Space
Flight Center in Huntsville, Alabama overrode the engineers' recommendation. They proceeded with the launch, violating safety rules about passing
information up the hierarchy in the process. An outraged Commission concluded that the disaster was not simply a technical failure but a managerial
failure of tragic proportions.

In the months following the disaster, the Commission's published report
and myriad media accounts led the public to believe that managerial wrongdoing was behind the launch decision. Based primarily on the Presidential

Commission's findings about production pressures at the space agency and repeated safety rule violations by Marshall managers responsible for the Solid Rocket Booster Project, the historically accepted explanation was this: These NASA managers, warned that the launch was risky, succumbed to production pressures and violated safety rules in order to stick to the schedule. I began this research with a hypothesis that the cause of the disaster was organizational misconduct. However, as my research went beyond readily accessible information into primary data stored at the National Archives, Washington, DC, I discovered that I was wrong. Many of my most important assumptions about the case were mistaken. Some examples are the following:

• In the years preceding the *Challenger* teleconference, NASA had repeatedly launched with known flaws in the Solid Rocket Boosters. The public impression was that NASA managers had a history of overriding engineering concerns. Yet the same Thiokol engineers that protested the *Challenger* launch were the ones responsible for risk assessments and launch recommendations during those controversial years. Despite technical problems, they had repeatedly recommended that NASA managers accept risk and fly.

• On the eve of the *Challenger* launch, not all engineers were opposed. Some were in favor, but only the ones who were opposed were called to testify. Moreover, all but two agreed that the engineering analysis was flawed and unlikely to convince managers.

• Rumor had it that NASA managers needed to launch because of a planned hook-up between the crew and then U.S. President Ronald Reagan who was making his State of the Union address the evening of the launch. However, NASA never allowed outside communication with the crew during the first 48 hours in orbit because the crew were too busy. Moreover, every launch had two launch windows, morning and afternoon. If NASA managers truly believed they were taking an exceptional risk, they could have launched *Challenger* in the afternoon when the temperature was predicted to reach between 40 and 50 °F and still maintained their targeted launch date.

• In the history of decision making on the Solid Rocket Boosters, 1977 to 1985, and on the eve of the launch, I discovered NASA managers abided by every NASA launch decision rule. With all procedural systems in place, they had a failure.

From the preceding insights, two things became clear to me. First, outsiders viewing what happened at NASA retrospectively saw key incidents and events very differently than insiders who were diagnosing problems as they occurred. In large part, this was due to retrospection. Equally important, outsiders had not been able to fully grasp NASA culture: the rules, procedures, and language that were key to understanding engineering decisions. Culture was even a barrier to the understanding of the Presidential Commission, which spent 3 months and enormous resources finding out what had happened. I concluded that only by learning NASA culture and only by reconstructing events chronologically to see how insiders assessed risk as the Solid Rocket Booster problems unfolded would I be able to understand the launch decision.

From my research, I concluded that the disaster was a mistake, not misconduct. However, the answer to why this mistake occurred is complex. Most important, this was no anomaly, something peculiar to NASA, as the managerial wrongdoing theory suggested. It was a mistake that could happen in any organization, even one pursuing its tasks and goals under optimal conditions. Typically, accident investigations focus on individuals and the decisions they make, excluding other important factors that are harder to see: how organizations are shaped by history and politics and how those external factors alter the organization itself, affecting individual decision making. Organizational and environmental factors help explain why individuals in organizations behave as they do. The *Challenger* disaster was a product of NASA's political environment, organization history, and organization culture and structure. First, I consider how these factors affected the history of decision making prior to 1986. This history, and the factors that shaped it, were crucial to the disaster because past decisions shaped the sensemaking of many who were present at the *Challenger* teleconference.

THE HISTORY OF DECISION MAKING: NORMALIZING DEVIANCE

Why, in the years preceding *Challenger*, did NASA continuing launching with a Solid Rocket Booster that repeatedly was having problems? In formal risk assessments in the years preceding the *Challenger* launch, managers and engineers in the work group responsible for formal risk assessments of the boosters continually normalized the technical anomalies (deviations from performance predictions) that they found on the boosters after a mission. By *normalized*, I mean that in official decisions, information they initially viewed as a signal of potential danger—evidence that the design was not performing as predicted—was, after consideration, reinterpreted as acceptable and nondeviant. Based on engineering analysis showing that if the primary O-ring failed a second O-ring would back it up, they continued to believe it was safe to fly. In other words, the work group developed a cultural belief—a shared worldview—that the situation was safe despite mounting evidence that indicated otherwise.

What is astonishing is that the work group gradually expanded the boundary of what was to them an acceptable risk. The critical decision was the first one when, expecting no damage to the O-rings, damage occurred and they found it acceptable. This started the work group on a slippery slope. As one manager told me, "Once you've accepted that first lack of perfection, that first anomaly, it's like you've lost your virginity. You can't go back." That initial precedent served as a basis for engineering risk assessments for subsequent launches. Gradually, in formal risk assessments, engineers accepted more and more risk. Each of these decisions, taken singly, seemed correct, routine, and indeed insignificant and unremarkable, but the decisions had a cumulative directionality of which the decision-making participants themselves were unaware—except in retrospect. Even as engineering concerns began to grow in 1985, the engineers assigned to the Solid Rocket Booster Project—including those who protested the *Challenger* launch—continued to come forward in official flight readiness decisions with recommendations that the boosters were an acceptable flight risk.

Why, if engineers were having concerns that they expressed to each other informally and in internal memos, did they continually normalize the technical deviation in official decisions? There were three reasons: (a) information and its context affecting the work group's definition of the situation; (b) the NASA organization and its political environment altered the culture of the organization, reinforcing decisions to continue launching; and (c) structural secrecy, which prevented people outside the work group from intervening and halting the work group's incremental descent into poor judgement.

Signals of Potential Danger: Information and Its Context

The context mattered to the interpretation of information. First, the Solid Rocket Booster work group was working in an organization in which having technical problems was expected and taken for granted; therefore, to have problems was not itself a signal of potential danger. In fact, having problems was normal for several reasons. The Space Shuttle System was a large-scale technical system with many interactive parts. The design of the vehicle was unprecedented. Further, the many component parts were made by different manufacturers and had to be put together by NASA. Putting them together was completely uncharted territory: Each had to be crafted to fit with the others, and no one knew how they would work when assembled. Finally, the shuttle was designed to be reusable. Despite engineering laboratory tests, field tests, and calculations, engineers could never predict and prepare for all the forces of the environment that the shuttle would experience once it left the launch pad. They knew it was going to come back with damage that required new analysis and correction before it could be launched again.

Taking this into account, in 1981, NASA created a document titled "The Acceptable Risk Process" in which the agency acknowledged that after they had done everything that could be done, the shuttle would still contain residual risks (Hammack & Raines, 1981). That residual risk had to be analyzed prior to each flight to determine whether or not it was acceptable. The document articulated, in broad strokes, the directions that the Acceptable Risk Process must take prior to each flight. Therefore, in the NASA culture, both having problems and taking risks was routine. Documents were full of the language stating acceptable risk, acceptable erosion, anomalies, and discrepancies. To outsiders after the disaster, this looked like rationality gone wild. To insiders, it was normal, everyday talk.

The second important aspect of context that affected the interpretation of signals of potential danger about the Solid Rocket Boosters was the pattern of information as problems began to occur. What, in retrospect, seemed to outsiders afterward to be clear signals that should have halted shuttle flight were interpreted differently by NASA personnel at the time. Because signals of potential danger occurred within a stream of information and decisions, the pattern of information made these warning signs look mixed, weak, or routine as individual decisions were made.

Mixed Signals. A mixed signal was one in which a signal of potential danger was followed by signals that all was well, convincing the managers and engineers in the work group that the situation was safe. To illustrate, when returning flights showed anomalies on the booster joints—a signal of potential danger—engineers analyzed and corrected the problem (a piece of lint on an O-ring was enough to cause damage to an O-ring). Subsequently, a number of flights showed no problems—signals that all was well.

Weak Signals. A weak signal was one that was hard to decipher or, after analysis, seemed such an improbable event that the work group believed there was little probability of it recurring. To illustrate, a launch in January 1985—a year before *Challenger*—showed the worst O-ring damage to that point. Cold temperature was thought to be a factor because the vehicle was on the launch pad through 3 consecutive days of 19 to 20 °F overnight Florida temperatures. Knowing that *Challenger* was affected by the cold, people saw this as a strong signal. However, at the time, engineers had no evidence that temperature was responsible for the damage they found— many factors had been causing problems—and they believed such a run of cold temperatures was unlikely to happen again. There was, in the words of Morton Thiokol engineer Roger Boisjoly, "no scramble to get temperature data" because no one expected a recurrence. The vehicle was tested and designed to withstand extremes of heat, not cold. Cold temperature was, to them, a weak signal—until the eve of the launch.

Routine Signals. In mid-1985, O-ring erosion began occurring on every flight. After the disaster, outsiders were incredulous that flight continued. For insiders, however, multiple instances of erosion indicated not danger but assurance that they correctly understood the problem. This increase in the frequency of technical deviations from design predictions was acceptable and normal to them because they had instituted a new procedure that guaranteed that the O-rings would be properly positioned. This procedure increased the probability of erosion, but because they believed the O-rings were redundant, erosion was not viewed as a problem. Better they assure redundancy by getting the rings in proper position than worry about erosion, which was, in fact, occurring exactly as they predicted. What outsiders saw as signals of potential danger were to them routine signals showing the joint was operating exactly as they expected.

I shift now from the microlevel interaction to the macrolevel contingencies that combined to reinforce the work group's cultural belief in acceptable risk, which kept them recommending launch in the years preceding *Challenger.*

Organization Culture:
NASA and Its Political Environment

After the disaster, analysts unanimously concluded that politics had altered the organization culture. They were correct. NASA's relationships with other organizations in its environment—Congress, the White House, contractors—altered the organization culture, which was an important factor that explained the normalization of deviance. Contradicting the conventional wisdom that decision making at NASA was governed by a monolithic, production-oriented culture, however, I found three cultural imperatives that drove the normalization of technical deviation in the work group's risk assessments: the original technical culture, political accountability, and bureaucratic accountability. The main lesson here is about the trickle-down effect: how political bargains and decisions made by top administrators attempting to negotiate power and resources in the space agency's external environment altered both the structure and the culture of the workplace, impacting the decision making of the people in the Solid Rocket Booster work group.

The Original Technical Culture. The standards of engineering excellence that were behind the successes of the Apollo era made up the original technical culture. That culture required that risk assessments be guided by scientific principles and rigorous quantitative analysis. Hunches, intuition, and observation, so essential to engineering, had a definite place in laboratory work. However, when it came to decisions about whether to proceed with a launch or not, the subjective and intuitive were not allowed: Flawless engineering analysis based on quantitative data was required. This original technical culture still existed at NASA during the shuttle program, but it was struggling to survive amidst two other cultural mandates.

Political Accountability. During the Apollo era, Congress gave NASA a blank check. When Apollo was over, the consensus for space exploration was diminished. NASA barely got the shuttle program endorsed but did so by selling it to Congress as a program that would to a great extent be self-funded. The space shuttle would be like a bus, ferrying people and objects back and forth in space. It could carry commercial satellites and at the projected launch rate, could produce enough income a year to support the program. Thus, the shuttle would survive as a business, and production pressures were born. Meeting the schedule became the key to continued funding from Congress. Consequently, performance pressures and political accountability lived side by side with the original technical culture.

Bureaucratic Accountability. The agency became "bureaupathological." After Apollo, the growing NASA–contractor structure meant increased rule following was required to simply put together and launch this very complex shuttle vehicle. In addition, the Reagan administration required increased accountability of all government agencies. As a consequence of both these developments, working engineers spent much more time doing desk work and filling out forms, and the entire launch decision process, normally guided by rigid rules for procedural accountability, was joined with burgeoning paperwork of another sort. For each launch, 60 million components and thousands of countdown activities had to be processed. With the accelerated launch schedule, managers and engineers were working evenings and weekends just to turn around all the paperwork.

The original technical culture still existed, but engineers struggled to adhere to its tenets under the production pressures and pressures for bureaucratic accountability. All three aspects of NASA culture contributed to the normalization of deviance, so even when engineers began to develop deep concerns, they continued to make formal risk assessments that recommended launching. Culture mediated between the environment, the organization, and individual choice, as follows.

Political accountability resulted in production pressures and scheduling concerns that became normal and taken for granted to engineers and managers assigned to the hardware. No one had to tell them the schedule was important; they knew. The original technical culture required that rigorous, scientific, quantitative engineering arguments back up engineering recommendations. What this meant was that as long as the managers and engineers in the Solid Rocket Booster work group had strong quantitative data showing that the hardware was safe to fly (which they did), they could not interrupt the schedule to do tests to see why it was operating as it was (which they did not know). Production pressures suffocated the intuition and subjective concerns that were basic to the research and development (R&D) organization NASA was during the Apollo era. Finally, bureaucratic accountability contributed to the normalization of deviance in a surprising way. The sensemaking and cognitive beliefs of managers and engineers were affected by the fact that they followed all the rules. They had confidence that if they followed all the rules and all the procedures, then they had done everything they could to assure safety. This included following the rules and procedures of the original technical culture as well as the rules and procedures of the NASA organization. Although in the aftermath of the disaster, many believed that deviance and rule breaking was behind decisions to continue launching with anomalies in those years preceding the *Challenger* teleconference, it was really conformity. As they tried to make sense of signals of potential danger, managers and engineers alike were conforming to all three dominant cultural mandates that governed their workplace.

Structural Secrecy

Information and its context shaped the work group's definition of the situation, and the interaction of the NASA organization in its environment altered the culture, propelling decision making along by maintaining and reinforcing the work group's belief that to launch under the conditions they faced was appropriate and normal, but there was a third important factor: Structural secrecy prevented those outside the work group from intervening. *Structural secrecy* refers to a very common problem in organizations: how organizational structure—division of labor, hierarchy, complexity, geographic dispersion of parts—systematically undermines the ability of people situated in one part of an organization to fully understand what happens in other parts. As organizations grow large, actions in one part are invisible to the others. Knowledge about tasks and goals becomes segregated, compartmentalized within subunits of the organization, and further cloaked by the division of labor within a subunit. Specialization increases the problem: People in one department or division lack the expertise to understand the work in another. Structural secrecy is particularly a problem between highly specialized experts working at the bottom of the hierarchy and administrators at the top. Although always mechanisms are in place to record, store, and exchange information, these efforts often have the ironic consequence of adding to the problem. The more information that accumulates on individual desks, the less is read and mastered.

Structural secrecy contributed to the normalization of deviance by concealing the seriousness of the O-ring problem from regulators and administrators outside the work group, preventing them from intervening and altering the pattern of accepting more and more anomalies and risk. Launch decisions were made in a formal, four-tiered, hierarchical decision-making structure known as Flight Readiness Review. The purpose was to assure that every shuttle component was flight ready. Managers and engineers in each work group assigned to a shuttle component were responsible for initiating launch recommendations based on engineering analysis and calculations of risk acceptability. These work groups were at the bottom of the Flight Readiness Review structure, reporting upward. The engineering analysis and recommendation of each work group was examined, criticized, and challenged at every level of Flight Readiness Review.

A critical factor contributing to structural secrecy was information dependency. Those regulators and administrators in other parts of the organization were dependent on the people in the work groups for information. Outsiders could not duplicate the research engineers had done; they were limited to criticizing it. Specialization, position and role in the organization, and division of labor among other participants in Flight Readiness Review restricted them to asking general questions (but difficult questions, as everyone noted) based on general engineering principles rather than the kind of the detailed questions people working on the same task might ask. Flight Readiness Review worked: Many technical problems were corrected prior to launching, and many engineering risk assessments were made more thorough by the process. However, because it was a bottom-up review procedure, if managers and engineers in work groups made a fundamental error but continued to bring forward launch recommendations backed by tight engineering analysis, structural secrecy prevented others from recognizing the mistake, intervening, and acting. In the case of the Solid Rocket Boosters, information dependencies combined with structural secrecy, and therefore, the technical deviation occurring in the Solid Rocket Boosters joints continued unabated.

In combination, information and its context, the intersection of organization and environment, and structural secrecy perpetuated the normalization of deviance, creating and maintaining the work group's belief in acceptable risk. In the years before the *Challenger* launch decision, the belief in acceptable risk had become a cultural understanding, institutionalized in the history of the organization and recorded in its documents. On the eve of the *Challenger* launch, this history of decision making and the belief in acceptable risk of the Solid Rocket Boosters was the all-important context against which new signals of potential danger were weighed. That night, the boundary of acceptable risk was expanded one more time. Once again, patterns of information, the organization culture, and structural secrecy combined to transform an anomaly—the unprecedented cold temperature—into an acceptable risk, normalizing yet another signal of potential danger prior to launching.

THE EVE OF THE LAUNCH

The decision to launch the Space Shuttle *Challenger* on January 28, 1986 was the outcome of a 2-hr teleconference held the night before between 34 people gathered around tables at Morton Thiokol in Utah, Marshall Space Flight Center in Alabama, and Kennedy Space Center in Florida. Uncertainty was extraordinary because the situation was unprecedented in three ways: (a) the predicted cold temperature for launch time was below that of any previous launch; (b) although teleconferences were routine at NASA, a launch decision had never before been made by teleconference; (c) engineers had never before come forward with a no-launch recommendation on the eve of a launch. Production pressure notwithstanding, the launch decision cannot be explained by the actions of amorally calculating managers who threw caution to the wind and succumbed to production pressures to meet the launch schedule, as many postdisaster analysts believed. Production pressures had a tremendous impact but on all participants and not in the way most postdisaster analysts thought. Moreover, other factors were equally important. As in the past, information and its context, the politically altered, three-faceted organization culture, and structural secrecy shaped the sensemaking of individual participants and thus the final outcome. Here, in abbreviated form, is how these factors mingled to create a disaster.

Concern about the cold temperature arose earlier in the day on January 27, 1986. Contractor engineers at Morton Thiokol in Utah were contacted. Political accountability, in the form of production pressures, operated early and invisibly on cognition: Thiokol engineers automatically set a deadline for a teleconference discussion to begin at 8:15 p.m. eastern standard time. They were used to working in a deadline-oriented culture deeply concerned about costs. They knew that if they could make a decision before midnight when the ground crew at Kennedy Space Center in Florida began putting fuel into the External Tank, they could avoid the costly de-tanking if the decision was "No-Go." NASA always de-tanked in the event a launch was canceled. De-tanking was an expensive, time-consuming operation. As a consequence of this self-imposed deadline, the engineers had to scramble to put together their engineering charts containing their risk assessments. They divided up the work of chart making, neglecting to look over the charts collectively before the teleconference began. Some people were putting together the final recommendation chart without seeing the data charts the engineers were creating. As each chart was completed, it was faxed to people in the other two locations.

As it turned out, the Thiokol charts assessing risk contained mixed, weak, and routine signals. Thus, the engineering analysis was filled with inconsistencies that violated the standards of NASA's original technical culture. The original technical culture required quantitative, scientific data for every engineering recommendation. Thiokol engineers argued that NASA should not launch unless the temperature was 53 °F or better because that was the previous coldest launch and the one that had suffered the most O-ring damage. However, the charts included data indicating the O-rings would hold at 30 °F and data indicating they had had problems at 75 °F, the warmest launch (mixed signals). Some of the charts were pulled from previous engineering presentations in which the same data had been used to recommend launches. Those charts were routine signals because they had been seen before and viewed as associated with safety, not increased risk. Finally, the recommendation chart said, "Do not launch unless the temperature is equal to or greater than 53 degrees," a conclusion based on observational data, not quantitative data. Within the strictures of the original technical culture, the engineering evidence was a weak signal, insufficient to overturn the preexisting, positivistic understanding of how the joint worked that Thiokol engineers successfully had presented in risk assessments in the preceding years.

Production pressures appeared a second time in the angry voices of Marshall managers who challenged these engineering arguments, intimidating the engineers. Marshall managers would be the ones who would have to carry forward the launch recommendation and defend the engineering analysis to top administrators in a system in which schedule was important. They had stopped launches before, but this time it appeared they were going to have to stop it with engineering analysis that was, within the original technical culture, not only flawed but based on observational data and an intuitive argument that was unacceptable: not real science. Moreover, production pressures were at work in another way. This particular 53 °F limit would stand as a rule, so all shuttles hence could not go unless the temperature were 53 °F—an awesome complication in a system required to meet a tight schedule. Under these circumstances, a tight engineering argument seemed particularly essential.

Structural secrecy also had a devastating effect on the discussion. In three locations, people could not see each other; therefore, words and inflections were all-important. Midway in the teleconference, the people assembled at Morton Thiokol in Utah held a caucus off the teleconference line. All locations were disconnected, so no one could hear what was transpiring at other facilities. During the offline caucus at Thiokol, an administrator who knew little about booster technology took charge, repeating the challenges of the Marshall managers. Without any new data to support their arguments, the engineers could not build a stronger data analysis.

Four Thiokol managers reversed the original engineering recommendation, going back on the teleconference hookup to announce that Thiokol had reexamined their data, reversed the decision, and recommended launch. When Marshall managers asked, "Does anybody have anything more to say?," no one spoke up; therefore, people in the other two locations did not know that the Thiokol engineers still objected. Ironically, because of structural secrecy Thiokol engineers did not know that during the caucus people at the other two locations believed the launch was going to be canceled. In fact, the top Marshall decision maker was making a list of people to call to stop the launch.

Bureaucratic rules and the culture's bureaupathology also played a critical role in the outcome. In an unprecedented situation, all participants invoked the usual rules about how decisions are made. These rules were designed to assure safety. They included vigorous adversarial challenges to engineering risk assessments, insistence on scientific, quantitative evidence, and allegiance to hierarchical procedures and norms about the roles of managers and engineers in engineering disagreements. Teleconference participants once again conformed, invoking all the usual rules in an unprecedented situation in which the usual rules were inappropriate. Although quantitative evidence was required for a go launch decision, engineering concerns and hunches should have been enough (seen in retrospect) in a no launch situation. Adversarialism was important to tighten up engineering analyses and insure there were no flaws, but in a situation of uncertainty with little available data, perhaps a cooperative, democratic, sleeves-rolled-up, what can we make of all this collectively decision-making session would have produced a different outcome. Finally, bureaucratic accountability prevailed by silencing people in other locations who had potentially useful information. They did not enter it into the conversation because organization norms and rules about who was empowered to speak inhibited them from talking. As history, culture, political environment, and organization characteristics intersected to normalize signals of potential danger in the past, so also were they repeated on the eve of the launch, normalizing signals of potential danger but this time resulting in a tragedy.

THE AFTERMATH: ROUNDING UP THE USUAL SUSPECTS

The 1986 Presidential Commission's report concluded with a series of recommendations that targeted the causes the Commission identified. The Commission did make the connection between cause and strategies for control, but they did not go far enough, failing to address the social context of decision making. In common with most postdisaster rituals, the investigation following *Challenger* focused attention on the individuals immediately responsible for the launch decision and on the technology. Their strategy for risk reduction targeted (a) the Solid Rocket Booster joints determined as the technical cause of the accident, (b) the middle-level managers at Marshall Space Flight Center, and (c) the decision-making process. In light of these findings, the Commission's recommended changes seemed appropriate and straightforward: fix the technology, replace the responsible individuals, and alter decision rules. Left untouched were important aspects of the social context of decision making: Important contributing causes in the environment and organization remained unchanged.

A SITUATED-ACTION APPROACH TO RISK REDUCTION

Those searching for preventive strategies should pay attention to the connection between cause and control. The *Challenger* case is an example of a situated-action research approach that takes into account the connection between environment, organization, and individual choice and action in a much-publicized failure. The chapters in this volume affirm the importance of environmental elements (such as scarcity, competition, and regulatory activities), organization characteristics, and culture in risky work. One logical question is how one might draw from these analyses to improve safety and reduce the risk that is inherent in complex technologies that have the potential for harmful outcomes. Despite the uniqueness of this case, it can be used as an example of what a situated-action approach suggests for risk reduction in other kinds of organizations concerned about safety. When engineering decisions about the Solid Rocket Boosters are analyzed as situated action, a complex causal picture emerges. Aspects of NASA's political environment and organization shaped the interpretation, meaning, and action of the specialists in the work group who worked on the boosters daily, conducted risk assessments, and initiated launch recommendations. Although strategies for safety must necessarily address individual performance and improve the technology, more could have been done at NASA to reduce risk of future failures. The following suggestions are representative of a situated-action approach to risk reduction. They target all three levels of analysis that comprise situated action: environment, organization, and individual. Moreover, they acknowledge the role of culture as a mediating link between the following three elements:

1. Responsibility at the top. An organization's environment presents both opportunities and constraints for organizational survival. In responding to environmental contingencies, top administrators in organizations must take

responsibility for safety by remaining alert to how their decisions affect the actions of people at the bottom who do the risky work. In the years preceding the *Challenger* disaster, NASA top administrators made decisions that contributed to the tragedy.

Negotiations with Congress had a major impact. At NASA, budget cuts by Congress and the White House curtailed resources available for hardware, research, testing, and personnel. Previously resource rich and fully supported by the national budget, NASA was able to have a superb Apollo program. However, the lack of funding for the Space Shuttle program made the space agency vulnerable to politics; therefore, NASA administrators had to try to do more with less to convince Congress that funding should be continued. Also, Congress passed bills requiring increased accountability for all federal agencies. This decision increased bureaucratic requirements for paperwork; therefore, NASA and contractor personnel spent much time completing reports and sending in requisitions that could have been spent on training and/or improving risk assessment strategies that would enhance understanding of the technology.

Administrative decisions significantly affect the organization culture, which in turn impacts decision making of people doing the hands-on work. To secure the program, top NASA administrators promised Congress that NASA could launch a certain number of shuttles each year with less resources. In making this pact, Congress and NASA converted an R&D organization into one that operated like a business. Although NASA administrators stressed safety as a goal, other actions and decisions convinced workers that schedule was the priority. Thus, production concerns and cost efficiency invaded the culture, affecting the content of work, its pace, and decision making in the Solid Rocket Booster work group.

Policy must bring goals and the resources necessary to meet them into alignment. At NASA, goals (launching a specific number of shuttles each year) were impossible to meet with available resources. Resources include personnel, training, equipment, and all other factors necessary to effect safety. If current organizational resources are inadequate to meet established goals, then either the resources need to be increased or goals must be brought in line with capacity. After the *Challenger* accident, NASA was given money to rebuild the lost shuttle, but the budget battle with Congress (the leading environmental causal factor in the fatal outcome) was not altered. High performance and an internationally competitive space program was still expected, but sufficient resources were not provided. The new NASA Administrator, Daniel Goldin, tried to bring goals into alignment with budgetary allotments. "Faster, Better, Cheaper" was the new agency motto. Subsequently, NASA reduced the number of manned missions of the shuttle, emphasizing instead smaller missions with scientific experiments. The second approach was to downsize: reduce the number of personnel and alter the organization structure. Initially, the revised NASA mission and goals seemed to be successful until a series of very expensive public mistakes—the Hubble telescope, the failed missions to Mars—repeated the lessons of the *Challenger*. Faster and cheaper were incompatible with better.

Administrative decisions to alter the structure of an organization can affect safety both negatively and positively. Because structural complexity is associated with increased probability of accidents in organizational-technical systems, reducing structural complexity would seem a logical step. However, changes must be undertaken cautiously. Proposals to downsize, merge, or eliminate a subunit should not be undertaken without research evaluating the impact on safety. Structural change that alters interaction among personnel results in loss of institutional memory and confusion about rules and division of labor, both of which increase risk during the transition period. When risky work is the issue, top administrators should evaluate the repercussions of proposed changes in advance by bringing in consultants: outside experts trained to analyze structure and process, such as organizational behavior specialists, anthropologists, and sociologists, and inside experts—experienced personnel at all hierarchical levels.

2. Culture. Administrators should actively cultivate foresight by doing diagnostics that could prevent failures of foresight. Conducting research on organization culture can yield discoveries that suggest ways to reduce risk. A finding in the *Challenger* research was that the culture was more complicated and its effects on decision making more subtle and hard to detect than even insiders realized. Members of an organization are sensitive to certain aspects of the culture but ignorant of others. Some become taken for granted, and therefore, cultural dictates are unquestioningly followed without workers realizing exactly what the culture is or its effects.

At NASA, a can-do attitude prevailed in the culture. Prior to the accident, this was considered a good thing, but this can-do culture had a dark side that contributed to the tragic consequences of the accident. Top administrators' can-do attitude differed greatly from that of their managers and engineers. Top administrators possessed a can-do hubris that all things were possible and shuttle success was taken for granted, a policy-affecting attitude decidedly out of touch with the risks of NASA's developmental technology. Being out of touch with the technological realities of the shuttle was behind administrative decisions to take nonastronauts on missions for political purposes. In contrast, work groups doing the risky work also had a can-do attitude, but it was based on strict adherence to rules and procedures, going by the book, and a belief in the long experience of personnel. The difference was based on daily contact with the technology and all its ambiguities, which gave the people doing the hands-on work a sincere appreciation for and fear of their uncertain explosive technology and all the unknowns in every launch condition. The lesson here is to find ways to keep administrators in touch with the technology and working conditions of their hands-on people. Had this cultural difference been known in advance, a corrective that kept top administrators in touch with the technology and working realities recognized by their engineers and managers could have been put in place. Perhaps such awareness might have altered policy decisions, in particular, top NASA administrative decisions to take civilians on shuttle missions.

Rules—and whether to obey them or not—are part of an organization's culture. Research on both conformity to rules and norms as well as deviations from them would be important in deciding if safety rules are effective. Research could examine both rule-following and rule-violating behavior in both routine and crisis decision-making conditions in organizations. Although research on rules—both conformity and violative behavior—is a sensitive subject and susceptible to distortion, historically ethnographic methods have consistently produced information on the rules of an organization and people's willingness and ability to follow them (see, e.g., Bensman & Gerver, 1963; Blau, 1955; Burawoy, 1979; Chambliss, 1996; Dalton, 1959; Hawkins, 1984; Perrow, 1984; Snook, 2000; Vaughan, 1996; Weick, 1993). Administrators seeking to understand the dynamics of their organizations to avoid failures of foresight could rely on ethnographically trained social scientists for these insights. Another possibility is an anonymous questionnaire exploring how extensive rule violations are and why people violate them. People violate rules for numerous reasons: A rule may be complex and thus is violated out of lack of understanding; a rule may be recent, and thus, people are unaware of it; a rule may be vague or unclear and thus is violated because people don't see that it applies to the situation they face; a rule may be perceived as irrelevant to the task at hand or in fact an obstacle to accomplishing it, and thus, the rule is ignored; a rule may conflict with norms about how to behave to assure safety. Perhaps the most challenging problem the *Challenger* tragedy raises for organizations with risky technologies is how to simultaneously instill rules to assure safety so they are followed automatically when people are under great pressure and simultaneously preserve the ability to innovate, to be creative, and to recognize the situation for which no rules exist and for which the existing rules do not apply (see especially Weick, 1993; Vaughan, 1996).

3. Normalizing deviance: Signals of danger and interpretive work. The *Challenger* disaster must be seen as one decision in a chain of decisions that show how the people responsible for risk assessment made an incremental descent into poor judgment. NASA's failure of foresight matches Turner's (1978) finding, which was that typically accidents and disaster have had long incubation periods with signals of potential danger that were either ignored or misinterpreted. At the space agency, the failure of foresight was a consequence of the normalization of deviance: Environmental and organizational

factors combined to affect individual cognition, blinding people to the significance of information and the possible harmful consequences of their actions. Points 1 and 2 previously suggest some strategies directed toward the environmental and organizational factors that contributed to the normalization of technical deviation at NASA. Now I turn to the third factor: decision making, signals of danger, and normalizing deviance.

Organizations have the advantage of being able to devise rules, train individuals, and create systems of linguistic, visual, and technologically produced signals. In situations in which critical information is being exchanged routinely, it is important to evaluate systems of communication, providing a repertoire of clear signals. Bureaucratic language used routinely tends to lose its power to communicate danger. For example, at NASA, the word *catastrophe* was neutralized by appearing as a failure effect on the description of every item labeled *Criticality 1* on a list of items that were not redundant that appeared on NASA's Critical Items List. When engineers were deeply concerned about the possibility of an accident, the bureaucratic language and rules of the organization provided no way for them to convey it in formal reports and risk assessment, and mandates about hierarchical relationships prevented them from conveying it informally to those above them in the hierarchy.

Organizations can train individuals to be alert to themselves as signal givers. People can be made aware that what they say affects how others interpret the riskiness of the situation and therefore how those others respond. Not only words and actions, but inflection, gestures, and body language affect how others make sense of what is happening. Individuals who do risky work can be trained to give accurate and clear signals in a manner appropriate to the understood hazards of a situation. They can be trained to avoid giving mixed signals in a critical situation and to use strong signals in situation-appropriate ways.

Significant in the eve-of-launch decision were the many missing signals: People who had important information did not enter it into the conversation. Administrative actions can be taken to minimize missing signals. Subordinates, tokens, and newcomers in all organizations often have useful information or opinions that they don't express. Democratic practices and respectful practices empower people to give signals, and everything possible should be done to develop such practices. One should be aware, however, that some kinds of information will still be hard to pass on in a work situation in which people are trained to suppress individuality to the collective well-being, to conform to rules, and to follow the commands of a leader. More difficult even than passing on information contradicting what appears to be the leader's strategy or the group consensus is saying no when everyone else appears to want to go forward. This is the equivalent of two engineers continuing to argue "don't launch" when all around them appear to want to go.

274   VAUGHAN

REFERENCES

Anheier, H. (Ed.). (1998). *When things go wrong*. Thousand Oaks, CA: Sage.

Bella, D. (1987). Organizations and systematic distortion of information. *Professional Issues in Engineering, 113*, 360–370.

Belli, R. E., & Schuman, H. (1996). The complexity of ignorance. *Qualitative Sociology, 19*, 423–430.

Bensman, J., & Gerver, I. (1963). Crime and punishment in the factory: The function of deviancy in maintaining the social system. *American Sociological Review, 28*, 588–598.

Blau, P. M. (1955). *The dynamics of bureaucracy*. Chicago: University of Chicago Press.

Bosk, C. (1979). *Forgive and remember: Managing medical failure*. Chicago: University of Chicago Press.

Burawoy, M. (1979). *Manufacturing consent*. Chicago: University of Chicago Press.

Carroll, J. (1995). Incident reviews in high-hazard industries. *Industrial and Environmental Crisis Quarterly, 9*, 175–197.

Carroll, J., & Perin, C. (1995). *Organizing and managing for safe production* (Rep. No. NSP95–005). MIT Center for Energy & Environmental Policy Research. Cambridge, MA: Massachusetts Institute of Technology.

Chambliss, D. (1996). *Beyond caring: Hospitals, nurses, and the social organization of ethics*. Chicago: University of Chicago Press.

Clarke, L. (1992). Context dependency and risk decision making. In L. Clarke & J. F. Short (Eds.), *Organizations, uncertainties, and risk* (pp. 27–38). Boulder, CO: Westview.

Clarke, L. (1993). The disqualification heuristic. In W. Freudenberg & T. Youn (Eds.), *Research in social problems and public policy* (Vol. 5). Greenwich, CT: JAI.

Clarke, L. (1999). *Mission improbable*. Chicago: University Chicago Press.

Clarke, L., & Perrow, C. (1996). Prosaic organizational failures. *American Behavioral Scientist, 39*, 1040–1056.

Dalton, M. (1959). *Men who manage*. New York: Wiley.

DiMaggio, P. J. (1997). Culture and cognition. *Annual Review of Sociology, 23*, 263–287.

Eisenhardt, K. (1993). High reliability organizations meet high velocity environment. In K. H. Roberts (Ed.), *New challenges to understanding organizations*. New York: Macmillan.

Emerson, R. M. (1983). Holistic effects in social control decision-making. *Law and Society Review, 17*, 425–455.

Emerson, R. M., & Messinger, S. (1977). The micro-politics of trouble. *Social Problems, 25*, 121–134.

Giddens, A. (1984). *The constitution of society*. Berkeley: University of California Press.

Hammack, J. B., & Raines, M. L. (1981, March 5). *Space Shuttle Safety Assessment Report*. Johnson Space Center, Safety Division, National Archives, Washington, DC.

Hawkins, K. (1984). *Environment and enforcement: Regulation and the social definition of pollution*. New York: Oxford University Press.

Heimer, C. (1988). Social structure, psychology, and the estimation of risk. *Annual Review of Sociology, 14*, 491–519.

Hughes, E. (1951). Mistakes at work. *Canadian Journal of Economics and Political Science, 17*, 320–327.

Landau, M. (1969). Redundancy, rationality, and the problem of duplication and overlap. *Public Administration Review*, July/Aug, 346–357.

Landau, M., & Chisholm, D. (1995). The arrogance of optimism. *Journal of Contingency and Crisis Management, 3*, 67–78.

Lanir, Z. (1989). The reasonable choice of disaster. *Journal of Strategic Studies, 12*, 479–493.

LaPorte, T. R. (1982). On the design and management of nearly error-free organizational control systems. In D. Sills, V. B. Shelanski, & C. P. Wolf (Eds.), *Accident at Three Mile Island*. Boulder, CO: Westview.

LaPorte, T. R. (1994). A strawman speaks up. *Journal of Contingency and Crisis Management, 2,* 207–211.

March, J. G., Sproull, L. S., & Tamuz, M. (1991). Learning from samples of one or fewer. *Organizational Science, 2,* 1–13.

Marcus, A. (1995). Managing with danger. *Industrial and Environmental Crisis Quarterly, 9,* 139–151.

Marcus, A., McAvoy, E., & Nichols, M. (1993). Economic and behavioral perspectives on safety. In S. Bacharach (Ed.), *Research in organizational behavior* (Vol. 15). Greenwich, CT: JAI.

McCurdy, H. (1989). The decay of NASA's technical culture. *Space Policy,* 301–310.

Paget, M. (1988). *The unity of mistake.* Philadelphia: Temple University Press.

Pate-Cornell, M. E. (1990). Organizational aspects of engineering safety systems. *Science, 250,* 1210–1216.

Perin, C. (1995). Organizations as contexts. *Industrial and Environmental Crisis Quarterly, 9,* 152–174.

Perrow, C. (1983). The organizational context of human factors engineering. *Administrative Science Quarterly, 28,* 521–541.

Perrow, C. (1984). *Normal accidents.* New York: Basic Books.

Presidential Commission on the Space Shuttle Challenger Accident. (1986). *Report to the President by the Presidential Commission on the Space Shuttle Challenger Accident* (Vols. 1–5). Washington, DC: U.S. Government Printing Office.

Reason, J. (1997). *Managing the risks of organizational accidents.* Aldershot, England: Ashgate.

Roberts, K. (1989). New challenges in organizational research: High reliability organizations. *Industrial Crisis Quarterly, 3,* 111–125.

Rochlin, G. I. (1993). Defining high-reliability organizations in practice: A taxonomic prologomena. In K. H. Roberts (Ed.), *New challenges to understanding organizations.* New York: Macmillan.

Rochlin, G. I., LaPorte, T. R., & Roberts, K. H. (1987). The self-designing high-reliability organization. *Naval War College Review, 40,* 76–90.

Romzek, B. S., & Dubnic, M. J. (1987). Accountability in the public sector: Lessons from the *Challenger* launch. *Public Administration Review, 47,* 227–238.

Sagan, S. (1993). *The limits of safety.* Princeton, NJ: Princeton University Press.

Sagan, S. (1994). Toward a political theory of organizational reliability. *Journal of Contingency and Crisis Management, 2,* 228–240.

Schulman, P. R. (1989). The "logic" of organizational irrationality. *Administration & Society, 21,* 31–33.

Schulman, P. R. (1993). The analysis of high reliability organizations. In K. H. Roberts (Ed.), *New challenges to understanding organizations.* New York: Macmillan.

Short, J. F., Jr., & Clarke, L. (Eds.). (1992). *Organizations, uncertainties, and risk.* Boulder, CO: Westview.

Snook, S. A. (2000). *Friendly fire: The accidental shootdown of U.S. Black Hawks over Northern Iraq.* Princeton, NJ: Princeton University Press.

Star, S. L., & Gerson, E. (1987). The management and dynamics of anomalies in scientific work. *Sociological Quarterly, 28,* 147–169.

Starbuck, W., & Milliken, F. J. (1988). Executives' perceptual filters. In D. C. Hambrick (Ed.), *The executive effect* (pp.). Greenwich, CT: JAI.

Turner, B. (1978). *Man-made disasters.* London: Wykeham.

Turner, B., & Pidgeon, N. (1997). *Man-made disasters* (2nd ed.). London: Butterworth-Heinemann.

U.S. Congression House. (1986). *Investigation of the Challenger Accident* [Hearings] (Vols. 1–2). Washington, DC: U.S. Government Printing Office.

Vaughan, D. (1992). Theory elaboration: The heuristics of case analysis. In C. C. Ragin & H. S. Becker (Eds.), *What is a case? Exploring the foundations of social inquiry* (pp. 173–202). Cambridge, England: Cambridge University Press.

Vaughan, D. (1996). *The Challenger launch decision: Risky technology, culture, and deviance at NASA*. Chicago: University of Chicago Press.

Vaughan, D. (1997). The trickle-down effect: Policy, decisions, and risky work. *California Management Review, 39*, 80–102.

Vaughan, D. (1998). Rational choice, situated action, and the social control of organizations. *Law & Society Review, 32*, 23–61.

Vaughan, D. (1999). The dark side of organizations: Mistake, misconduct, and disaster. *Annual Review of Sociology, 25*, 271–305.

Vaughan, D. (2002). Signals and interpretive work: The role of culture in a theory of practical action. In K. A. Cerulo (Ed.), *Culture in mind: Toward a sociology of culture and cognition* (pp. 28–54). New York: Routledge.

Weick, K. E. (1987). Organizational culture as a source of high reliability. *California Management Review, 29*, 116–136.

Weick, K. E. (1993). The collapse of sensemaking in organizations: The Mann Gulch disaster. *Administrative Science Quarterly, 38*, 628–652.

Weick, K. E. (1995). *Sensemaking in organizations*. Thousand Oaks, CA: Sage.

Weick, K. E., & Roberts, K. A. (1993). Collective mind in organizations: Heedful interrelating on flight decks. *Administrative Science Quarterly, 38*, 357–381.

Weick, K. E., Sutcliffe, K. M., & Obstfeld, D. (1999). Organizing for high reliability. In B. Staw & R. Sutton (Eds.), *Research in organizational behavior*. Greenwich, CT: JAI.

19

Making Ethical Decisions in Professional Life

Iordanis Kavathatzopoulos
Uppsala University

The global world of today dominated by fast technological change, scientific innovations, communication, and openness implies new unanticipated ethical challenges. Not knowing the morally right way to act may have a serious impact on all kinds of activities, including professional life. Given the high pace of changes inside and outside organizations, it is impossible to foresee ethical problems and conflicts, and it is therefore very difficult to construct ethical guidelines with any practical value. Under such conditions, individual decision makers as well as groups and organizations must acquire high ethical competence and confidence in handling all possible ethical problems that may arise in everyday, real-life professional activities. What is needed is a psychological approach to ethical competence implying high ethical awareness, adaptive problem-solving and decision-making abilities at personal and organizational levels, effective ethical argumentation skills, and high ethical confidence. In this chapter I discuss development of assessment methods as well as construction and implementation of training methods for ethical decision making and problem solving. Application in real-life professional activities has shown that individuals, groups, and organizations can be trained to cope with difficult ethical problems.

Moral knowledge is necessary for the functioning of society. Humans need to be able to anticipate the actions of other people as much as other people need to know what we are going to do in a certain situation. Ethical rules convey moral knowledge and guide persons, groups, and organizations in their relationships to other people. This is true under stable conditions in society, but ethical rules and guidelines are even more important in the fast changing world of today.

THE DIFFICULTY OF HANDLING MORAL PROBLEMS IN REAL LIFE

The changing, open, and global world gives rise to ethical issues the resolution of which are of vital importance for the functioning of society. Information technology, which is the motor and channel of the new, fast-moving world, develops and changes more rapidly itself; scientific advancement in the areas of medicine and biology forces people into making decisions that they are not prepared to make but cannot impossibly avoid. People are confronted steadily with an increasing number of unanticipated ethical challenges without having the same chance as before, under stable and recurrent conditions, to use already existing moral knowledge or consult insightful moral authorities. The changing conditions make it hard to foresee actual or future moral problems and conflicts, a condition that results in making it increasingly difficult to construct ethical guidelines that can be used in real life. In this changing context, only skills can offer rescue because they are not connected to specific problems as ready answers are but are general methods applicable to all problems of the same kind. That means that professionals, as well as their organizations, have to acquire practical ethical skills. They need ethical competence and confidence in handling all possible moral problems that may arise in everyday, real-life, professional activities. Professionals need high ethical awareness, adaptive ethical problem-solving and decision-making abilities at personal and organizational levels, effective ethical argumentation capacity, and high confidence in their ability to cope with moral problems.

Just as people living and working together need moral rules, so do organizations. These rules tell professionals how to be with each other and how to cooperate and to coordinate activities. Moral rules help people to predict the behavior of others, guide their actions, and tell them what to expect from their own and other groups, organizations, and social institutions. This knowledge is absolutely necessary. Society cannot function without moral rules. It would be impossible for organizations to operate and for persons to live and work together with other people if they do not know what is expected of them and how relevant others will act under certain conditions.

The interesting issue today is what people do when they do not have access to such knowledge, that is, moral rules that can guide them right in the fast changing society of today. When people do not have this moral knowledge conveyed by adaptive moral rules, do they at least know how to acquire that knowledge? The answer is no, and the reasons for that are many.

People live in an open and fast-changing environment. Modern society is global, more unpredictable, and more complex than ever. That gives people more pressure to try to find out what the right answers and actions are. Moral authorities, ethical codes, and guidelines are not as functional as they used to be. What people are usually missing in front of a moral problem are pieces of good working advice; they need concrete hints and directions for satisfying solutions. Another way to describe this is to say that people need to know how to apply their personal or organizational principles to concrete moral problems. General moral principles are impossible to apply directly in real-life problems because people cannot automatically deduct the right course of action from them; that is usually the cause of moral problems. The conclusion is that old ways, habits, and traditions cannot provide the right answers to modern moral problems.

Organizations try to adapt to these new conditions by changing themselves. Hierarchical structures are suitable for stable and predictable conditions, but organizations decentralize to be more flexible. Responsibility of making decisions is pressed downward creating a more flat organizational structure. However, this demolition of organizational pyramids is not followed by the corresponding transfer of power and resources. At the same time, an increasing and intense public and media interest is directed toward everything persons and organizations do or fail to do. The press on decision makers to come up with satisfying answers to moral problems is immense, resulting in increasing uncertainty, insecurity, stress, and anxiety. A solution that attracts many people is to somehow get a ready answer, usually by relying on ethical guidelines or authorities. However, functioning ethical guidelines demand stable conditions, and this is not what the world looks like today. Applicable and satisfying solutions to moral problems are also generally difficult to construct because moral solutions are controversial in themselves. Moral problems are also difficult to process critically because of all the strong emotions involved and because the decision maker more easily focuses on authority and content than on process and method (Kavathatzopoulos, 2003; see also Griseri, 2002; Schwartz, 2000).

What should people do then? What should be the focus of business ethics research and what should be the focus of training programs for decision making in ethics? There is a consensus that decision makers, persons as well as organizations, need some capacity to cope with moral issues. They need to anticipate moral problems, recognize them, and take them seriously when confronted by them. They need to raise their level of ethical awareness and become more alert to moral issues. They need to be unconstrained by moral fixations. They need to know how to explain to others involved, but also to themselves, why their decision was the best possible and to do it convincingly. They also need to know how to lead others in their organization in their common effort to take care of actual moral problems as well as to construct ethical rules and guidelines (see, e.g. Stark, 1993).

Yet the dominating idea and educational practice is that all of the preceding goals can be achieved by the transmission of moral values and principles. Theoretical knowledge is at focus in business ethics education programs (see, e.g. Tullberg, 2003). Students learn theories of moral philosophy and train in philosophical argumentation. This educational method does not bring the success in professional life one would expect. Students may be successful in their academic business ethics courses but subsequently have great difficulty showing the same level of excellency in real professional life (Sims, 2002; Weber, 1990).

Nevertheless, it would be very difficult to be successful in achieving the goal of transmission of moral principles. People in responsible positions, leaders, and professionals are not easy to convince when they believe something is right. They usually have strong ideas about the matter. Accepting someone else's authority would mean a corresponding loss of their own, and that is hard for a professional to accept. Thus, what commonly happens in training programs is that the moral values offered are at such a general level that everybody can agree with them.

Handling moral problems in real life is not easy. Educators may very well be successful in transmitting moral values to participants in an education program, but this is not really any gain. These kinds of moral values are general in character and therefore not possible to apply automatically and directly in a concrete, real-life situation. Besides the fact that all people already possess general moral values, probably the same values as those educators try to transmit through education, the real problem is how to apply them in reality. It seems that knowing what is right and wrong at a general level is not that useful in solving concrete moral dilemmas. The difficulty is inherited in the nature of moral problems: Moral problems are about the conflict of valid moral values. Solutions to real-life dilemmas are both right and wrong. It appears then that what really matters is to know how to handle such conflicts, to acquire ethical competence.

If the goal of education is to acquire competence, training should focus on the components of this competence. Ethical awareness is one of those skills. It must be easy for decision makers to see when and where a moral problem may arise to prepare for dealing with it. If they do not anticipate an upcoming problem, they can do nothing, and it can grow in an uncontrolled way causing greater damage. Education can train decision makers' ability to see what is important for customers, employees, the public, and so forth. They can also be informed of what issues are important for different stakeholders.

The next step is to acquire the skill to cope satisfactorily with moral problems. Confronted with a moral problem, decision makers need to know how to think, how to analyze the problem, and how to reach a solution. In a corresponding way, making satisfying moral decisions at an organizational level implies the adoption of appropriate routines and processes. The organization must be able to find solutions and to create ethical rules according to its basic values and the values of its stakeholders. It must also have the ability to adjust its ethical activities continuously to existing conditions. Persons and organizations need also to be able to argue, motivate, and defend their moral decisions convincingly. Confidence in one's own ethical ability, as well as readiness to execute difficult and controversial moral decisions, are very important, too.

Thus, ethical competence contains ethical awareness, personal decision-making and problem-solving skills, appropriate organizational processes, communication abilities, and ethical confidence. In that way the definition of ethical competence is based on the processing and treatment of moral problems rather than on the normative aspects of problems or the moral qualities of end solutions to the problems. However, it is not that easy for people to acquire and use this competence because of the controversy of moral issues and the extra difficulty to adopt independent and unconstrained thought processes in handling moral problems.

Therefore, ethics education of professionals, or future professionals, should focus on the acquisition and use of ethical problem-solving skills. Everybody would agree that the kind of method one uses to handle a problem is important. Consequently, one should learn methods rather than rehearse ready answers. However, this approach is rarely applied in ethics education or in ethical assessment (Holt, Heischmidt, Hammer Hill, Robinson, & Wiles, 1997/1998; Rossouw, 2002; Vitell & Ho, 1997).

In psychological research, there are, however, some efforts to focus on the processes of ethical problem solving. Studies have been conducted on how people reason about moral problems and how they solve such problems, what methods are adopted in moral problem solving, and what effect these methods have on achieving moral goals (Piaget, 1932; see also Kurtines & Gewirtz, 1995). It is known from this research that people's moral thinking is often constrained by certain moral principles or authorities, by uncontrolled decisions, by automatic reactions, and by responsibility avoidance. What is needed is an ability to focus on the concrete moral problem, an ability to perceive the relevant and significant aspects of the problem, and an ability to start the process of unconstrained analysis. This is the skill of ethical autonomy, which is supposed to be the basis of ethical competence.

Based on that assumption psychologists and educators could construct assessment methods to describe professionals' different ways of handling moral problems as well as to develop educational programs for learning to use what is supposed to be a better way to attack a moral problem. Some caution is however needed here to avoid the risk of confounding normative or moral content aspects in the psychological, ethical, decision-making and problem-solving process (such as the early work of Kohlberg, 1981, 1984; see also Brown, Debold, Tappan, & Gilligan, 1991; Gilligan, 1982; Jaffee & Hyde, 2000; Miller, 1994; Walker, 1995). Training and assessment of ethical decision making has to be based on the psychological hypothesis of autonomy that is supposed to constitute the critical factor for ethical competence (Kohlberg, 1985; Piaget, 1932).

ASSESSING ETHICAL COMPETENCE IN PROFESSIONAL LIFE

According to the preceding hypothesis, it is possible to describe ethical competence by assessing the psychological construct of autonomy, that is, the different ways people handle moral problems. This has been previously done through interviews (Kavathatzopoulos, 1993; Piaget, 1932), but later research has shown that this can be done by using paper-and-pen questionnaires. These questionnaires contain six to nine moral dilemmas, and they have been formulated so as to be very similar to real and ordinary moral problems in different professional activities (Kavathatzopoulos, 1994b; Kavathatzopoulos, 2000; Kavathatzopoulos & Rigas, 1998).

The task of participants answering the questionnaire is to place themselves in the position of the main agent in each story, accept the dilemmas as their own, and attempt to solve the problems. The items are short stories about some professional dilemma followed by four alternatives representing different aspects to be considered before any decision is made. These alternatives express different ways of thinking according to the hypothesis of autonomy. Two of the alternatives represent a constrained, automatic, and authoritarian way of thinking. The other two alternatives represent thinking that is focused on the concrete conditions of the problem, considering pertinent values involved in the situation. Here is a sample:

> You are the president of a major bank, and you have discovered that one of the oldest and most trusted employees in the organization systematically uses a computer-routine to transfer client capital to accounts of his own. He is a highranking executive and is seen as one of the bank's well-known profiles with the public. Will you press charges or discretely settle the matter with him?
>
> Which of the following alternatives are, in your opinion, the most and second most important to consider before making your decision? (Mark with 1 and 2 respectively)
>
> (a) He has betrayed, deceived and hurt a lot of people

(b) Damages claimed will be high

(c) It is important to protect the bank's good reputation

(d) The temptation is great; anyone could have done the same thing

Four alternatives are given (two heteronomous, a and d, and two autonomous, b and c) to make the choice independent of the preferred solution of the dilemma, for example, to press the charges or not. One of the heteronomous (d) and one of the autonomous (b) imply a preference for not pressing the charges, whereas the other two (a and c) are close to the opposite solution. By neutralizing the normative content of the solution, the focus is concentrated clearly on the psychological process of making a decision.

Different versions of the questionnaire, adapted to different professional activities (business and working life [Ethical Autonomy Questionnaire–Working Life and Business], and politics [Ethical Competence Questionnaire–Political]), have already been tested on more than 1,000 persons working at different organizations and at different organization levels (Kavathatzopoulos, 2000; Kavathatzopoulos & Rigas, 1998). The results have shown satisfying reliability coefficients: Cronbach's alpha coefficient varied between 0.59 and 0.74, which indicates that the scale has sufficient homogeneity; the stability coefficient varied from 0.74 to 0.78. Confirmatory factor analyses have also shown that the items in all versions of the questionnaire assess ethical autonomy.

The results also showed that in all scales there was a positive correlation between hierarchy level and autonomy score, indicating the validity of the instrument. Decision makers at higher levels of organizations scored higher than persons at lower levels (see Table 19.1). The effect size of the difference between the scores of higher level and lower level decision makers was medium to high (Cohen's d varied from 0.54 to 1.16). Furthermore, it has been shown that professionals scored higher on autonomy compared to people with no experience in the same professional activities.

TABLE 19.1
Scoring on Ethical Autonomy: Professional Decision Makers
at Higher Organizational Levels Score Higher;
Professionals Score Higher than Nonprofessionals

Profession	n	High Level	Medium Level	Low Level	Control
Business	1,234	5.3	4.6	4.0	3.9
Politics	291	5.3	n/a	4.4	n/a

Note. Autonomy score varied from 0 to 10. Standard deviations varied from 2.4 to 4.8. n/a = not applicable. From Kavathatzopoulos (2000). Adapted with permission from Arktéon. From "A Piagetian Scale for the Measurement of Ethical Competence in Politics" by Kavathatzopoulos and Rigas (1998), *Education and Psychological Measurement, 1998, 54*, 791–803, copyright 1998 by Sage Publications. Adapted with permission.

There are, however, certain limitations connected to the medium level of homogeneity, but given the relatively low number of items in each version of the scale, as well as the high variation of the item content, homogeneity is acceptable. Other limitations are common to all paper-and-pen questionnaires such as the rigidity of fixed alternatives in each item and their weakness in expressing and capturing the differences of ethical thinking when compared to the flexibility of the interview method. The questionnaires (Ethical Autonomy Questionnaire–Working Life and Business, and Ethical Competence Questionnaire–Political) in their present form have sufficient reliability, but the addition of new items would increase the reliability of the instrument, facilitating its use in applied settings. It is, however, possible to use it for evaluations of training programs or mapping of ethical skill needs, as well as for other similar purposes. The strength of this instrument is that it tries to avoid mixing in or linking normative content or moral principles to the psychological process of ethical decision making. It represents a promising alternative, avoiding many of the problems that can be found in other ethical or moral tests, and therefore, it is more easily applicable and acceptable in many different professional organizations.

TRAINING FOR ETHICAL DECISION MAKING AND PROBLEM SOLVING

Professional decision makers need the ability to handle moral issues satisfactorily, and this means that they need a high level of ethical competence. Ethical competence consists of a number of skills at personal and organizational levels such as high ethical awareness, personal ability to handle and solve moral problems, appropriate organizational processes and routines, argumentation and communication skills, and ethical confidence. Ethical competence is obviously something that professionals can acquire without any external help in the form of training and education, as the scoring results on autonomy scales have already shown. However, education could still contribute to the acquisition of ethical competence, at least as a complement to real-life experience or as a means to accelerate the development of ethical competence. In that case, the goal of education in ethical decision making should be the previously mentioned skills, and the focus should be on the psychological function of ethical autonomy.

Indeed, an education program (Kavathatzopoulos, 1994a, 1994b, 2000; Liiti, 1998) based on the preceding principles has been developed and tested on decision makers coming from different organizations and professional activities. Evaluations by the different versions of autonomy assessment instruments have produced very positive results. After training, participants showed higher scores on autonomy shortly after the course, as

TABLE 19.2

Scoring on Ethical Autonomy: Training Promotes the Acquisition
of Ethical Autonomy in Different Professions

Profession	n	Before Training	Directly After Training	1 Month After Training	4 Months After Training	2½ Years After Training
Private Sector	17	3.4	8.6	7.5	n/a	n/a
Public Sector	49	3.8	n/a	n/a	5.0	n/a
Politics	32	4.2	n/a	6.2	n/a	6.1

Note. Autonomy score varied from 0 to 10. Standard deviations varied from 0.75 to 4.2. n/a = not applicable. From Kavathatzopoulos (1994a). Adapted with permission from Uppsala University. From Kavathatzopoulos (1994b), *Journal of Business Ethics*, 1994, *13*, 379–386. Copyright 1994 by Kluwer Academic Publishers. Adapted with permission. From Kavathatzopoulos (2000). Adapted with permission of Arktéon.

well as up to 2½ years later (Table 19.2). It has therefore been clearly shown that participation in those training programs stimulated autonomy in ethical decision making.

Ethical competence and its component skills were assessed by using self-report questionnaires and interviews (Kavathatzopoulos, 1994a, 1994b, 2000; Liiti, 1998). The results showed clearly that the participants used their new skills in their real professional life and that they were very satisfied. After training, they had higher ethical awareness, it was easier to handle ethical problems at a personal level, ethical argumentation and communication was more effective, and their ethical confidence was higher. However, organizational handling of moral issues was not similarly satisfying. One explanation may be the orientation of these education programs, which was focused on the training of personal and small-group skills. Organizational structure and processes are undoubtedly a decisive factor in the use of ethical competence. Without proper adaptation of organizational structures as well as training of all people in an organization, it would be difficult to extend the positive effects of ethical autonomy education to the whole organization.

These education programs are 2-day or 3-day classical workshops with at least a 1-day follow-up approximately a month later (Kavathatzopoulos, 1994a, 1994b, 2000). They focus primarily on the difference between heteronomous and autonomous thinking that participants have to learn by practicing on a number of moral problems. Participants work though six blocks of exercises covering all aspects of ethical competence as well as its application in real life.

After a short introduction during which autonomous thinking is demonstrated, participants are placed in small groups to work together on moral problems. They are encouraged to identify real problems from their own professional life: problems they feel are important or problems they are concerned about. Practicing autonomy on one's own real-life moral problems is a presupposition for learning. As I have already discussed previously, practicing on hypothetical problems at university courses does not lead to ethical competence in professional life (Sims, 2002; Weber, 1990). Furthermore, learning is facilitated if instructions are adapted to the zone of proximal development (Vygotsky, 1962, 1978) or the extension of cognitive schemata (Piaget, 1962), and this happens when instructions are about real problems. Practicing on one's own real moral problems gives participants the chance to experience directly the value of autonomy in decision making and problem solving. Experiencing the value of instructions is the necessary precondition for learning and using what is learned in real life.

Autonomy training is based entirely on participants' personal and organizational moral values. Autonomous ethical thinking implies the consideration of significant values involved in a concrete moral problem rather than speculation of general or unrelated moral principles. During training, participants learn how to use autonomy to identify pertinent values and weigh them against possible actions. Autonomy is a tool for the satisfaction of the most important values participants or their organizations possess. This is the focus of autonomy education and the criterion of the usefulness of autonomy skill. In a corresponding way, the use of autonomy to create ethical guidelines for an organization or profession is focused on expressing fundamental organizational and personal values in everyday situations rather than importing irrelevant principles or simply making a list of abstract moral values.

The aim of autonomy training is to provide participants with a cognitive tool to use in ethical problem solving and decision making of real problems and not with a method to handle moral philosophical issues, to participate in general moral discussions, or to contemplate privately hypothetical moral problems. What is important is the way people reason in front of concrete moral problems; therefore, education is concentrated exclusively on psychological processes. Training is expected to result in an improved ability to cope with personal and organizational moral problems as well as to obtain satisfying solutions. At an organizational level, the use of autonomy implies the construction of usable ethical guidelines and the implementation of needed and necessary processes for handling moral issues.

Participants in those workshops (Kavathatzopoulos, 1994a, 1994b, 2000) are trained to use the autonomy method, which implies the opening up of independent and unconstrained decision making about real moral problems. The autonomy method in solving professional ethics problems means that decision makers focus their attention on the concrete problem rather than being constrained by certain moral values or moral authorities. They identify significant values, interests, feelings, and principles pertinent solely to the moral problem at stake. They investigate the relations of influence and dependency among alternative solutions and stakeholder values. Based on this sovereign, down-to-earth analysis, they make their decision.

CONCLUSION

The psychological approach to ethical decision making avoids many of the problems inherited in other efforts to assess and train ethical competence, efforts that are not independent of fixations to particular normative aspects and certain moral principles. Focus on autonomy as a pure psychological process allows the reliable description of ethical competence as well as the construction of education methods that promote the acquisition of ethical problem-solving and decision-making skills applicable in real professional life.

REFERENCES

Brown, M., Debold, E., Tappan, M., & Gilligan, C. (1991). Reading narratives for conflict and choice for self and moral voices: A relational method. In W. M. Kurtines & J. L. Gewirtz (Eds.), *Handbook of moral behavior and development: Vol. II. Research* (pp. 25–62). Hillsdale, NJ: Lawrence Erlbaum Associates.

Gilligan, C. (1982). *In a different voice.* Cambridge, MA: Harvard University Press.

Griseri, P. (2002). Emotion and cognition in business ethics teaching. *Teaching Business Ethics, 6,* 371–391.

Holt, D., Heischmidt, K., Hammer Hill, H., Robinson, B., & Wiles, J. (1997/1998). When philosophy and business professors talk: Assessment of ethical reasoning in a cross disciplinary business ethics course. *Teaching Business Ethics, 3,* 253–268.

Jaffee, S., & Hyde, J. S. (2000). Gender differences in moral orientation: A meta-analysis. *Psychological Bulletin, 126,* 703–726.

Kavathatzopoulos, I. (1993). Development of a cognitive skill in solving business ethics problems: The effect of instruction. *Journal of Business Ethics, 12,* 379–386.

Kavathatzopoulos, I. (1994a). *Politics and ethics: Training and assessment of decision-making and problem-solving competency* (Uppsala Psychological Reports No. 436). Uppsala, Sweden: Uppsala University, Department of Psychology.

Kavathatzopoulos, I. (1994b). Training professional managers in decision-making about real life business ethics problems: The acquisition of the autonomous problem-solving skill. *Journal of Business Ethics, 13,* 379–386.

Kavathatzopoulos, I. (2000). *Autonomi och etisk kompetensutveckling: Utbildnings- och utvärderingsverktyg för personer och organisationer* [Autonomy and the development of ethical competence: Education and evaluation tools for individuals and organizations]. Uppsala, Sweden: Arktéon.

Kavathatzopoulos, I. (2003). The use of information and communication technology in the training for ethical competence in business. *Journal of Business Ethics, 48,* 43–51.

Kavathatzopoulos, I., & Rigas, G. (1998). A Piagetian scale for the measurement of ethical competence in politics. *Educational and Psychological Measurement, 58,* 791–803.

Kohlberg, L. (1981). *Essays on moral development: Vol. I. The philosophy of moral development.* San Francisco: Harper & Row.

288   KAVATHATZOPOULOS

Kohlberg, L. (1984). *Essays on moral development, Vol. II: The psychology of moral development.* San Francisco: Harper & Row.

Kohlberg, L. (1985). The just community: Approach to moral education in theory and practice. In M. Berkowitz & F. Oser (Eds.), *Moral education: Theory and application* (pp. 27–87). Hillsdale, NJ: Lawrence Erlbaum Associates.

Kurtines, W. M., & Gewirtz, J. L. (1995). *Moral development: An introduction.* Boston: Allyn & Bacon.

Liiti, T. (1998). *Etisk kompetens: Utvärdering av en utbildningsmetod* [Ethical competence: Evaluation of an education method]. Stockholm: Försäkringskassan i Stockholm.

Miller, J. G. (1994). Cultural diversity in the morality of caring: Individually oriented versus duty-based interpersonal moral codes. *Cross Cultural Research: The Journal of Comparative Social Science, 28*, 3–39.

Piaget, J. (1932). *The moral judgment of the child.* London: Routledge & Kegan Paul.

Piaget, J. (1962). *Comments on Vygotsky's critical remarks.* Cambridge, MA: MIT Press.

Rossouw, G. J. (2002). Three approaches to teaching business ethics. *Teaching Business Ethics, 6*, 411–433.

Schwartz, B. (2000). Self-determination: The tyranny of freedom. *American Psychologist, 55*, 79–88.

Sims, R. R. (2002). Business ethics teaching for effective learning. *Teaching Business Ethics, 6*, 393–410.

Stark, A. (1993 May/June). What's the matter with business ethics? *Harvard Business Review, 71*, 38–48.

Tullberg, J. (2003). *Etik i ekonomundervisningen* [Ethics in business education]. Unpublished manuscript, Stockholm School of Economics at Stockholm.

Vitell, S. J., & Ho, F. N. (1997). Ethical decision making in marketing: A synthesis and evaluation of scales measuring the various components of decision making in ethical situations. *Journal of Business Ethics, 16*, 699–717.

Vygotsky, L. S. (1962). *Thought and language.* Cambridge, MA: MIT Press.

Vygotsky, L. S. (1978). *Mind and society.* Cambridge, MA: Harvard University Press.

Walker, L. J. (1995). Sexism in Kohlberg's moral psychology? In W. M. Kurtines & J. L. Gewirtz (Eds.), *Moral development: An introduction* (pp. 83–108). Boston: Allyn & Bacon.

Weber, J. (1990). Measuring the impact of teaching ethics to future managers: A review, assessment, and recommendations. *Journal of Business Ethics, 9*, 183–190.

ADVANCES IN NATURALISTIC DECISION-MAKING METHODOLOGY

CHAPTER

20

A Software Program to Trace Decision Behavior in Lending and an Empirical Example on How Professionals Make Decisions*

Patric Andersson
Center for Economic Psychology
Stockholm School of Economics

A key issue in naturalistic decision making (NDM) is to study experienced decision makers for the reason that "only those who know something about the domain would usually be making high-stakes choices" (Klein, 1999, p. 4). NDM researchers study preferably how experts make decisions, but given the difficulties in identifying persons who may be considered as expert decision makers (cf. Shanteau, Weiss, Thomas, & Pounds, 2002), it is sufficient to study experienced or proficient decision makers (cf. Lipshitz, Klein, Orasanu, & Salas, 2001). A proficient decision maker is someone who has acquired relevant experience or knowledge through extensive practice within a certain domain.

When investigating experienced decision makers, NDM researchers often rely on a technique called cognitive task analysis. Simply put, this technique aims at providing an understanding of the foundations of expertise as well as suggesting methods to elicit the knowledge and cognitive processing on which the experts rely (cf. Klein, 1999). The findings obtained from the cognitive task analysis may result in the construction of a simulated world where

*Revised conference paper submitted to the book based on the Fifth Conference on Naturalistic Decision-Making, Stockholm, May 26–28, 2000. The original title of the conference paper was *The decision simulator P1198: A software program to trace decision processes in loan decisions to small business enterprises* and was presented in two sessions chaired by Professor Henry Montgomery (business decision-making) and Professor Ranan Lipshitz (observing decision-making) respectively.

decision behavior can be formally tested. Advanced computer technology has made it possible to construct simulated worlds involving many of the complex features in the real world, such as time pressure, distractions, and uncertainty. Examples of simulated worlds are flight simulators and computerized micro-worlds like NewFire and MORO (cf. Brehmer, 1999).

To capture decision behavior in simulated worlds (or laboratories), variants of verbal protocols (Ericsson & Simon, 1993) may be employed. Alternatively, a device for tracking cognitive behavior could be added to the construction of the simulated world, so that decision behavior is stored in log-files. Due to its many methodological advantages, computerized process-tracing software should be considered. Some of its major advantages are: (a) decision behavior can be reliably and accurately measured (Biggs, Rosman, & Sergenian, 1993); (b) complex and interactive environments can be implemented (Brehmer, 1999); and (c) multiple participants can be investigated simultaneously (Brucks, 1988). In conjunction, one must note that computerized process-tracing may be associated with two forms of invalidity as discussed by Biggs et al. (1993). First, different computer practice among participants may threaten internal validity. Second, the inability of participants to view information simultaneously on several screens may make the task dissimilar to real-life practice, so that the external validity may be questioned. However, these methodological objections must be viewed in light of people's increased use of computers as well as the increased use of Internet as a tool to buy and sell products. As a result, it might be argued that there are better possibilities for theories to reflect in reliable and valid ways how people actually make decisions.

The present chapter has two goals. The first goal is to describe a software program that is designed to capture professionals' decision behavior in a real business context. The participants acquire information from the software that provides the parameters for deciding whether a loan proposition should be granted or rejected. One might argue that this process-tracing software is like a simulated world. The second goal is to illustrate the type of data on decision behavior that the software collects by reporting results from a pilot study where two experienced loan officers participated.

The remainder of the chapter is organized in four sections. First, the rationale for developing the software is described. Second, the software is described. Third, an example of collected data is presented. Fourth, the chapter ends with a short discussion.

RATIONALE FOR DEVELOPING THE SOFTWARE

The process-tracing software called P1198 has been designed to investigate how experienced loan officers make decisions about loan propositions concerning small firms. (The name P1198 originates from the fact that the first step toward the development of the software was taken in November 1998.) Besides being designed as a process-tracing device, the software was designed as a training tool for loan officers. These professionals are expected to make better decisions by gaining insights into their decision behavior.

The design was motivated by two circumstances. First, the creditworthiness of small firms is hard to assess mainly because of the following factors: (a) the traditional analyses of financial ratios perform poorly in reflecting performance of small firms (Keasey & Watson, 1987), (b) loan officers, thus, have to rely on other kinds of information that are ambiguous (Argenti, 1976), and (c) empirical research on bankruptcy cannot be generalized to small firms due to their special characteristics such as high variance in expected returns (Hall, 1992). Second, insights about how experienced loan officers make lending decisions are crucial for developing training programs for novice loan officers and constructing expert systems (cf. Shanteau & Stewart, 1992). Such insights may also improve economic efficiency (cf. Camerer, 1995) in that the number of incorrect decisions might be decreased; implying a more efficient capital allocation to small firms (cf. Cressy & Olofsson, 1997).

Another motive for the design was the aforementioned fact that computerized process tracing gives accurate and reliable insights into decision behavior. Reviews of various literature databases showed that there existed (at that time) three main process-tracing software programs: Search Monitor (Brucks, 1988), MouseLab (Payne, Bettman, & Johnson, 1993), and DSMAC (Saad, 1998). Analyses of the applications of those software programs showed that they were designed to deal with artificial and simple decision tasks involving a small number of cues ($n < 12$). For that reason, they were deemed inappropriate to employ in a project investigating how loan officers really make decisions (cf. Andersson, 2001a). P1198 was, therefore, designed to improve the applicability of process tracing to more complex decision-making tasks.

Similar to the other process-tracing programs the P1198 assumes that decision making is a sequential process with four stages (cf. Einhorn & Hogarth, 1980): acquiring information, evaluating the acquired information, deciding to acquire additional information, and making the final decision. The sequence of the stages does not necessarily reflect the way decision makers actually make decisions. March (1994) has argued that the decision maker acquires information, makes the decision, and then acquires more information in order to confirm the made decision. Thus, the P1198 software is flexible, so the participant can acquire information and make decisions in any sequence.

In contrast to the aforementioned programs, P1198 has strong ecological or face validity because it has been designed to reflect the real world. Viewpoints from experienced loan officers and empirical research findings on lending (Andersson, 2001; Argenti, 1976; Hedelin & Sjöberg, 1995; Maines,

1995) ensured a good fit between the experimental world and the real world (cf. Beach, 1992).

DESCRIPTION OF THE SOFTWARE

The P1198 software runs in the web-browser (e.g., Internet Explorer 5), which must be connected to a special server (i.e., Lotus Domino). The source code was Lotus Notes and the software consists of two interacting systems: a database of three loan propositions and a user interface. Collected data on decision behavior are stored in log-files, which provide blueprints of the cognitive processing used by the participant (cf. Biggs et al., 1993).

The Loan Proposition Database

The database includes three loan propositions with information about three real but anonymous small firms. The three real small firms were selected from a small sample ($n < 25$) that was randomly drawn from the stratified sample of 465 small firms used by Wiklund (1998) from data collected in 1996. One of these three companies constitutes a warm-up case used for practice purposes. The other two companies represent the propositions used to capture decision behavior. The first proposition concerns a survival small firm producing lifting devices with turnover about $3.5 million for the last year (1996) and an applied amount of $294,000. The second proposition concerns an insolvent small firm producing metallurgy products with a turnover about $1.6 million for the last year and an applied amount of $235,000.

Table 20.1 describes the information available to the participant. It should be emphasized that all these cues represent the information that is accessible for loan officers in the real world and that the selection has been done in accordance with research findings as well as discussions with loan officers.

The financial information consisting of balance sheets, income/cost reports, and financial ratios was taken from credit reports certified by a Swedish credit rating agency. The financial cues cover three subsequent years. As the credit reports had abundant information, some simplifications were made to decrease the number of financial cues. For each loan proposition the database provides the participant with a total of 57 financial cues. The 17 non-financial cues were derived from two main sources. First, the credit rating agency provided some data. Second, some responses to the questionnaires in Wiklund (1998) were used. The questionnaires included several questions about the company, the board of directors, the decision style of management, the industry, and the business environment. Re-

TABLE 20.1
The 74 Financial and Nonfinancial Cues Available to the Participant

Category of Information	Content of Cues		
Financial Information: Balance Sheet (21 Cues)	Fixed assets	Liquid assets	Total shareholders' equity
	Land, buildings and improvements	Total current assets	Total untaxed reserves
	Plant and equipment	Total Assets	Long-term debt
	Total fixed assets	Share capital	Accounts payable-trade
	Inventories	Legal reserves	Other current liabilities
	Trade accounts receivable	Retained earnings	Total liabilities
	Other accounts receivable	Net income for the year	Total shareholders' equity and liabilities
Financial Information: Income/Cost Report (11 Cues)	Sales	Financial income	Special adjustments
	Cost of goods sold	Financial expenses	Taxes
	Depreciation	Income from ordinary activities	Net income for the year
	Earnings before interest and taxes	Non-recurring income and expenses	
Financial Information: Financial Ratios (25 Cues)	Return on equity	Cash-flow-to-total-debt-ratio	Ratio of average inventory to annual sales
	Return on total assets	Rate of self-financing	Ratio of current liabilities to annual sales
	Average interest rate on debt	Inventory turnover rate	Average collection period
	Gross margin	Annual sales per employee	Average supplier credit period
	Profit margin	Ratio of personnel costs to annual sales	Current ratio
	Net margin	Ratio of working capital to annual sales	Acid test ratio
	Interest coverage ratio	Ratio of accounts receivable to annual sales	Equity-asset ratio
	Working capital turnover		Capitalization rate
	Cash-flow-to-total-cost-ratio		Debt-equity ratio
Nonfinancial or Qualitative Information (17 Cues)	Business	Competitors	Accountant
	The purpose of the proposition	Products	Collateral
	Managing director	Development of industry	Auditor's report
	Board of directors	Suppliers	Employees
	Customers	One year prognosis	Salaries
		Five year prognosis	Credit rating

Note. The cues in each category of information appear from left to right in the sequence they appear on the screen. Additionally, the financial cues are re-ported in the traditional sequence. All accounting cues are numbers. Most of the nonfinancial cues are written as very short statements.

sponses to those questions are essential for accurate credit assessments. To facilitate reading and understanding, these statements were written in a straightforward manner with short sentences and simple words. For instance, the range of the numbers of words for the non-financial information is between a minimum of 2 words (i.e., "certificate auditor") to a maximum of 41 words (i.e., the description of the manager).

User Interface

The user interface is window-driven and consists of the following interacting components: (a) an introductory screen, (b) a loan proposition menu, (c) four screens for the information categories, (d) a decision menu, and (e) a pop-up window. In principle, the components interact through the sequential stages as illustrated in Fig. 20.1. Note that the participant can jump back and forth between the second, the third, and the fourth stage.

In the first stage, the P1198 software starts with an introductory screen informing the participant about the purpose of the research project and the fact that mouse clicking navigates the software. The participant is briefly told about the research project and he or she is asked to make decisions about two loan propositions. The participant is also told that he or she has various kinds of information reported by an assumed colleague and that a warm-up case will be used for practical purposes. Before this case is presented, the participant is asked to respond to some short questions about age, education, and experience. It is mentioned that the responses will be treated confidentially.

The second stage displays the loan proposition screen that describes in a few words (i.e., in 37–59 words) the company and the requested amount of loan. In addition, this screen informs the participant that "by means of the available information your task is to make a decision about this loan proposition." At the top of the screen there is a toolbar for the following components: (i) the loan proposition screen, (ii) the four screens for each

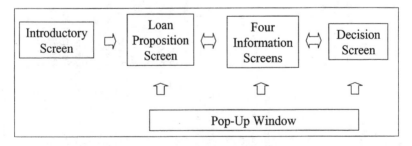

FIG. 20.1. Simplified flowchart on how the components of the P1198 software interact. Note that the pop-up window occurs in intervals of a predetermined number of minutes.

category of information, and (iii) a decision screen. The participant can jump back and forth between these six screens in any sequence.

In the third stage, the participant turns to each of the four screens representing the four information categories (see Table 20.1). All cues are hidden, so that information acquisition behavior can be captured. By mouse clicking the participant activates the cues one by one. For instance, to view the financial cue "Return on equity," the participant clicks on a button and the figures for three subsequent years are shown. The other cues are activated in a similar way. Once a cue has been activated it remains so.

While the participant considers the loan proposition and the information screens, a pop-up window appears at intervals of a predetermined time period (e.g., 2 minutes). To the best of my knowledge, the pop-up window is unique for the P1198 software, but parallels can be drawn to the experimental design employed by Russo, Meloy, and Medvec (1998) in their research on the pre-decisional distortion. The pop-up window aims to investigate the ongoing decision process by asking the participant at intervals to state his or her current judgment on a seven-point verbally anchored scale ranging from *loan is absolutely denied* (0) to *loan is absolutely granted* (6). The scale has a neutral point (3) implying that the participant is dubious. Some criticisms can be raised toward the use of repeatedly asking the participant about their ongoing judgment for three major reasons. First, a participant might neglect diagnostic information and thus be a victim of the pre-decisional distortion bias (Russo et al., 1998). Second, he or she might search for confirmatory information and thus be a victim of the well-known confirmation bias (Plous, 1993). Third, cognitive dissonance might occur resulting in efforts to reduce that dissonance by neglecting or boosting information (Festinger, 1957). Moreover, in the real world the participant is seldom asked to repeatedly state their ongoing decision process. Nevertheless, it is argued that the use of pop-up windows has limited influence and gives valuable insights into how credit decisions are made, as will be reported in the next section.

The fourth stage concerns the decision screen. Once the participant has made up his or her mind, he or she moves to this screen and is asked to grant or deny the loan proposition. If a loan is granted, the participant is asked to state the interest rate and the number of years of amortization, as well as to write an explanation why the loan is granted. On the other hand, if a loan is denied, the participant is asked to motivate why he or she rejects the loan proposition and what complementary information is required. For both kinds of decisions, the participant is asked to rate his or her confidence in the decision on a seven-point verbally anchored scale ranging from *absolutely unsure* (0) to *absolutely sure* (6).

To proceed, the participant clicks on a button. Then the next loan proposition appears on the screen and the participant returns to the aforementioned second stage. This procedure is repeated until he or she has made decisions about all three loan propositions.

AN EXAMPLE OF DATA COLLECTED
BY THE P1198 SOFTWARE

This section reports results on tentative analyses of collected data concerning the decision behavior of two experienced loan officers. The data have been extracted from the log-files registering the behavior of those two participants. Because the P1198 software supplies a vast amount of data stored in log-files, it is important to know in advance what kind of analyses to perform. It is otherwise easy to get lost in the data. Put bluntly, the researcher may be choked by the haze in this minefield of data.

Table 20.2 summarizes the decision behavior of the two loan officers. The two participants made identical decisions and rejected the two propositions. There was variability in confidence levels as well as the time spent on making decisions: Loan officer 1 tended to be more confident in his decision making than his colleague, who on the other hand, made his decisions quicker. Minor differences in amount of acquired information were found.

TABLE 20.2

Example of Data Collected by the Software P1198: Decision Behavior
of Two Experienced Loan Officers: Amount of Acquired
Information and Depth of Acquisition

Information Acquisition Behavior	Total Available Cues	Loan Proposition 1		Loan Proposition 2	
		Loan Officer 1	Loan Officer 2	Loan Officer 1	Loan Officer 2
Decisions		Deny	Deny	Deny	Deny
Confidence		Very Sure	Absolutely Sure	Absolutely Sure	Somewhat Sure
Amount (depth) of acquired information					
Total acquired cues in Balance sheet	21	17 (81%)	13 (62%)	18 (86%)	15 (71%)
Total acquired cues in Income/Cost report	11	10 (91%)	7 (64%)	7 (64%)	5 (45%)
Total acquired cues in Financial ratios	25	8 (32%)	9 (36%)	9 (36%)	4 (16%)
Total acquired qualitative cues	17	12 (71%)	17 (100%)	10 (59%)	17 (100%)
= Total acquired cues	74	47 (64%)	46 (62%)	44 (59%)	41 (55%)

Note. Depth of acquisition refers to the ratio between each category of acquired cues and the total available cues in each category.

As regards the first and the second proposition, loan officer 1 acquired a total of 47 and 44 cues, respectively. Loan officer 2 acquired almost the same amount of information: 46 and 41 cues, respectively. Across the two loan propositions, there were consistent differences in terms of the depth of information acquisition. Loan officer 1 tended to acquire more financial cues but less qualitative cues than did loan officer 2. It follows that loan officer 2 was prone to rely on non-financial cues (qualitative cues) to a large extent.

The graphs in Fig. 20.2 visualize how the decisions made by the two loan officers evolved. The pop-up window makes it possible to get insights into the participants' processing during their decision making. With regard to the first loan proposition, the following observations could be made. Loan officer 1 was uncertain about the proposition in almost 12 minutes although he had acquired 50% of the available cues. He then acquired an additional 12% of the cues (i.e., financial ratios and the cue about collateral) and arrived at a rejecting decision. Loan officer 2, on the other hand, was positive in about 8 minutes, but after acquiring cues from the balance sheet and financial ratios he rejected the proposition. In addition, it could be noted that the loan officers acquired the cues in a similar sequence ($rs = 0.64$) and that they motivated their decisions similarly by stressing low profitability and low equity/asset ratio.

As regards the second proposition, loan officer 1 had an indifferent attitude. He was initially dubious, but while acquiring cues from the "income/cost report" he leaned toward rejecting the proposition. Once he had viewed the cues from the balance sheet, he turned to be dubious again. Then (like the first proposition) he acquired financial ratios and decided to reject. In contrast, loan officer 2 followed almost the same pattern as he did in the first proposition. He was positive while viewing the qualitative cues, but when he had considered some financial ratios he chose to reject the proposition. The loan officers also acquired the cues in almost the same sequence ($rs = 0.67$). Moreover, each loan officer had a high degree of internal consistency ($rs > 0.98$) measured as the sequence in which they acquired information.

In conclusion, the tentative analyses of the reported data on the two loan officers suggest several implications. The observed high levels of within-subjects and between-subjects agreement fit with the behavioral conditions on expert judgment as claimed by Einhorn (1974). The participants' relatively low amounts of acquired information correspond to Shanteau's (1992) assumption that experts do not rely on much information when making decisions. The observations from Fig. 20.2 relate to the ideas of regarding the decision process as a search for a cognitive structure where one option dominates (cf. Montgomery, 1989) and as a framework with phases of differentiation and consolidation (cf. Svensson, 1996). Parallels can also be drawn to other descriptive models of decision making such as the recogni-

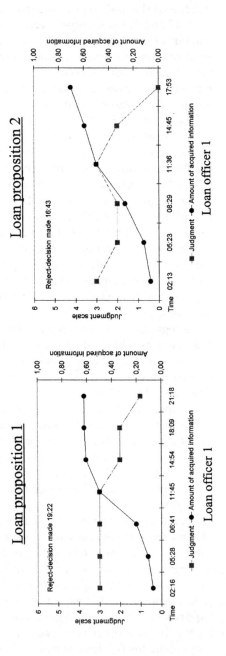

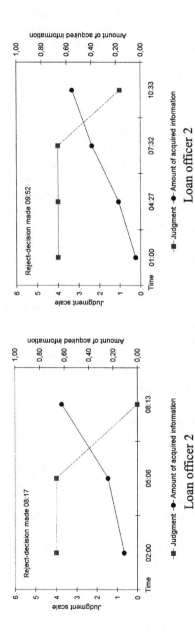

Reject-decision made 08:17

Loan officer 2

■ Judgment   ● Amount of acquired information

Reject-decision made 09:52

Loan officer 2

■ Judgment   ● Amount of acquired information

FIG. 20.2. Decision processes and amount of acquired information concerning the two loan officers.

Note. The graphs show the first impression of the propositions and how that impression is influenced by information acquisition. The horizontal axis indicates minutes. The vertical axis has two values: the judgment scale and the amount of acquired information (defined as the number of acquired cues divided by the total number of available cues). Participants were asked at intervals of circa 3 minutes to state their judgment on the following seven-point verbally anchored scale:

Judgment scale

Deny absolutely	Deny probably	Deny maybe	Dubious	Grant maybe	Grant probably	Grant absolutely
0	1	2	3	4	5	6

Due to minor technical problems some intervals deviated from three minutes. The observant reader may note that the time for the final decision deviates from time spent on decision making as reported in Table 2. The reason is that the stated time in Table 2 also includes the time the participants spent on motivating their decision.

301

tion-primed decision model (see Klein, 1999; Lipshitz et al., 2001) and the image theory (Beach, 1990) to name a few. However, given that the data only concerned two participants and that a preliminary version of the software was employed, caution may be taken with regard to the implications.

DISCUSSION

Despite its focus on a narrowly defined decision domain, the software reported in this chapter should be of interest for NDM researchers. The advanced design of this software may stimulate future projects on simulated worlds. For instance, the large scope of information and the use of pop-up windows may be some features from the P1198 software that may be considered in such projects. Furthermore, the reported (tentative) analyses of log-files and particularly the graphs in Fig. 20.2 that visualize how decision making evolves should be of interest. To the best of my knowledge, such visualization of decision making has not been done in the past research on judgment and decision making.

So far, the software P1198 has been used to investigate how experience influences decision behavior in lending; see Andersson (in press). Among other things, this study showed that loan officers with more than 10 years of practice tended to acquire a very large amount of information. To evaluate the validity of the P1198 software, a questionnaire was handed out to the experienced participants. Because they stated that the included information was realistic and sufficient, it is reasonable to expect that observations provided by the present software have external validity.

Like other simulated worlds, the present software has some limitations. The main limitation is that the participant is restricted to sequentially consider the cues reported in the software. As a result, the participant may become a passive information collector (see Raynard & Williamson in this volume). In real life, loan officers encounter the clients, demand additional information, consider the revenues of granting loan, and may rely on portfolio thinking. In other words, the real world has dynamic features that are hard to reproduce in the simulated world.

Nevertheless, the software should have external validity because in real life loan officers are confronted with the majority of the cues in the present format. Besides, one might speculate that in the future banks would implement procedures where subordinates electronically send documents concerning loan propositions to their senior loan officer, who then makes the final decisions. Process-tracing devices such as the present software could in this case easily be set up. The decision behavior stored in the log-files would not only attract the interest of researchers in decision making, but also internal auditors.

ACKNOWLEDGMENTS

This research has been enabled by grants from the Ruben Rausing's Foundation for Research on New Firm Creation and Innovation as well as Sparbankernas Research Foundation. I wish to thank Mr. Magnus Dahlberg, who did the computer programming, and Dr. Jan Edman for valuable advice. Viewpoints from professors Lennart Sjöberg, Ingolf Ståhl, Henry Montgomery, Robyn Dawes, and Alice Isen are gratefully acknowledged. I am obliged to Dr. Johan Wiklund for giving me access to his empirical material. I also thank Professor Ranan Lipshitz and conference participants for helpful comments. The majority of the chapter was written during my stay as a visiting researcher at the London School of Economics and Political Science. I am grateful for the Tom Hedelius Scholarship from the Handelsbanken Research Foundation for making this stay possible.

REFERENCES

Andersson, P. (2001). *Expertise in credit granting: Studies on judgment and decision-making behavior.* Published doctoral dissertation, Stockholm School of Economics, Stockholm.

Andersson, P. (in press). Does experience matter in lending? A process-tracing study on experienced loan officers' and novices' decision behavior. *Journal of Economic Psychology.*

Argenti, J. (1976). *Corporate collapse: The causes and symptoms.* London: McGraw-Hill.

Beach, L. R. (1990). *Image theory: Decision making in personal and organizational contexts.* London: Wiley.

Beach, L. R. (1992). Epistemic strategies: Casual thinking in expert and non-expert judgment. In F. Bolger & G. Wright (Eds.), *Expertise and decision support* (pp. 107–127). New York: Plenum.

Biggs, S. F., Rosman, A. J., & Sergenian, G. K. (1993). Methodological issues in judgment and decision-making research: Concurrent verbal protocol validity and simultaneous traces of process. *Journal of Behavioral Decision Making, 6,* 187–206.

Brehmer, B. (1999). Reasonable decision-making in complex environments. In P. Juslin & H. Montgomery (Eds.), *Judgment and decision making: Neo-Brunswikian and process-tracing approaches* (pp. 9–22). Mahwah, NJ: Lawrence Erlbaum Associates.

Brucks, M. (1988). Search monitor: An approach for computer-controlled experiments involving consumer information search. *Journal of Consumer Research, 15,* 117–121.

Camerer, C. (1995). Individual decision making. In J. H. Kagel & A. E. Roth (Eds.), *Handbook of experimental economics* (pp. 587–703). Princeton, NJ: Princeton University Press.

Cressy, R., & Olofsson, C. (1997). European SME finance: An overview. *Small Business Economics, 9,* 87–96.

Einhorn, H. J. (1974). Expert judgment: Some necessary conditions and an example. *Journal of Applied Psychology, 59,* 562–571.

Einhorn, H. J., & Hogarth, R. M. (1980). Behavioral decision theory: Processes of judgment and choice. *Annual Review of Psychology, 32,* 53–88.

Ericsson, K. A., & Simon, H. A. (1993). *Protocol analysis: Verbal reports as data.* Cambridge, MA: MIT Press.

Festinger, L. (1957). *A theory of cognitive dissonance.* Stanford, CA: Stanford University Press.

Hall, G. (1992). Reasons for insolvency amongst small firms—A review and fresh evidence. *Small Business Economics, 4,* 237–250.

Hedelin, L., & Sjöberg, L. (1995). Bankers' judgment of loan applicants in regard to new ventures. In B. Green (Ed.), *Risky business* (pp. 67–94). Stockholm: Stockholm University.

Keasey, K., & Watson, R. (1987). Non-financial symptoms and the prediction of small company failure: A test of Argenti's hypothesis. *Journal of Business, Finance & Accounting, 14,* 335–354.

Klein, G. (1999). *Sources of power: How people make decisions.* Cambridge, MA: MIT Press.

Lipshitz, R., Klein, G., Orasanu, J., & Salas, E. (2001). Taking stock of naturalistic decision-making. *Journal of Behavioral Decision-Making, 14,* 331–352.

Maines, L. A. (1995). Judgment and decision-making research in financial accounting: A review and analysis. In R. H. Ashton & A. H. Ashton (Eds.), *Judgment and decision-making in accounting and auditing* (pp. 76–101). New York: Cambridge University Press.

March, J. G. (1994). *A primer on decision making: How decisions happen.* New York: Free Press.

Montgomery, H. (1989). From cognition to action: The search for dominance in decision-making. In H. Montgomery & O. Svenson (Eds.), *Process and structure in human decision-making* (pp. 23–49). Chichester, UK: Wiley.

Payne, J. W., Bettman, J. R., & Johnson, E. J. (1993). *The adaptive decision maker.* New York: Cambridge University Press.

Plous, S. (1993). *The psychology of judgment and decision-making.* New York: McGraw-Hill.

Russo, J. E., Meloy, M. G., & Medvec, V. H. (1998). Predecisional distortion of product information. *Journal of Marketing Research, 35,* 438–452.

Saad, G. (1998). The experimenter module of the DSMAC (Dynamic Sequential MultiAttribute Choice) interface. *Behavior Research Methods, Instruments, & Computer, 30,* 250–254.

Shanteau, J. (1992). How much information does an expert use? Is it relevant? *Acta Psychologica, 81,* 75–86.

Shanteau, J., & Stewart, T. R. (1992). Why study expert decision-making? Some historical perspectives and comments. *Organizational Behavior and Human Decision Processes, 53,* 95–106.

Shanteau, J., Weiss, D. J., Thomas, R. P., & Pounds, J. C. (2002). Performance-based assessments of expertise: How to decide if someone is an expert or not. *European Journal of Operational Research, 136,* 253–263.

Svenson, O. (1996). Decision-making and the search for fundamental psychological regularities: What can be learned from a process perspective. *Organizational Behavior and Human Decision Processes, 65,* 252–267.

Wiklund, J. (1998). *Small firm growth and performance: Entrepreneurship and beyond.* Published doctoral dissertation, Jönköping International Business School, Jönköping, Sweden.

Conversation-Based Process Tracing Methods for Naturalistic Decision Making: Information Search and Verbal Protocol Analysis

Rob Ranyard
Janis Williamson
Bolton Institute, United Kingdom

You have gone to an electrical superstore to buy a new washing machine—there is a bewildering array of options, but eventually you choose one. How do you make that final decision? Which factors are most important? After you have made your choice, the salesperson offers you the chance to buy an extended warranty on your chosen product. Do you take up this offer or not—what factors sway your decision? You are at a garage making a final decision about a secondhand car purchase. What factors would be important in making this decision? Would you buy the extended warranty or take the credit deal offered by the salesman? Purchasing decisions such as these are made every day by large numbers of consumers. We have developed a method for tracing people's thought processes as they think about decisions such as those just described. In this chapter, we briefly explain the main features of the method and discuss possible applications to naturalistic decision research. Our main objectives are to illustrate the kind of data obtained and to describe and discuss alternative ways of analyzing it. We draw on earlier reports that have evaluated the method and applied it to consumer risk management (Ranyard, Hinkley, & Williamson, 2001; Williamson, Ranyard, & Cuthbert, 2000a, 2000b).

In an attempt to develop new methodologies for studying real-world decisions, Huber and his colleagues (Huber, Beutter, Montoya, & Huber, 2001;

Huber, Wider, & Huber, 1997) examined the effects of information presentation conditions on realistic but fictitious decision scenarios. In standard laboratory decision tasks, it is usual to present a complete set of information.
However, Huber et al. (1997) argued that this is unlike real-world decision
tasks in which we have to search for information we consider to be relevant. If respondents use a piece of information in a complete presentation
of information situation, it does not necessarily mean they would have used
it in a real-world decision task. Huber et al. (1997) concluded that more naturalistic process tracing methods are needed for real-world decisions, and
therefore he developed the method of active information search (AIS). Here
the respondent is presented with a minimal description of the decision situation and has to ask questions to obtain further information supplied in
written form; in this respect, AIS is somewhat analogous to the information
board technique (Payne, Bettman, & Johnson, 1993; Westenberg & Koele,
1994) in that information is available but has to be acquired by the respondent. When this method was compared to a complete presentation of information condition, Huber et al. (1997) found that respondents in AIS seldom
asked for information about probability of events but would use it in conditions in which it was given to them. Clearly, supplying complete information
encourages respondents to behave differently than they do when left to formulate their own questions. Huber et al. (1997) concluded that if one wants
to study which information is genuinely relevant to the decision maker,
then AIS and not complete information should be used; he also recommended the addition of think-aloud instructions to the basic AIS method. In
this chapter, we describe a methodology based on the AIS technique but extended by using a conversational context.

A CONVERSATION-BASED PROCESS TRACING
METHOD

Many process tracing methods attempt to minimize interaction between
the researcher and respondent for fear of distorting thought processes.
However, Schwarz (1996) commented that psychologists' focus on individual thought processes has led to a neglect of the social context in which individuals do their thinking. Whereas researchers comply with conversational maxims outside the laboratory, they are much less likely to do so
within the research setting itself, and this may have a major contribution to
the type of data obtained. Therefore, to address this issue, our method uses
a more natural conversational setting and tries to facilitate the elicitation of
respondents' questions in a familiar and comfortable social context. The
three main components of the method are the basic AIS technique, think-
aloud instructions, and postdecision summaries.

The Basic AIS Method

Respondents' questions about the decision task are responded to orally rather than in written form by the interviewer who takes the role of a helpful consultant. He or she is careful only to give the specific information requested and no more. Prompts can be given ("Don't forget you can ask me anything you want"), and the question asking, or AIS stage, ends when the respondent makes his or her decision. This leads to a protocol containing the following types of conversational sequence: question-specific answer; question-clarification; prompt-question-specific answer. The content, order, and organization of such sequences can be used as evidence of the respondents' evaluation strategies in a similar way to information board research except that it is more certain that respondents have asked only for information they are interested in and have not used it simply because it was there.

AIS + Think-Aloud Instructions

An additional component to the method is the use of think-aloud instructions. In the basic method as outlined previously, respondents merely ask questions and receive answers. Although this tells us about what aspects or attributes of a topic people are interested in, it is less useful in explaining how they are using the information they ask for in reaching a decision. In formal think-aloud methods for use in controlled conditions, the researcher does not respond to the speaker in any way to try to avoid distortion of thought processes (Ericsson & Simon, 1980, 1993). However, this is unnatural; people do not like speaking into an empty space, and therefore our technique is an adaptation of the normal think-aloud method for use in field settings. Interviewers take the role of nonjudgmental listeners and retain the formal technique's emphasis on achieving good descriptive validity of verbal data as evidence of thought processes while at the same time creating a more natural interactive social setting. Instructions adapted from Russo, Johnson, and Stephens (1989) form the basis of the technique, but the main differences between the conversational and formal method are that (a) the instructions to think aloud stress that the respondent is speaking directly to a listener rather than to a tape recorder and (b) the interviewer responds to the thinking aloud using signals that they are listening and interested. These signals include the nonverbal such as head nods, direct gaze, and attentive interested expression and verbal acknowledgments and invitations to continue, such as "I see . . . yes. . . . OK . . . uhuh." We believe that in general these signals do not influence the course of information processing (i.e., that they will not produce reactivity) but rather encourage the respondent to fully express their thoughts. The protocols resulting from this approach will contain sequences of uninterrupted statements and/or statement acknowledgment sequences that can be analyzed in the same way as other think-aloud data.

Postdecision Summary

The method also uses a postdecision interview in which respondents are asked to summarize how they made each of their decisions. It consists of an open-ended question: "Can you say in your own words how you made your decision?" More detail could be sought using the prompt "Anything else?"; otherwise, the interview is nondirective. The postdecision summary is therefore a retrospective report and as such is useful for checking the validity or internal consistency of data elicited during the decision task. This technique may have merits of its own; there is no major memory issue because the summary is close in time to the decision, and the respondent is not doing two tasks concurrently, so they can attend fully to the reporting task. It would be especially useful in situations in which think-aloud instructions were not being used.

We note one important limitation on the applicability of the method at this point. Because novices often do not know what questions to ask when presented with a minimal description of a decision-making task, the method is mainly appropriate for the study of experienced decision makers (although not necessarily experts). In the next sections, we look in more detail at the nature of the data collected. We illustrate and discuss alternative ways of analyzing it.

ANALYZING INFORMATION SEARCH

Componential Analysis: Basic Search Sequences and Attributes

Two of the key components of information search are the type of search sequence, attribute based versus alternative based, and the actual attributes of the alternatives that are inspected. The basic analysis of the type of search sequence can be illustrated via excerpts from respondents' protocols in the baseline process tracing technique: conversational AIS. Recall that here respondents engage in question asking to obtain the information they need to reach a decision and then provide a postdecision summary. They are not required to think aloud. Examples of the two basic types of questioning sequence are illustrated in Table 21.1: (a) *attribute based*, or choosing a particular attribute and asking about it in relation to all three options (Lines B1 to B6); or (b) *alternative based*, or choosing one of the alternatives and focusing the questions on that (Lines C13 to C20). For the

TABLE 21.1
Excerpts From Protocols Illustrating Attribute-
and Alternative-Based Questioning

B: Car task, attribute-based questioning
  1. R: Okay, what's the price of the Golf?
  2. I: The price of the Golf is £3,995.
  3. R: The Rover?
  4. I: the Rover is £4,300
  5. R: And the Tipo?
  6. I: £3,590
  7. R: How many owners had the Rover had?
  8. I: The Rover's had one owner, a male.
  9. R: And how many owners for the Golf?
  10. I: The same, one male owner
  11. R: I'll ask the same question for the Tipo?
C: Washer task, alternative-based search
  13. R: The AEG, the price on that please?
  14. I: £429.99
  15. R: The maximum load?
  16. I: Five kilos
  17. R: Where was it built?
  18. I: Germany
  19. R: Does it have half load?

Note. R = respondent; I = interviewer. Adapted with permission from Williamson et al., 2000a, p. 220.

product decisions in our study, attribute-based questioning was by far the more common strategy, with around 75% of the questions in this format.

Now we consider the basic analysis of attributes searched. Because attributes are selected by the respondent rather than by the researcher, they need to be identified by applying content analysis to the questions asked. The attributes respondents ask about can be of interest for a number of reasons. For example, we were interested in the extent to which people were concerned with product reliability when choosing a car and a washer (Williamson et al., 2000b). In relation to the washer, 18% of the questions concerned reliability in some direct sense, although there were other questions that might also be seen as assessing reliability in some way. For example, those could be about the particular manufacturer or about the number of programs—a commonly expressed view was that the more programs there were the more chance there was of something going wrong. In relation to the car, there were less direct questions about reliability; however, many of the questions could be seen as trying to assess reliability in more indirect ways—asking about the service history, number of previous owners, accidents, garage reports, and so forth. Our content analysis of questions asked thus gave an indication of how the risk of product failure concerned consumers at the point of purchase. These two componential analyses give an insight into what information respondents seek, the order in which they do so, and the extent of their interest in particular attributes.

Detailed Search Descriptions and Useful Search Indexes

Of course, more detailed analysis of search patterns can be carried out. Essentially, the careful methods developed for the information board paradigm can be applied (Payne et al., 1993; Westenberg & Koele, 1994). The main difference is that an alternative-by-attribute information matrix has to be constructed from each protocol rather than being predetermined by the researcher. Table 21.2 illustrates two matrices constructed from one respondent's questions while deciding on the type and source of credit to be used for the two products. In each case, three actual credit options were available: fixed installment credit from the store or garage (store or garage credit), fixed installment credit from a major bank (bank loan), and a flexible repayment credit source (credit card). The rows represent the alternatives interrogated, and the columns represent the attributes about which information was requested. For the washer credit, a 3 × 4 matrix was constructed indicating that the respondent asked questions about all three alternatives and about four attributes. For the car, a 3 × 3 matrix was produced. The matrices, therefore, summarize the problem space actually used by the respondent for each decision. The second main function of the matrices is to summarize the sequence of information search and acquisition. The order of rows, top to bottom, indicates the initial sequence of acquisition of information about al-

TABLE 21.2
Information Search Matrices Constructed From a Respondent's Questions
While Choosing a Credit Option for the Car and the Washer

Credit	APR	TI ($400)	Discount	TI ($500)
Washer				
Store	1	7	6	
Credit card	2	5*		
Bank	3	4		8

Credit	APR	MI^1	MI^2
Car			
Bank	1	4	5
Garage	2	6	7
Credit card	3*		

Note. APR = Annual percentage rate of interest (true rate of interest); TI($400) = Total interest charged on a 2-year loan of $400; Discount = Does the store offer a discount on the product if a loan is taken?; TI($500) = Total interest charged on a 2-year loan of $500; * = Information inferred by respondent; MI^1 = Monthly interest charges, 2-year loan, cheaper car; MI^2 = Monthly interest charges, 2-year loan, expensive car.

ternatives. Similarly, the order of columns, left to right, shows the initial search sequence of attributes. The numbers in the body of the matrix represent the order of acquisition of items of information, and an asterisk indicates an item of information inferred by the respondent.

The matrix for the washer credit decision shows that the respondent first asked about the annual percentage rate (APR) of each option, that is, the store card, the credit card, and the bank loan. In fact, he asked for "all the APRs"; therefore, in this case, the numbers 1 to 3 indicate the order in which they were given rather than that in which they were requested. He then asked for the total interest charged on a £400 loan from the bank and followed this up by inferring the total charges for a £400 credit card loan. (Incidentally, his calculations were quite inaccurate.) He then turned his attention to the store option, first asking whether they were offering any discount on the washer and then asking about the total interest charges for the £400 loan. Finally, he returned to the bank option, checked the total charges for a £500 loan, and decided to take it. In terms of the problem search space, the decision was quite straightforward because the selected option dominated the other two.

The respondent's search pattern for the car loan decision displayed both similarities and differences to that for the washer loan. The first attribute searched was again the APR, although this time it was requested for each option in turn, starting with the bank. The APR for the credit card was inferred (accurately) using the minimal description of the task, which had given the range of APRs. Next, attention returned to the bank loan, even though the garage loan had the lower APR. First, monthly interest charges were requested for a 2-year loan to buy the cheapest car, and then the same information was requested for the most expensive car. Finally, the garage option was searched in the same way. Interestingly, the dominant alternative was not selected this time, although there was only a small difference in interest charges between the bank and garage options.

A number of aspects of search patterns like these can give useful information relevant to understanding respondents' decision strategies. First, the order of search of attributes may indicate their perceived importance. Second, early and/or thorough inspection of an alternative may indicate a higher degree of interest in it. Third, early termination of inspection of an alternative may indicate that it has been eliminated. In addition to simple features of the search pattern, the indexes developed using the information board paradigm, which summarize important and more complex features of search patterns, may be useful. For example, the general direction of search, attribute based versus alternative based, can be summarized by the Payne (1976) index. Also, the degree of compensation between attribute values of different alternatives can be assessed using Koele and Westenberg's (1995) compensation index.

ANALYZING VERBAL REPORTS

In the previous section, we show that the baseline conversational AIS technique is a valuable process tracing tool, which in this basic form can give
useful insights into information search and acquisition strategies. It does,
however, provide only limited insight into how respondents thought about
the information they asked for or the part it played in their final choices,
and this is where think-aloud instructions may have much potential. Think-
aloud statements are invaluable in that they reveal important information
about the decision process that is not available from any other source.

Harte, Westenberg, and van Someren (1994) described a framework for
the analysis of think-aloud data that is appropriate for the conversational
AIS with think-aloud procedure. They classified analytic techniques according to the completeness with which the decision process was modeled: the
analysis of fragments (or components), sequences, or complete process descriptions. We have found the analysis of components of the decision process to be fruitful as we illustrate in the following subsection.

Componential Analysis: Defusing Operators
and Recognition Primed Decisions (RPD)

As indicated earlier, one aim of our study was to investigate how people
manage the risks of mechanical breakdown at the point of purchase of
products such as a new washing machine or a secondhand car (Williamson
et al., 2000b). We looked at the extent to which they were prepared to purchase an extended warranty or to use alternative risk defusing operators.
We therefore sought to identify in the protocols certain component processes derived from theory, that is, alternative risk defusing operators such
as the construction of worst case plans or attempts to control the probability of breakdown. Each protocol and postdecision summary was scanned
for statements referring to the use of defusing operators, and this set provided the basis for content analysis. For the initial analysis, two coders independently classified a sample of statements into four mutually exclusive
categories. However, the reliability of this coding was relatively low, and
therefore, two categories were merged. This resulted in a reliable coding
system that was used to classify all risk defusing statements. The findings
were summarized by a table of percentages of respondents using each type
of defusing operator together with examples of each type, as illustrated in
Table 21.3.

A less formal analysis of components can also be useful. For example,
the protocol evidence in relation to both accepting and rejecting the warranty suggested that some respondents seemed to apply an RPD heuristic

TABLE 21.3

Examples From Respondents' Protocols of Alternative Defusing Operators

---

Long-term plans/new alternatives

    Yes, I wouldn't consider that [extended warranty] at all—I could put that towards another washing machine. (Resp. 42)

    I'd pay money into an account every month, instead of paying for the warranty, that would build up to the warranty amount over the year. (Resp. 88)

Worst case plans

    I know a few people who know things about cars, so if anything was to go wrong I'd be ok. (Resp. 64)

    I have a certain amount of mechanical knowledge and would try to repair the machine myself. Or if I couldn't do I would have one or two people I've used in the past for repairs. So that's the route I'd go down if something went wrong. (Resp. 44)

    £480 over two years is about £20 a month, I can't see this car having that many problems. Anyway I can afford to pay a fair whack if something major went wrong with it. (Resp. 2)

    If anything does go wrong, I've got a friend who actually used to work for [names electrical superstore] and I would just call him out. (Resp. 95)

Control probability of negative event

    I would get it serviced regularly and hopefully wouldn't need the warranty. (Resp. 46)

    If I don't overload it, and use it sensibly, do the usual things to keep it running reasonably . . . I shouldn't need the warranty. (Resp. 73)

    I would take a mechanic with me to look the car over, and anyway I get it serviced every year which covers quite a lot of things. (Resp. 79)

Note. Resp. = respondent. Number in parentheses is respondent number. Adapted with permission from Williamson et al., 2000b, p. 32.

(Klein, Calderwood, & MacGregor, 1989). The following protocol segments contain some evidence of the use of RPDs that illustrate both positive and negative orientations toward the warranties on offer (Williamson et al., 2000b, p. 32).

... on previous experience, all electrical goods. ... I always take extended warranties.

I've always taken it out, I've never actually had to use it, but I've always taken it on.

[It's] never been my preference to do so ... because it's quite expensive over a period of time.

I always say no because you know, you've already paid for something.

Hence, for many respondents, insurance decisions were made with reference to their previous related insurance decisions. The RPD heuristic allows insurance choices to be made with minimal effort and information processing.

Constructing Process Descriptions

The preceding analyses of defusing operators and recognition statements illustrated how the content analysis of statements relating to specific components of the decision process can inform specific theoretical issues. A more detailed analysis of verbal reports and information search can be carried out using Harte et al.'s (1994) approach to the construction of complete process descriptions. Ranyard, Hinkley, and Williamson (2003) carried out a similar, but less formal analysis of the credit decision process based on the 32 participants in the AIS + think-aloud condition of our consumer choice study. Space does not allow a full account of this analysis, but see Ranyard et al. (2003) for further details. We conclude with a brief discussion of the reliability and validity of the conversation-based process tracing method and its applicability to other naturalistic decision domains.

DISCUSSION

Reliability

As with any interview, the first step in establishing the reliability of the analysis is to make sure the speech of both parties is recorded with adequate clarity and that the written transcription is prepared to professional standards. The reliability of the coding and interpretation of the written protocol can be approached in a number of ways. The foundation is a clear, explicit, and transparent coding procedure. Our componential analyses applied the established intercoder reliability measure for content analysis, Cohen's kappa. Less formal coding procedures can achieve good reliability, but at least two coders should independently analyze a set of protocols during their development.

Validity

In the full report of our evaluation study, we described the interviewing technique that should be adopted (Williamson et al., 2000a). To ensure the validity of the responses, it is important that interviewers are trained in the use of this style of interviewing and execute it to a high professional standard. The next stage of our work related to this is the production of a more detailed manual to facilitate interviewer training. A major issue concerning validity is whether the procedures used to elicit concurrent and retrospective verbal reports lead to significant changes to the decision process. Ericsson and Simon (1993) argued against our procedure of creating a conversational context for thinking aloud on the grounds that it leads people to construct reasons for their behavior, thoughts they would not have otherwise had. Our main counterargument is that the search for reasons is often a core decision process, regardless of think-aloud conditions. Nevertheless, our recommendation is always to split the sample into two groups, with and without think-aloud instructions, and then to test for effects of the procedure on both decision processes and final decisions. Similar considerations are relevant to the procedure eliciting postdecision summaries. Previous research has shown that predecision processes can be affected by requirements to justify a decision after it has been made (e.g., Huber & Seiser, 2001). Thus, prior knowledge of a requirement to summarize a decision may produce justification biases in predecision processing. To avoid this, we recommend that participants are not asked to practice giving postdecision summaries and are not informed about this part of the task until after the final decision has been made.

Concluding Remarks

In our initial studies using conversation-based process tracing, participants were explicitly and deliberately provided with optimal conditions for decision making. We argue that it is important to obtain evidence of how people think and behave under optimal conditions because theories of naturalistic decision making (NDM) need to be able to account for decision making in both optimal and suboptimal conditions. Consequently, our respondents were not under time pressure and the available information was not distorted by, for instance, sales personnel with their own agenda. Because our use of the question asking technique (AIS) provided respondents with a supportive environment in which they could use information that they had specifically asked for, insight was gained into what information was most important to them in making their decisions. The addition of think-aloud instructions gave valuable insight into why the information requested was important, and this is an extremely beneficial component of the method. It also gave insight into the extent to which decisions were based on prior experience rather than information acquired during the task. Without think-aloud information, this would have been much more difficult to evaluate. The postdecision summary was also very useful for accessing respondents' reasons for making their choice, and as it was close in time to the decision process, it could be regarded as a reasonably accurate record. Overall, our application of conversation-based process tracing applied to good effect the recommendation by Ericsson and Simon (1980, 1993) that both concurrent and retrospective reports should be collected whenever possible.

Of course in the real world, we also need to understand the effects of time pressure, information that has not explicitly been asked for (advertising campaigns for instance), and the other major aspects of NDM discussed by Lipshitz, Klein, Orasanu, and Salas (2001). We do not claim that our method is applicable to all naturalistic decisions. In particular, it is not relevant to the initial stages of complex decisions in which a large number of alternatives are available or situations in which options are created during the decision process. Nevertheless, previous naturalistic research, notably that in the context of image theory (see Beach, 1997, for a review), has found that in many situations people seek information about and carefully consider a short list of promising alternatives identified after a screening stage. Our next development of the conversation process tracing method will focus on this selection stage of the decision process in realistic, suboptimal decision contexts. As Lipshitz et al. (2001) pointed out, the use of simulations, thinking aloud, and information search monitoring are becoming increasingly important tools in NDM research. We hope that the conversational processing method described in this chapter will prove to be a powerful addition to the toolkit.

ACKNOWLEDGMENTS

Janis Williamson died on July 18th, 2000, at the age of 42.

Lisa Cuthbert and Eileen Hill assisted with this study, which was supported by the Economic and Social Research Council of the United Kingdom, Award L211 25 2051.

Correspondence should be addressed to Rob Ranyard, Psychology Subject Group, Bolton Institute, Deane Rd., Bolton, Lancs, BL3 5AB, UK. Send e-mail to rr1@bolton.ac.uk

REFERENCES

Beach, L. R. (1997). *The psychology of decision making.* Thousand Oaks, CA: Sage.

Ericsson, K. A., & Simon, H. A. (1980). Verbal reports as data. *Psychological Review, 87,* 215–251.

Ericsson, K. A., & Simon, H. A. (1993). *Protocol analysis: Verbal reports as data* (Rev. ed.). Cambridge, MA: MIT Press.

Harte, J. M., Westenberg, M. R. M., & van Someren, M. (1994). Process models of decision making. *Acta Psychologica, 87,* 95–120.

Huber, O., Beutter, C., Montoya, J., & Huber, O. W. (2001). Risk-defusing behaviour: Towards an understanding of risky decision making. *European Journal of Cognitive Psychology, 13,* 409–426.

Huber, O., & Seiser, G. (2001). Accounting and convincing: The effect of two types of justification on the decision process. *Journal of Behavioral Decision Making, 14,* 69–85.

Huber, O., Wider, R., & Huber, O. (1997). Active information search and complete information presentation in naturalistic decision tasks. *Acta Psychologica, 95,* 15–29.

Klein, G. A., Calderwood, R., & MacGregor, D. (1989). Critical decision method for eliciting knowledge. *IEEE Systems, Man and Cybernetics, 19,* 462–472.

Koele, P., & Westenberg, M. J. M. (1995). A compensation index for multiattribute decision strategies. *Psychonomic Bulletin and Review, 2,* 398–402.

Lipshitz, R., Klein, G., Orasanu, J., & Salas, E. (2001). Taking stock of naturalistic decision making. *Journal of Behavioral Decision Making, 14*, 331–352.

Montgomery, H., & Svenson, O. (1989). *Process and structure in human decision making.* Chichester, England: Wiley.

Payne, J. (1976). Task complexity and contingent processing in decision making: A replication and extension. *Organizational Behavior and Human Performance, 11*, 366–387.

Payne, J., Bettman, J., & Johnson, E. (1993). *The adaptive decision maker.* Cambridge, England: Cambridge University Press.

Ranyard, R., Hinkley, L., & Williamson, J. (2001). Risk management in consumers' credit decision making: A process tracing study of repayment insurance decisions. *Zeitschrift Für Sozialpsychologie, 32*, 152–161.

Ranyard, R., Hinkley, L., & Williamson, J. (2003). *Tracing people's thoughts about consumer credit decisions: Information search and evaluation strategies.* Unpublished manuscript.

Russo, J. E., Johnson, E., & Stephens, D. (1989). The validity of verbal protocols. *Memory and Cognition, 17*, 759–769.

Schwarz, N. (1996). *Cognition and communication: Judgmental biases, research methods and the logic of conversation.* Mahwah, NJ: Lawrence Erlbaum Associates.

Westenberg, M., & Koele, P. (1994). Multi-attribute evaluation processes: Methodological and conceptual issues. *Acta Psychologica, 87*, 65–84.

Williamson, J., Ranyard, R., & Cuthbert, L. (2000a). A conversation-based process tracing method for use with naturalistic decisions: An evaluation study. *British Journal of Psychology, 91*, 203–221.

Williamson, J., Ranyard, R., & Cuthbert, L. (2000b). Risk management in everyday insurance decisions: Evidence from a process tracing study. *Risk, Decision and Policy, 5*, 19–38.

22

Modeling Naturalistic Decision-Making Cognitive Activities in Dynamic Situations: The Role of a Coding Scheme

Jean-Michel Hoc
CNRS and University of Nantes, IRCCyN, France

René Amalberti
IMASSA, Brétigny-sur-Orge, France

The study of naturalistic decision making (NDM) typically involves research into complex situations considered from an ecological perspective. This approach has the potential to bring into question numerous psychological theories that have their basis only in laboratory studies. However, the NDM perspective very often shows some weakness in methodology, meeting ecological criteria but jeopardizing the preconditions of any empirical proof of theories, even though the latter are neat and plausible. For several years now, we have been dealing with this problem, trying to draw as straight a line as possible from theory (Hoc, Amalberti, & Boreham, 1995) to methodology (Amalberti & Hoc, 1998; Hoc & Amalberti, 1999) and empirical studies (Amalberti, 1996, 2001; Amalberti & Deblon, 1992) in the domain of dynamic situation management (DSM; blast furnace control, air traffic control, aircraft piloting).

Our theoretical approach to dynamic situations is in line with the cognitive engineering perspective (Rasmussen, Pejtersen, & Goodstein, 1994) but takes a stronger (human) cognitive ergonomic orientation. This approach has several commonalities with NDM research:

- The prominence of case recognition over deep reasoning or calculation stressing the role of affordances (Gibson, 1979/1986).
- The management of a satisfying level of performance instead of the optimal performance put forward by control theorists. Given the complexity of the situations with which the human operator is confronted, the

importance of context, and the existence of tolerance margins, the operator is managing a trade-off between several factors to result in an acceptable performance (Amalberti, 1996).

- As with most NDM approaches, the goal of our research is twofold: First, to conduct basic psychological research with the notion that field studies in complex dynamic situations have the potential to gain access to the dynamic control of cognition, and second, to conduct applied research with regard to situated ergonomics in aviation, blast furnace control, nuclear power plant (NPP) industry, and medicine.

From a methodological point of view, our target is not only explicit behavior but also the implicit part of cognitive activity that controls behavior. Using spontaneous verbal protocols, we try to avoid using invasive techniques wherever possible. This leads us to favor the symbolic level of control of the activity and to make use of inference when reconstructing the activity from verbal and nonverbal behaviors. This perspective emanates from well-established approaches to the study of problem solving (Duncker, 1945; Ericsson & Simon, 1984) and from more recent attempts to support implicit cognitive process elicitation from individual protocols (Sanderson et al., 1994).

In the first section of this chapter, we summarize the main features of dynamic situations. We devote the second section to the presentation of our theoretical framework—DSM. In the third section, we develop the coding scheme, derived from this architecture, to describe the elementary cognitive activities. The coded protocol is a necessary step from raw data to cognitive modeling. In the fourth section, we take stock of the benefits of this approach but also open up a discussion on the drawbacks that have to be overcome, as is the case in most of the NDM approaches.

DYNAMIC SITUATIONS

In dynamic situations (as opposed to static ones), human operators do not fully control their environment. Many natural situations in human–machine system studies belong to this category, in industry (e.g., nuclear power plants), transportation, or medicine (e.g., anesthesiology). In static situations, the participants bring about all the transformations within their environment. In dynamic situations, human actions are combined with technical process dynamics to produce any transformation. For example, an action on the helm of a ship is combined with wind strength and direction of the current, with the ship's inertia, and so on before bringing about a change in the ship's trajectory. In technical processes, within an industrial plant or aircraft, for example, automation has introduced autonomous machines with which human control (or supervision) is shared.

The research community has conducted extensive studies within this context over a long period of time (Bainbridge, 1978; Hoc, Cacciabue, & Hollnagel, 1995; Hollnagel, Mancini, & Woods, 1988; Rasmussen et al., 1994). Confronted with a wide variety of cognitive theories and models in this domain, we have tried to define a general cognitive architecture capable of generating these models on a coherent basis (Hoc, et al., 1995). This coherence is necessary to enable researchers to compare results and identify invariant properties as well as differences across situations. Among these invariant properties, some of them appear to be more significant than others.

The management of dynamic situations requires the acquisition of a large amount of technical knowledge. The operators must understand process changes that are not fully or directly related to their interventions. Most of the time, case recognition can be sufficient, but some problem-solving activity must develop in unusual situations. The current representation must be regularly updated to integrate events not directly caused by the operator's action.

Dynamic situations are partly uncertain and demand a regular control of performance and risk. External risk concerns possible incidents or accidents and their effect on persons or equipment. Internal risks are often neglected when one adopts a purely technical point of view. Human operators do not only manage the technical processes under control but also their own resources such as work load, competency, vigilance, and so on. The major risk for a human operator is the loss of control of the situation, which is an essential determinant of performance. Thus, maintaining optimal performance can be dangerous if this strategy leads to the overloading of cognitive resources and to subsequent loss of control, with no reserves to deal with a possible complex state. This is why human operators maintain safety margins, not only from the technical system's point of view but also from their own cognitive system's point of view.

Dynamic situations very often require the management of several goals simultaneously. Several technical subprocesses can develop in parallel and must be managed concurrently. Even when a single process is considered, there may be several temporal spans with possible contradictory criteria. Thus, the process must be maintained within acceptable limits in the short term without jeopardizing medium or long-term objectives. Finally, conflicting criteria such as production, safety, cost, and efficiency must be prioritized.

DSM Cognitive Architecture

The DSM cognitive architecture distinguishes between three feedback loops in terms of abstraction level and temporal span of the decision (Fig. 22.1). It is inspired largely by Rasmussen's (1986) step ladder model of diag-

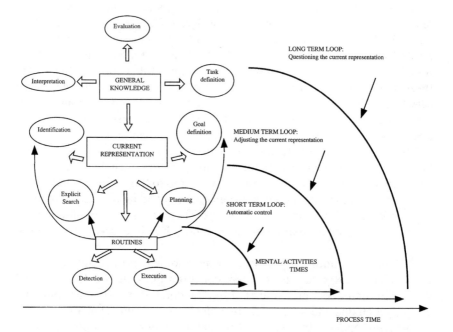

FIG. 22.1. The Dynamic Situation Management (DSM) cognitive architecture. The DSM model expands the Rasmussen (1986) step ladder model to provide a comprehensive view on how cognition acts dynamically to make compromises that result in both efficient short-term and long-term control of the process. Indeed, the cognition faces contradictory requirements: on one hand to effectively control the process and on the other hand to invest time and resources to understand any unexpected or bizarre events that are likely to occur in all human activity (surprises, unexpected results, deviations, etc.). When resources are limited, a compromise is needed. Time must be freed from the short term to allow the understanding of surprises, measuring the potential loss of relevance of the ongoing mental model and possibly replanning and changing this model. The DSM architecture assumes that the required dynamic compromise results from the activation of three semi-independent and parallel (short-term, medium-term, and long-term) loops. Most of the time, priority is given to the short-term control loop, which is based on affordances and routines. However, a continuous flow of unexpected events passes to the medium- and long-term loops and piles up, requiring further analysis and adjustment of the mental representation. The DSM architecture justifies the fact that many surprises can remain cognitively unexplored, or at least poorly explored, because of the impossibility of freeing sufficient resources from the medium- and long-term loops to enable the long-term loop to operate. Sometimes, uncertainty is so high that a decision between several different explanations could not be made before the occurrence of a long-term feedback.

nosis and decision making, at least when one considers the elementary activity modules such as detection, explicit information search, and planning. However, it devotes more consideration to feedback and to synchronization between cognitive processes and the technical process dynamics.

This architecture introduces the current representation of the process state and evolution as a prominent entity for an inner determination of the activity. The current representation not only integrates the technical process state and evolution (as was the case in the concept of situation awareness described, e.g., by Endsley, 1995) but also the operators' metaknowledge of their own activity (internal risks, resources, and plans). Priority is always given to the short-term control of the technical process, which is more likely to be performed by automatic processes bearing on subsymbolic representations and short-term feedback, supervised by a symbolic plan at the symbolic level of the current representation. However, some desynchronization between deep cognitive processes and the technical process evolution can occur. The current plan can reveal itself to be poorly adapted and may need some local adjustment. Sometimes, local adjustment is not sufficient and more in-depth replanning is necessary, requiring both more time and a long-term loop to get its feedback. These three control levels must operate in parallel or at least on a time sharing basis.

This conception of DSM cognitive processes has several implications, consistent with NDM research findings.

Diagnosis, prognosis, and planning are crucial ingredients for decision making. However, these activities are not performed for themselves (e.g., diagnosis exclusively for comprehension purposes) but as a means of establishing action decisions (e.g., diagnosis to access the best decision considering the circumstances and, in particular, temporal constraints). This is the role of the current representation of the situation that corresponds to a level of understanding just sufficient to act adequately. Obviously the evaluation of adequacy is subjective. It depends on the operators' personality, motivation, trade-off between contradictory goals, competency, and so on.

When the process is slow and subject to a low level of uncertainty, the three control loops can operate in series. However, when the process is rapid and uncertain, they must operate in parallel as shown in fighter aircraft piloting (Amalberti & Deblon, 1992). When there is really no time to replan, contingency preplanning is needed. In this case, plans can integrate a number of possible events to prepare adequate responses. More generally, the architecture can generate diverse cognitive models (implementation) in relation to circumstances. It can be predictive when the mechanisms of such implementations are known.

Decisions are made regularly in relation to an overall plan and to affordances obtained from the environment in real time. When the plan is identified as incorrect, it is not immediately abandoned, but affordances play a major role in its adaptation to deal with unexpected circumstances.

Decisions are more often suboptimal than optimal. A major objective is maintaining an acceptable performance level and at the same time maintaining an acceptable reserve of resources (not only now, but also in the future). The main risk is the loss of control of the situation. The current representation does not only include information on the external environment (the technical process under control) but also on cognitive resources (metaknowledge).

This architecture poses several research questions. The proposed coding scheme has only solved some of them. Its main objective is to prove the existence and the importance of some mechanisms and to compare different situations with evaluation purposes in mind. In diverse studies, our main concern was creating a tool to meet reproduction criteria. In our domain, replication of experiments is very difficult. However, we would like to reach a systematic method of protocol encoding with some guarantee of reproducibility by coders sharing the same explicit rules. We would like to escape from the sole use of excerpts of protocols for illustration purposes and to devise a systematic encoding that enables us to evaluate the importance of phenomena across situations and participants in a statistical way. We also tried to avoid overly domain-specific descriptions, stressing instead general cognitive processes in terms of a clear cognitive architecture.

A CODING SCHEME AND ITS APPLICATIONS

At the individual level, our coding scheme identifies three main classes of activities (Fig. 22.2) described by a predicate and arguments structure (Sanderson et al., 1994). Predicates code activities, and arguments specify these activities and representations that are to be processed. These activities are supposed to develop at the symbolic level because they are inferred mainly from spontaneous verbal protocols.

The coding scheme (see Fig. 22.2) is divided into three main categories: information elaboration (before deep interpretation), diagnosis (interpretation relevant to action, integrating prognosis), and decision making (integrating feedback evaluation) plus a specific coding for metacognitive activities. In the case of metacognitive activities, the representations processed do not concern the external environment (technical process) but the mental or internal process. In our approach to human cognition in dynamic environments, we (Hoc & Amalberti, 1999) have stressed three aspects of metacognition: the management of internal risks, the management of internal costs, and metaplanning. Internal risk management considers the risk as "the representation individuals have for an event—with a subjective probability and cost—to which they refuse to adapt their plan for reasons of re-

INFORMATION ELABORATION	DIAGNOSIS	DECISION MAKING
IG : Information Gathering A piece of information is gathered (in the sense of its consideration in the activity) without deep interpretation or information that has been already interpreted is transmitted to the operator.	**CREC: Case recognition** An activity that aims to categorise a variable value in a way that is meaningful to action, with certainty	**SCH-DEC: Schematic decision :** Decision can be schematic (abstract plan)...
IGG: Information gathering goal An intention of gathering a piece of information is verbally reported, sometimes with a justification.	**HYPGEN: Hypothesis generation** Hypothesis generation is also a categorisation activity, but with an element of uncertainty. In dynamic situations, uncertainty is linked to a lack of information during diagnosis	**PRCS-DEC : Precise decision** ...Or precise (full-specified plan).
IRM: Information recovery from memory Several pieces of information, gathered at different times, are grouped together at a certain time, because they are considered to be related. Such groups give access to the content of the current representation, as recaps.	**TEST: Hypothesis testing** The prominent feature of a hypothesis being uncertainty, any hypothesis calls for testing. Although hypothesis testing is often expressed in an all or none way in the verbal protocols, it basically consists of modifying the level of uncertainty of a hypothesis. Several tests of the same hypothesis are often necessary, along time, to entirely reduce uncertainty, except in formal reasoning as it is widely studied in the laboratory.	**ACTION: Decision implementation :** Decision making does not always result in action implementation. A long history of decision elaboration can take place before an action, integrating hypothesis generation and testing, decision evaluation, and so forth. Secondly, an action can be implemented without a corresponding decision-making cue in the verbal protocol (e.g., in routine episodes).
		DECI-EVAL: Decision evaluation This elementary activity is coded when it is clearly linked to decision making. The cognitive system being adaptive, feedback evaluation is crucial. Such an evaluation can be performed before or after implementation (anticipative vs. reactive evaluation).

FIG. 22.2. Coding scheme (predicates). Example in ATC: IG(<OBJECT TYPE>,<OB-JECT>,<VARIABLE>,<VALUE>,<CONDITION>,<AIM>) IG(aircraft,SU999,beacons,NOR-PTV,trans-rc,assume). The object is an individual aircraft as opposed to a group of several conflicting aircraft. Its call sign is SU999. The operator (radar controller) is interested by the entry (NOR) and exit (PTV) beacons. This information is gathered when the strip (written information about the aircraft) is just transmitted by the planning controller (in charge of the coordination with adjacent sectors and of the radar controller's work load). The aim of this activity is the integration of this aircraft. The value trans-rc of the argument condition gives a reactive status to this IG activity as opposed, for example, to the value IGG that corresponds to an anticipative strategy. In this case, the radar controller previously set the goal of gathering this information before it was available.

325

sources, cognitive cost, and motivation" (Hoc, 1996, p. 155). Internal cost management can also be implicit or explicit. Last, risk and cost management result in metaplanning that can also be implicit. This type of activity is more salient in self-confrontation protocols when operators evoke alternative plans that were excluded on this basis. Thus, although metaplanning is situated at a very high level of control of the activity, it can remain implicit.

Coded activities do not include subsymbolic processes, although the latter can play a prominent role in some episodes. Our scheme enables us to account for the microstructure of a protocol but only at the symbolic level. Those arguments that code relations between individual activity occurrences (especially conditions and goals of the individual activity) enable us to access part of the macrostructure. In dynamic situations in which the (macro) structure of the protocol is not only determined by the strategy but also by the unexpected evolution of the technical process under control, automatic pattern identification from the series of elementary activities is not very useful.

PROTOCOL ENCODING

The scheme codes cognitive activities that can remain implicit for the coder, although they are supposed to be explicit for the participant (processed at the level of attention) even if verbal protocols are collected. Thus, inference is necessary to identify these activities. Inference is controlled on the basis of three types of knowledge:

1. Cognitive Architecture. The cognitive architecture ensures that the same activity is encoded in the same way at different locations in the protocol and from one application domain to another. Obviously, from one domain to another and according to the main objective of the analysis, the coding scheme can put local emphasis on some categories of predicates, for example, on cooperation, diagnosis, or error management.

2. Behavior and Context. Verbal and nonverbal behavior is collected and synchronized with successive states of the technical process under control (not only depending on the behavior but also on the spontaneous process trend and on unexpected external factors). This enables the coder to situate an elementary activity within a context.

3. Domain Knowledge. The coder's level of knowledge in the domain is very important in guiding inference of psychological mechanisms. When this knowledge is poor, inference could be impossible or restricted to surface characteristics of behavior.

The reliability of this encoding strategy requires mutual control between different coders. Indeed, there are several techniques for evaluating consistency between coders. However, we are more interested in discussions between coders when local disagreements occur, often resulting in a consensus on explicit rules following an enlargement of the analysis framework.

AN EXTENDED EXAMPLE OF THE USE OF THE DSM METHOD

In this example, we describe a PhD student's (Lien Wioland) experimental contribution to a 10-year research program conducted at IMASSA on error management and ecological safety management (Amalberti, 1996, 2001; Wioland, 1997; Wioland & Amalberti, 1996). This example is an illustration of how one can draw a straight line from methodology to NDM theory by the means of the proposed framework.

Purpose of the Study. The purpose of the study was to describe the components of human error management and how they are dynamically organized.

Experimental Protocol. This experiment made use of a microworld Terminal Radar Approach CONtrol (TRACON) simulating a fairly realistic air traffic control situation. Two groups of participants were considered: one made up of controllers and the other of naïve participants (to assess the learning curve of error management). For the purposes of this example, we only present the results from professional controllers. Controllers had to manage en route and airport traffic, with instructions to be as safe and efficient as possible. High-workload periods, system failures, miscommunications, and poor weather conditions were present in half of the scenarios.

Participants. Eight participants took part in nine simulations.

Data Collection. The microworld TRACON used a Macintosh Quadra 840AV. For each scenario, traces were recorded from the replay function (technical data, aircraft speed, etc.), controller behavior was video recorded, and think-aloud verbal reports were produced.

Data Analysis (See Fig. 22.3). For each scenario, the verbal reports, key information from the replay, and participants' actions were written down on a MacSHAPA spreadsheet (Sanderson et al., 1994). Data were stored in separate columns synchronized by a time base. The DSM coding scheme we present in this chapter was used to code data in a third column as shown in the excerpt in Fig. 22.3. Once this coding was completed, a last column was added to code error management activities (error production, error detection, error recovery).

THINK-ALOUD PROTOCOL	CONTEXT (INFORMATION GATHERED ON THE INTERFACE (REPLAY) AND FROM THE VIDEO (SUBJECT 'S ACTIONS)	DSM CODING	ERROR MANAGEMENT CODING
00:07:39:37 this aircraft should go down to Flight level 16	00:07:39:37-00:07:41:12 Aircraft N85Q: stable Flight level 30, final approach to the airport	15 SCH-DEC (Schematic decision) (N85Q, Descent FL16, when?)	
00:07: 40: 20 Yes…I I had forgotten it was already at 220 Knots. I thought I should check the speed before transferring because I remembered I had been given the instruction to deliver the aircraft to the approach controller at a speed of 250 Knots maximum. I had transferred the aircraft already, so I didn't have the strip under my eyes anymore	00:07: 40: 02 Message from en-route AF701 reacting to the fact that the controller has given an instruction to reduce speed to 250 Knots, but the pilot has already announced that he was at 220 Knots The controller clears AF 701 for 220Kts	16 IRM (Information recovery from memory) (AF701, conflict transfer speed, confirmation 220)	Human Error Nature : Slip Detection based on deleterious outcome Passive recovery
00:07: 40: 40 Well, back to the approach, I clear AC31 for the final descent, FL16	00:07: 40: 40 the demand of final descent regards AC31 and not N85Q. AC 31 is approaching the same airport from a different route.	17 ACTION (descent FL16, AC31)	
00:07: 40:50 Ho, I mismanaged the order, should have cleared first N85Q slow down N85Q	00:07: 40:50 AC 31 is let first/.N85Q becomes second in the approach	DECI-EVAL (inversion SCH-DEC 15 between AC31 and N85Q; error recovery)	Human error Nature : Slip Detection based on deleterious outcome Partial recovery

FIG. 22.3. Excerpt from data analysis, Terminal Radar Approach CONtrol experiment (Wioland, 1997).

Main Results

A total of 174 errors were recorded during a period of about 22 hr of simulation, with an average frequency of about 8 errors per hour. Slips were the most frequent (68%). The detection rate ranged from 68% for slips to 57% for mistakes.

More interesting for NDM was the percentage of detected but not recovered errors, which reached about 32% of the total number of detected errors. Most of these detected, nonrecovered errors had few or no consequences. Some were recovered with a significant delay after identification; others were detected and just ignored. This result is a good illustration of the DSM architecture as well as of the NDM theory. The controllers gave priority to short-term control and were careful to avoid spending too much time in the activation of a thorough analysis of the process. They often waited for evidence (frank deleterious outcome) before making the detection and even more often did nothing after it or postponed recovery. Another frequent strategy consisted in adapting the plan to the error, therefore avoiding the undo process.

To sum up for this specific example, the DSM coding scheme has allowed us to objectively characterize and aggregate cognitive processes, hence recombining information to systematically find out the underlying mechanisms of error management. Without this preliminary step, for example, if only referring to a macroanalysis from raw data, many error management mechanisms would have been misinterpreted; even more would not have been identified if simply relying on a tenuous cognitive evaluation.

DISCUSSION

Natural behaviors in complex work environments often differ from ostensibly identical behaviors in laboratory testing. Cognitive psychology has much to learn in the study of these natural, ecological situations because they offer a unique opportunity to access and model the complexity and richness of the dynamic control of cognition and its associated trade-offs (between speed and accuracy, performance and risk, etc.).

However, because of the importance of the contextual control of cognition in natural behaviors, the intrinsic variability of context in the field, and the difficulty in gathering a large panel of observations, results often take the form of case-based anecdotes and isolated examples, neither of which entirely satisfy usual scientific criteria of proof.

Should one access these mechanisms of control through a method of protocol coding that is sufficiently stable and refutable, the scientific validity of theories of natural psychology (a concept encompassing NDM as well as ecological safety and related approaches) would be vastly improved.

An ideal coding scheme has to capture three kinds of data. The first type of data is the characterization of the context and episodes as a piece of a real-life story with stakes, associated burdens, and simple solutions (the macrostructure, including macrostrategies). A second type of data is the screening and characterization of the ongoing microcognitive processes (search for information, problem solving, decision making, communication, etc.). A third type of data is gained through access to associated short-term routinized activities, medium-term revisitations of representation, and long-term activation of deep knowledge and the system of interactions between all the identified levels of cognition and external constraints (to validate a frame and contextual model of cognition).

The coding scheme we proposed in this chapter is satisfactory for gathering and formalizing data for the first and second level (macrostructure and microstructure), although some clarifications are still needed for some concerns. Much harder problems persist in gaining access to the third level. This discussion is therefore structured in two parts. In the first subsection, we question the remaining points that limit the coding of levels one and two. In the second subsection, we focus on the bottleneck to access the third level.

Coding the Context and the Cognitive Microstructure: An Almost Successful Method

The coding scheme has been refined so that it provides a direct visibility of the microstructure and of the context. Predicates are kept to a minimum and code the microstructure (the cognitive elementary operations). Arguments are more numerous and "tell the story," including the evaluation of results, such that any person can read the protocol explicitly at the two levels.

The reliability of the coding of participants' verbal reports is warranted by two mechanisms: One is a parallel coding of objective data (on the process and the context); another is the coder's knowledge in the domain. Limitations remain threefold.

Stability/Duplication of Results. Whatever the quality of observations, protocols can only give periodic snapshots of a continuous physical and mental process. Most data remain unrecorded. This means that processes must be inferred from local insights; hence, the control of inference becomes central to the quality of coding. Through experience, we tend to consider that the control of inference mostly depends on the coder's expertise in the domain in combination with his or her expertise of psychological models. In other words, the results among coders and even within coders over an extended period are subject to variation, especially for the value of arguments. To reduce this bias, we use a double-blind coding approach with coders that have significant experience in the domain and share the same framework in terms of cognitive architecture.

Reference for Performance Evaluation. We said previously that arguments code most of the value of performance but that this is only possible if there is a recognized reference for performance. The problem is that there are typically at least three different references of performance. The first reference is the expected ideal/optimal professional performance as described in manuals, with an expected level of end result plus a set of policies and procedures to respect when carrying out the job. The second reference is the local contract between the worker and the superiors: What is the very envelope of intangible constraints for task completion, not in general but now, for this specific case? The third reference is the worker's individual reference of what is to be done to avoid anything risky for the individual and of what is simple to complete; this last reference greatly depends on the participant's knowledge and metaknowledge. We must confess that, in most cases, the only reference we can get (although sometimes not very easily) is the first one (the ideal), which tends to assess standards protocols with excessive negative values. Should the model of natural psychology be true, including the permanent participant's intention to free margins and resources resulting in a suboptimal performance, the use of the second or third reference could become better than the first reference. The problem lies in characterizing these second and third references. We have no magic solutions, but it may be eased by undertaking interviews with workers just after completion of the work to assess the quality of the domain knowledge of the analyst.

Recoding Contextual Cognitive Modes of Control From Local Processes. Until now, the analysis is rather satisfactory in capturing the story and the participant's current cognitive processes. A complete approach should be able to infer from the microstructure, and with a reproducible procedure, the local control level of cognition. Although this procedure seems accessible to the recorded data, more work is needed to perform this translation precisely.

Accessing the Related Hidden Levels of Cognition: The Bottleneck Itself

As already stated in the introduction to this discussion, this is the final bottleneck faced by almost all coding schemes.

Verbal protocols and objective contextual data are effective methods for capturing evidence of the conscious, dominant activity of cognition. Nobody should affirm, however, that it is the total of cognitive activities. As assumed by the cognitive architecture we presented in the first section of this chapter, routine activities, as well as slow revisions of representation and mobilization of deep knowledge, are underconsidered. The limitation does not only concern what is going on at these two levels but also the underconsideration of the serialism/parallelism between levels. Comprehension activities can develop with or without a clear link to immediate action. Diverse levels of comprehension must be defined in relation to diverse planning levels.

One must acknowledge that most of the episodes coded as conscious, symbolic, cognitive processes on the basis of objective change in the situation probably carry out an important part of routinized activities, even at the level of decision making.

The problem is that most of the methods that could identify subsymbolic activities are intrusive in task completion (interview techniques, physiological recording set, etc.). Focusing on routines often changes the task, and the microstructure and macrostructure of protocols. Conversely, without any specific technique, routine activities tend to be mute.

Access to the slow pace of change in representation and related moves of activation of deep knowledge in cognition raises a different problem. For these long-term changes, there are very few occurrences to observe in each protocol. In addition, these macrolevel changes are likely to be partly accessible to awareness. The small number of protocols that characterize field experiments aggravates this low number of data to be found within each protocol. We think that this level will only become accessible to psychology with a much larger number of observations to be able to relate the long-term changes to local events. Hence, such an approach should demand the setup over a long period of time of an observatory of cognitive work in a given complex situation. To our knowledge, this does not yet exist.

CONCLUSIONS

In this chapter, we have presented a coding scheme that is applicable to dynamic, complex, and natural work activities in line with a cognitive architecture for these activities. The cognitive architecture captures some basic ideas of Rasmussen's (1986) step ladder model and adds a permanent possibility of parallelism between subsymbolic activities, several kinds of symbolic activities, and the activation and updating of deep knowledge. Results from several domains show that the coding scheme is relatively easy to use and robust according to scientific criteria. Some difficulties still exist, however, in accessing the parallelism of cognitive activities and the dynamic integrated control and adaptation of cognition that allow participants to exhibit a limited suboptimal performance and nevertheless reach their goals conveniently.

This kind of approach has another type of limitation. It does not explicitly address performance as the result of cognitive activity in the terms of the work domain. It provides researchers and designers with an internal evaluation of performance (from the participant's point of view). This kind of evaluation is crucial for identifying design problems in terms of internal criteria such as cognitive cost, risk, and so forth. However, it is only one side of the coin. The other side is external evaluation of performance from the work domain's point of view (Long, 1996). Obviously, it is not reasonable to support strategies that are ineffective from this external point of view, although they could be considered as effective with appropriate support from the internal point of view (Vicente, 1999). The work domain's constraints and objectives allow some degree of freedom to choose (and assist) effective and compatible strategies from both points of view. Thus, our analysis should also integrate criteria coming from a work domain model.

ACKNOWLEDGMENT

This work was undertaken while the authors were members of LAMIH, CNRS laboratory at the University of Valenciennes, Valenciennes, France.

REFERENCES

Amalberti, R. (1996). *La conduite de systèmes à risques* [Risky process control]. Paris: Presses Universitaires de France.

Amalberti, R. (2001). The paradoxes of almost totally safe transportation systems. *Safety Science, 37*, 109–126.

Amalberti, R., & Deblon, F. (1992). Cognitive modelling of fighter aircraft's process control: A step towards an intelligent onboard assistance system. *International Journal of Man–Machine Studies, 36*, 639–671.

Amalberti, R., & Hoc, J. M. (1998). Analyse des activités cognitives en situation dynamique : pour quels buts ? comment ? [Cognitive activity analysis in dynamic situations: Why and how?]. *Le Travail Humain, 61*, 209–234.

Bainbridge, L. (1978). The process controller. In W. T. Singleton (Ed.), *The study of real skills: Vol. 1. The analysis of practical skills* (pp. 236–263). St Leonardgate, UK: MTP.

Duncker, K. (1945). On problem-solving. *Psychological Monographs, 58*(Whole No. 270).

Endsley, M. (1995). Toward a theory of situation awareness in dynamic systems. *Human Factors, 37*, 32–64.

Ericsson, K. A., & Simon, H. A. (1984). *Protocol analysis: Verbal reports as data.* Cambridge, MA: MIT Press.

Gibson, J. J. (1986). *The ecological approach to visual perception.* Hillsdale, NJ: Lawrence Erlbaum Associates. (Original work published 1979)

Hoc, J. M. (1996). *Supervision et contrôle de processus: La cognition en situation dynamique* [Process control and supervision: Cognition in dynamic situation]. Grenoble, France: Presses Universitaires de Grenoble.

Hoc, J. M., & Amalberti, R. (1999). Analyse des activités cognitives en situation dynamique : D'un cadre théorique à une méthode [Cognitive activity analysis in dynamic situations: From a theoretical framework to a method]. *Le Travail Humain, 62*, 97–130.

Hoc, J. M., Amalberti, R., & Boreham, N. (1995). Human operator expertise in diagnosis, decision-making, and time management. In J. M. Hoc, P. C. Cacciabue, & E. Hollnagel (Eds.), *Expertise and technology: Cognition & human–computer cooperation* (pp. 19–42). Hillsdale, NJ: Lawrence Erlbaum Associates.

Hoc, J. M., Cacciabue, P. C., & Hollnagel, E. (Eds.). (1995). *Expertise and technology: Cognition & human–computer cooperation.* Hillsdale, NJ: Lawrence Erlbaum Associates.

Hollnagel, E., Mancini, G., & Woods, D. D. (Eds.). (1988). *Cognitive engineering in complex dynamic worlds.* London: Academic.

Long, J. (1996). Specifying relations between research and the design of human–computer interaction. *International Journal of Human–Computer Studies, 44,* 875–920.

Rasmussen, J. (1986). *Information processing and human-machine interaction.* Amsterdam: North-Holland.

Rasmussen, J., Pejtersen, A. M., & Goodstein, L. P. (1994). *Cognitive systems engineering.* New York: Wiley.

Sanderson, P., Scott, J., Johnson, T., Mainzer, J., Watanabe, L., & James, J. (1994). MacSHAPA and the enterprise of exploratory sequential data analysis (ESDA). *International Journal of Human–Computer Studies, 41,* 633–681.

Vicente, K. (1999). *Cognitive work analysis.* Mahwah, NJ: Lawrence Erlbaum Associates.

Wioland, L. (1997). *Étude des mécanismes de protection et de détection des erreurs, contribution à un modèle de sécurité écologique* [Study of error protection and detection mechanisms]. Unpublished doctoral dissertation, University of Paris V.

Wioland, L., & Amalberti, R. (1996). When errors serve safety: Towards a model of ecological safety. In H. Yoshikawa & E. Hollnagel (Eds.), *Proceedings of the First Asian Conference on Cognitive Systems Engineering in Process Control (CSEPC 96)* (pp. 184–191).

23

The Knowledge Audit as a Method for Cognitive Task Analysis

Gary Klein
Laura Militello*
Klein Associates Inc.

The Knowledge Audit was designed to survey the different aspects of expertise required to perform a task skillfully (Crandall, Klein, Militello, & Wolf, 1994). It was developed for a project sponsored by the Naval Personnel Research & Development Center. The specific probes used in the Knowledge Audit were drawn from the literature on expertise (Chi, Glaser, & Farr, 1988; Glaser, 1989; Klein, 1989; Klein & Hoffman, 1993; Shanteau, 1989). By examining a variety of accounts of expertise, it was possible to identify a small set of themes that appeared to differentiate experts from novices. These themes served as the core of the probes used in the Knowledge Audit.

The Knowledge Audit was part of a larger project to develop a streamlined method for Cognitive Task Analysis that could be used by people who did not have an opportunity for intensive training. This project, described by Militello, Hutton, Pliske, Knight, and Klein (1997), resulted in the Applied Cognitive Task Analysis (ACTA) program, which includes a software tutorial. The Knowledge Audit is one of the three components of ACTA.

The original version of the Knowledge Audit is described by Crandall et al. (1994) in a report on the strategy that was being used to develop ACTA. That version of the Knowledge Audit probed a variety of knowledge types: perceptual skills, mental models, metacognition, declarative knowledge, analogues, and typicality/anomalies. *Perceptual skills* referred to the types of perceptual discriminations that skilled personnel had learned to make. *Mental models* referred to the causal understanding people develop about

---

*Ms. Militello is currently a human factors psychologist at the University of Dayton Research Institute.

how to make things happen. *Metacognition* referred to the ability to take one's own thinking skills and limitations into account. *Declarative knowledge* referred to the body of factual information people accumulate in performing a task. *Analogues* referred to the ability to draw on specific previous experiences in making decisions. *Typicality/anomalies* referred to the associative reasoning that permits people to recognize a situation as familiar, or, conversely, to notice the unexpected.

Some other probes were deleted because they were found to be more difficult concepts for a person just learning to conduct a Cognitive Task Analysis to understand and explore; others were deleted because they elicited redundant information from the subject-matter experts being interviewed. To streamline the method for inclusion in the ACTA package, eight probes were identified that seemed most likely to elicit key types of cognitive information across a broad range of domains.

CURRENT VERSION

The Knowledge Audit has been formalized to include a small set of probes, a suggested wording for presenting these probes, and a method for recording and representing the information. This is the form presented in the ACTA© software tutorial. The value of this formalization is to provide sufficient structure for people who want to follow steps and be reasonably confident that they will be able to gather useful material.

Table 23.1 presents the set of Knowledge Audit probes in the current version. These are listed in the column on the left. Table 23.1 also shows the types of follow-up questions that would be used to obtain more information. At the conclusion of the interview, this format becomes a knowledge representation. By conducting several interviews, it is possible to combine the data into a larger scale table to present what has been learned.

However, formalization is not always helpful, particularly if it creates a barrier for conducting effective knowledge elicitation sessions. We do not recommend that all the probes be used in a given interview. Some of the probes will be irrelevant, given the domain, and some will be more pertinent than others. Furthermore, the follow-up questions for any probe can and should vary, depending on the answers received.

In addition, the wording of the probes is important. Militello et al. (1997) conducted an extensive evaluation of wording and developed a format that seemed effective. For example, they found that the term *tricks of the trade* generated problems because it seemed to call for quasi-legal procedures. *Rules of thumb* was rejected because the term tended to elicit high-level, general platitudes rather than important practices learned via experience on the job. In the end, the somewhat awkward but neutral term *job smarts* was used to ask about the techniques people picked up with experience.

TABLE 23.1
Knowledge Audit Probes

Probe	Explanation and Follow-Up Questions
Past and future	Experts can figure out how a situation developed, and they can think into the future to see where the situation is going. Among other things, this can allow experts to head off problems before they develop.
	Is there a time when you walked into the middle of a situation and knew exactly how things got there and where they were headed?
Big picture	Novices may only see bits and pieces. Experts are able to quickly build an understanding of the whole situation—the big picture view. This allows the expert to think about how different elements fit together and affect each other.
	Can you give me an example of what is important about the big picture for this task? What are the major elements you have to know and keep track of?
Noticing	Experts are able to detect cues and see meaningful patterns that less experienced personnel may miss altogether.
	Have you had experiences where part of a situation just "popped" out at you, where you noticed things going on that others didn't catch? What is an example?
Job smarts	Experts learn how to combine procedures and work the task in the most efficient way possible. They don't cut corners, but they don't waste time and resources either.
	When you do this task, are there ways of working smart or accomplishing more with less—that you have found especially useful?
Opportunities/ improvising	Experts are comfortable improvising—seeing what will work in this particular situation; they are able to shift directions to take advantage of opportunities.
	Can you think of an example when you have improvised in this task or noticed an opportunity to do something better?
Self-monitoring	Experts are aware of their performance; they check how they are doing and make adjustments. Experts notice when their performance is not what it should be (this could be due to stress, fatigue, high workload, etc.) and are able to adjust so that the job gets done.
	Can you think of a time when you realized that you would need to change the way you were performing to get the job done?
Anomalies	Novices don't know what is typical, so they have a hard time identifying what is atypical. Experts can quickly spot unusual events and detect deviations. Also, they are able to notice when something that ought to happen doesn't.
	Can you describe an instance when you spotted a deviation from the norm or knew something was amiss?
Equipment difficulties	Equipment can sometimes mislead. Novices usually believe whatever the equipment tells them; they don't know when to be skeptical.
	Have there been times when the equipment pointed in one direction, but your own judgment told you to do something else? Or when you had to rely on experience to avoid being led astray by the equipment?

Turning to another category, the concept of perceptual skills made sense to the research community but was too academic to be useful in the field. After some trial and error, the term *noticing* was adopted to help people get the sense that experience confers an ability to notice things that novices tend to miss. These examples illustrate how important language can be and how essential the usability testing was for the Knowledge Audit.

The wording shown in Table 23.1 is not intended to be used every time. As people gain experience with the Knowledge Audit, they will undoubtedly develop their own wording. They may even choose their own wording from the beginning. The intent in providing suggested wording is to help people who might just be learning how to do Cognitive Task Analysis interviews and need a way to get started. In structured experimentation, researchers often have to use the exact same wording with each participant. The Knowledge Audit, however, is not intended as a tool for basic research in which exact wording is required. It is a tool for eliciting information, and it is more important to maintain rapport and follow up on curiosity than to maximize objectivity.

CONDUCTING A KNOWLEDGE AUDIT

We have learned that the Knowledge Audit is too unfocused to be used as a primary interviewing tool without an understanding of the major components of the task to be investigated. It can be too easy for subject-matter experts to just give speeches about their pet theories on what separates the skilled from the less skilled. That is why the suggested probes try to focus the interview on events and examples. Even so, it can be hard to generate a useful answer to general questions about the different aspects of expertise. A prior step seems useful whereby the interviewer determines the key steps in the task and then identifies those steps that require the most expertise. The Knowledge Audit is then focused on these steps, or even on substeps, rather than on the task as a whole. This type of framing makes the Knowledge Audit interview go more smoothly.

We have also found that the Knowledge Audit works better when the subject-matter experts are asked to elaborate on their answers. In our workshops, we encourage interviewers to work with the subject-matter experts to fill in a table, with columns for deepening on each category. One column is about why it is difficult (to see the big picture, notice subtle changes, etc.) and what types of errors people make. Another column gets at the cues and strategies used by experts in carrying out the process in question. These follow-up questions seem important for deriving useful information from a Knowledge Audit interview. In this way, the interview moves from just gathering opinions about what goes into expertise and gets at more details. For a skilled interviewer, once these incidents are identified, it is easy to turn to a more in-depth type of approach, such as the Critical Decision Method (Hoffman, Crandall, & Shadbolt, 1998). Once the subject-matter expert is describing a challenging incident, the Knowledge Audit probes can be used to deepen on the events that took place during that incident.

For example, Pliske, Hutton, and Chrenka (2000) used the Knowledge Audit to examine expert–novice differences in business jet pilots using weather data to fly. Additional cognitive probes were applied to explore critical incidents and experiences in which weather issues challenged the pilot's decision-making skills. As a result of these interviews, a set of cognitive demands associated with planning, taxi/takeoff, climb, cruise, descent/approach, and land/taxi were identified, as well as cues, information sources, and strategies experienced pilots rely on to make these difficult decisions and judgments.

FUTURE VERSION

One of the weaknesses of the Knowledge Audit is that the different probes are unconnected. They are aspects of expertise, but there is no larger framework for integrating them. Accordingly, it may be useful to consider a revision to the Knowledge Audit that does attempt to situate the probes within a larger scheme.

In considering the probes presented in Table 23.1, they seem to fall into two categories. One category is types of knowledge that experts have or "what experts know," and the second category is ways that experts use these types of knowledge or "what experts can do." Table 23.2 shows a

TABLE 23.2
Aspects of Expertise and Knowledge Audit Probe

What Experts Know	What Experts Can Do
• Perceptual skills/*noticing*	• Run mental simulations/*past and future*
• Mental models/*big picture*	To diagnose
• Sense of typicality and associations	To explain
• Routines/*job smarts*	To form expectancies
• (Declarative knowledge)	• Spot anomalies and detect problems/*anomalies*
	• Find leverage points/*opportunities/improvising*
	Perform workarounds
	• Manage uncertainty
	• Plan and replan
	• Assess complex situations
	• Manage attention
	• Take their strengths and limitations into account/*self-monitoring*

Note. Probes from the Knowledge Audit are in italics.

breakdown that follows these categories. Probes from the current version of the Knowledge Audit are included in italics next to the aspect of expertise each addresses.

The left-hand column in Table 23.2 shows different types of knowledge that experts have. They have perceptual skills, enabling them to make fine discriminations. They have mental models of how the primary causes in the domain operate and interact. They have associative knowledge, a rich set of connections between objects, events, memories, and other entities. Thus, they have a sense of typicality allowing them to recognize familiar and typical situations. They know a large set of routines, which are action plans, well-compiled tactics for getting things done.

Experts also have a lot of declarative knowledge, but this is put in parentheses in Table 23.2 because Cognitive Task Analysis does not need to be used to find out about declarative knowledge.

The right-hand column in Table 23.2 is a partial list of how experts can use the different types of knowledge they possess. Thus, experts can use their mental models to diagnose faults, and also to project future states. They can run mental simulations (Klein & Crandall, 1995). Mental simulation is not a form of knowledge but rather an operation that can be run on mental models to form expectancies, explanations, and diagnoses. Experts can use their ability to detect familiarity and typicality as a basis for spotting anomalies. This lets them detect problems quickly. Experts can use their mental models and knowledge of routines to find leverage points and use these to figure out how to improvise. Experts can draw on their mental models to manage uncertainty. These are the types of activities that distinguish experts and novices. They are based on the way experts apply the types of knowledge they have. One can think of the processes listed in the right-hand column as examples of macrocognition (Cacciabue & Hollnagel, 1995; Klein, Klein, & Klein, 2000).

Also note that the current version of the Knowledge Audit does not contain probes for all the items in the right column.

Table 23.2 is intended as a more organized framework for the Knowledge Audit. It is also designed to encourage practitioners to devise their own frameworks. Thus, R. R. Hoffman (personal communication) has adapted the Knowledge Audit. His concern is not as much with contrasting experts and novices as with capturing categories of cognition relevant to challenging tasks, such as forecasting the weather. Hoffman's categories are noticing patterns, forming hypotheses, seeking information (in the service of hypothesis testing), sensemaking (interpreting situations to assign meaning to them), tapping into mental models, reasoning by using domain-specific rules (e.g., meteorological rules), and metacognition.

One of the advantages of the representation of the Knowledge Audit shown in Table 23.2 is that the categories are more coherent than in Table

23.1. Although there is considerable overlap between the Knowledge Audit probes in Table 23.1 and the aspects of expertise in Table 23.2, we have not developed wording for all of the probes taking into account the new focus on macrocognition. Moreover, we have not yet determined the conditions under which we might want to construct a Knowledge Audit incorporating more of the items from the right-hand column of Table 23.2, the macro-cognitive processes seen in operational settings.

CONCLUSIONS

The Knowledge Audit embodies an account of expertise. We contend that any Cognitive Task Analysis project makes assumptions about the nature of expertise. The types of questions asked, the topics that are followed up, and the areas that are probed more deeply all reflect the researchers' concepts about expertise. These concepts may result in a deeper and more insightful Cognitive Task Analysis project, or they may result in a distorted view of the cognitive aspects of proficiency. One of the strengths of the Knowledge Audit is that it makes these assumptions explicit.

Because it distinguishes between types of knowledge and applications of the knowledge, the proposed future version of the Knowledge Audit is more differentiated than the original version. This distinction has both theoretical and practical implications. There are also aspects of expertise that are not reflected in the Knowledge Audit, such as emotional reactivity (e.g., Damasio, 1998), memory skills, and so forth. We are not claiming that the Knowledge Audit is a comprehensive tool for surveying the facets of expertise. Its intent was to provide interviewers with an easy-to-use approach to capture some important cognitive aspects of task performance.

There can be interplay between the laboratory and the field. Cognitive Task Analysis tries to put concepts of expertise into practice. It can refine and drive our views on expertise just as laboratory studies do. The Knowledge Audit is a tool for studying expertise and a tool for reflecting about expertise.

REFERENCES

Cacciabue, P. C., & Hollnagel, E. (1995). Simulation of cognition: Applications. In J. M. Hoc, P. C. Cacciabue, & E. Hollnagel (Eds.), *Expertise and technology: Cognition and human–computer co-operation* (pp. 55–73). Hillsdale, NJ: Lawrence Erlbaum Associates.
Chi, M. T. H., Glaser, R., & Farr, M. J. (Eds.). (1988). *The nature of expertise*. Hillsdale, NJ: Lawrence Erlbaum Associates.

Crandall, B., Klein, G., Militello, L., & Wolf, S. (1994). *Tools for applied cognitive task analysis* (Tech. Rep. Contract No. N66001–94–C–7008 for the Naval Personnel Research and Development Center, San Diego, CA). Fairborn, OH: Klein Associates Inc.

Damasio, A. R. (1998). Emotion and reason in the future of human life. In B. Cartledge (Ed.), *Mind, brain and the environment: The Linacre lectures 1995–1996* (pp. 57–71). Oxford, England: Oxford University Press.

Glaser, R. (1989). Expertise and learning: How do we think about instructional processes now that we have discovered knowledge structures? In D. Klahr & K. Kotovsky (Eds.), *Complex information processing: The impact of Herbert A. Simon* (pp. 269–282). Hillsdale, NJ: Lawrence Erlbaum Associates.

Hoffman, R. R., Crandall, B. W., & Shadbolt, N. R. (1998). Use of the critical decision method to elicit expert knowledge: A case study in cognitive task analysis methodology. *Human Factors, 40,* 254–276.

Klein, D. E., Klein, H. A., & Klein, G. (2000). Macrocognition: Linking cognitive psychology and cognitive ergonomics. In *Proceedings of the 5th International Conference on Human Interactions with Complex Systems* (pp. 173–177). Urbana-Champaign: University of Illinois at Urbana-Champaign, The Beckman Institute.

Klein, G. A. (1989). Recognition-primed decisions. In W. B. Rouse (Ed.), *Advances in man-machine systems research* (Vol. 5, pp. 47–92). Greenwich, CT: JAI.

Klein, G. A., & Crandall, B. W. (1995). The role of mental simulation in naturalistic decision making. In P. Hancock, J. Flach, J. Caird, & K. Vicente (Eds.), *Local applications of the ecological approach to human–machine systems* (Vol. 2, pp. 324–358). Hillsdale, NJ: Lawrence Erlbaum Associates.

Klein, G. A., & Hoffman, R. (1993). Seeing the invisible: Perceptual/cognitive aspects of expertise. In M. Rabinowitz (Ed.), *Cognitive science foundations of instruction* (pp. 203–226). Hillsdale, NJ: Lawrence Erlbaum Associates.

Militello, L. G., Hutton, R. J. B., Pliske, R. M., Knight, B. J., & Klein, G. (1997). *Applied cognitive task analysis (ACTA) methodology* (Tech. Rep. prepared for Navy Personnel Research and Development Center Contract No. N66001–94–C–7034). Fairborn, OH: Klein Associates Inc.

Pliske, R. M., Hutton, R. J. B., & Chrenka, J. E. (2000). *Weather information requirements for business jet pilots: A cognitive task analysis* (Final report under NASA Langley Contract No. NAS1–99158). Fairborn, OH: Klein Associates Inc.

Shanteau, J. (1989). Cognitive heuristics and biases in behavioral auditing: Review, comments and observations. *Accounting Organizations and Society, 14*(1–2), 165–177.

24

Recruitment and Selection in Hotels: Experiencing Cognitive Task Analysis

Julie Gore
Michael Riley
University of Surrey

Interest in the investigation of managerial cognition has been accruing across a range of research domains over the past few decades, offering an extension of researchers' understanding of managerial processes. Consequently, research communities are still seeking to develop and refine appropriate methodological tools to explore managerial cognition. Cognitive mapping instruments have been developed to assist our understanding of internal thinking processes (see Huff, 1994 for a useful review). Attributed to Tolman (1948), cognitive mapping is now more commonly used as an "umbrella term" that covers methodologies, which explore how individuals make sense of their world. Thus, assisting our understanding of cognition is perception, thinking, problem solving, and decision making.

In general, research enquiry about the United Kingdom hotel and hospitality industry and its participants is steadily accruing, although in a fairly haphazard fashion. Its development has often been reported as being hindered by a lack of understanding and communication between academics and practitioners. Gaining access to key participants within organizations is problematic, and using methodologies that are deemed useful for hotel practitioners are seen to be of key importance (Brotherton, 1999).

The success of methodological techniques such as cognitive mapping have been reported more recently by Eden and Spender (1998), as they offered innovative ways of aiding our understanding of complex management activities. They appear however, often difficult to use. Jenkins (1998) suggested that to overcome this problem, cognitive research design should take issues such as epistemology, validity, reliability, and practicability into account. In other words, the theoretical basis for representing cognition should be considered. The methodology should capture issues that are salient to the participant. As far as possible, both the researcher and participant should be free from systematic bias, and finally, the research methods used should be efficient and challenging rather than time consuming and irritating. Jenkins (1998) argued that these are important considerations, especially as they allow management researchers to build the sort of relationships that are needed within the management community.

The development of task analytic techniques, which have been crucial to the development of training, concur with the development of research focusing on managerial cognition. Essentially, cognitive task analysis (CTA) methods primarily attempt to identify how experts perform a cognitive task. Historically, however, CTA does not have one single and well-accepted definition. In addition, knowledge elicitation methods vary greatly. The most frequently used methods include structured and semistructured interviews, group interviews, verbal think-aloud protocols, retrospective verbal protocols, analysis of previous incidents, and observation of task performance. Each of the methods have had some success using realistic problem-solving and decision-making tasks, many experts and many tasks, or many different scenarios revolving around the same task. A key concern here is that CTA methods have been criticized, as they are often difficult to use, very time consuming, and result in problematic data analysis.

In direct response to the preceding difficulties, applied CTA (ACTA) has been developed by Klein Associates (Militello, Hutton, & Miller, 1997) who are interested in making cognitive mapping techniques more accessible to practitioners across a range of fields.

Militello and Hutton (1998) detailed the ACTA techniques and their development, which were part of a project funded by the Navy Personnel Research and Development Center. They suggested that as task analytic techniques have begun to focus on both cognitive and behavioral elements, they have become less accessible to practitioners. ACTA, therefore, was designed to assist professionals who may benefit from the use of CTA. Militello and Hutton (1998) also provided a useful evaluation of the ACTA technique, reporting its ease of use, flexibility, and provision of clear output.

Like CTA, the ACTA technique is intended to assist the identification of key cognitive elements required to perform a task proficiently. The cognitive requirements that CTA address are difficult judgments and decisions, attentional demands, identifying critical cues and patterns, problem-solving strategies, and other related topics. ACTA includes both knowledge elicitation and knowledge representation techniques. Knowledge elicitation techniques involve the use of interviews (and sometimes observation), whereas knowledge representation techniques are a means to depict cognitive information (cognitive mapping).

The ACTA techniques were developed to compliment each other, each tapping into different aspects of cognitive skill. The first technique, the task diagram interview, gives an overview of the task, highlighting cognitive difficulties that can be explored in greater detail later. This provides the interviewer with a broad overview of the task. The second technique, the Knowledge Audit, reviews the aspects of expertise required for an explicit task or subtask. The audit stems from the research literature on expert–novice differences and critical decision method studies. As the aspects of expertise are elicited, they are individually probed for further detail and concrete examples associated with the task are investigated. This technique also encourages the interviewee (subject matter expert) to identify why elements of the task may present a problem to inexperienced individuals. The Knowledge Audit has been developed to capture key aspects of expertise, improving and streamlining data collection and analysis. The third technique, the simulation or scenario, contextualizes the job/task that is difficult to obtain with the other techniques. It allows the interviewer to explore and probe issues such as situation assessment, potential errors, and how a novice would be likely to respond to the same situation. Last, a cognitive demands table is suggested as a means to merge and synthesize data. This is intended for practitioner use to focus the analysis on the project aims and outcomes.

Taking on board the associated difficulties highlighted by Jenkins (1998) about researching managerial cognition, and Brotherton's (1999) concerns about researching hospitality management, ACTA appears to be a realistic way forward. The ACTA techniques are concerned with epistemology, the study of expert knowledge and how it is formed; face validity also seems evident, as does practicability. The reliability of the techniques, however, have not been rigorously investigated. Having a firm theoretical base, the ACTA techniques, their purpose, and their use are clearly defined. The approach, therefore, seems to offer a systematic package for practitioners to elicit expertise and map and communicate cognition.

We anticipated that ACTA would be a challenging method, which would have credibility and face validity for its participants and researchers. Having little prior experience of CTA, we studied the compact disc ACTA multimedia training tool (Militello et al., 1997). This provided a comprehensive overview of CTA, detailing examples of the three ACTA methods and provided information about how to produce a cognitive demands table.

The purpose of this study was to experience and assess the usability of the ACTA techniques with hospitality practitioners. The task of recruitment and selection was chosen for analysis, as it is a complex management activity, and further systematic exploration of this area is of interest to academics and practitioners alike. It was envisaged that accessing experts experiences in this area would provide useful material for education and training purposes.

METHOD

Participants

Five hotel human resource professionals with an average of 14 years' domain experience participated in the investigation, following Militello et al.'s (1997) recommendations that three to five subject management experts (SME) are suitable for this type of task analysis.

Procedure

Each of the stages of the ACTA were tape-recorded, and the total structured interview time for each participant was approximately 3 hr. The human resource managers—the SMEs—completed the following:

1. A task diagram, that is, a perceptual map of the recruitment and selection process for the hotel industry.
2. A Knowledge Audit, that is, detailed the experts' perceptions of the recruitment process within the labor market environment.
3. A simulation interview, that is, a simulation/case scenario of a recruitment process, detailing cognitive information by looking at the task of recruitment in context.

Developing the Simulation Scenario

The development of a scenario for Stage 3 of the ACTA (see Table 24.1), which would be challenging for the participants, entailed detailed discussions with hotel Human Resource Directors, utilizing our knowledge and experience of the hotel labor market (Riley, 1991). We also piloted several scenarios with academics and practitioners linked to the United Kingdom hotel industry.

Developing an appropriate scenario was not as difficult or time consuming as we had envisaged. We were aware, however, of not answering the question that Hoffman, Crandall, and Shadbolt (1998) and Militello and Hutton (1998) have stressed, that is, we did not empirically verify the scenario knowledge base.

TABLE 24.1
Simulation Scenario

The environment in which today's policymakers in the hospitality industry operate is changing. The old adage "change is here to stay" epitomizes the workplace over the last three decades. Market forces have forced many organizations to radically review their employment strategies and the "flexible workforce" continues to be an integral part of the hospitality industry. Numerical flexibility has increased over the past 5 years, as has the use of contracting out, temporary agency workers, and the use of fixed-term and part-time employees.

The Human Resource Manager of a large hotel located at one of the United Kingdom's most profitable airports has recently been forced to implement a series of new recruitment initiatives. Obtaining and retaining housekeeping staff has been an ongoing and problematic concern. The work is very seasonal and coupled with zero levels of unemployment in the immediate area has meant a radical re-think of recruitment strategy. In the past few weeks, 40% of the staff have left, opting to work for 50 to 80 pence extra per hour in other hotels or retail outlets.

Imagine you are the HR Manager. What would you do?

FINDINGS

Figure 24.1 provides an illustration of one of the SME's overviews of the recruitment and selection process. The human resource directors each provided very similar overviews of the task, segmenting the task into between four and five subtasks. The cognitive map then provided a guide to focus on elements of the task that require cognitive skill. We found that the participants were eager to assist with the process, and they also felt that the production of the diagram really focused their attention.

Only 1 of the 5 participants found this first stage of the ACTA difficult, wanting to provide far more information than was required at this stage. Some time spent redirecting her, however, provided a clearer overview.

Table 24.2 provides an example of a knowledge audit. The Knowledge Audit provides an example of the recruitment process based on a real experience. As Militello and Hutton (1998) noted, this process is derived from the literature on expert–novice differences. We found that although the audits did not provide as much rich detail as other methods (as warned by

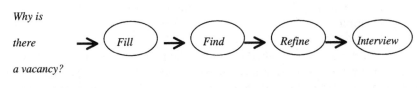

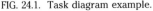

FIG. 24.1. Task diagram example.

TABLE 24.2
Example of a Knowledge Audit Table

Aspects of Expertise	Cues and Strategies	Why Difficult
Past and future		
For example, recruiting a waiter/waitress	Looking for the right attitude is essential	Hard to judge
Big picture		
Strategy, future long-term plan, jigsaw	Ask why, when, what's happening about it; contact person who is leaving and why	Must be able to adapt, difficult to step back
Noticing		
Body language, eye contact	Gut reaction "kind of know"—Rapport, communication, personality	Easier with experience, can get it wrong, intuition

TABLE 24.3
Example of a Simulation Interview Table

Events	Actions	Situation Assessment	Critical Cues	Potential Errors
Identifying how many have left and/or are needed	Decide where to recruit; identify standards (of room); think about redeployment	Look at projected business levels; assess hours; analyze rates of pay in the area	Look at peaks and troughs	Misjudge; not analyzing the job market enough; for example, How competitive are we?
Recruit/interview	Define person specification/job description	Maybe look at alternative markets	Appoint the right caliber of person	Lack of suitable applicants
Induct, train				

Militello & Hutton, 1998), for example, critical incident technique (Flanagan, 1954) or the critical decision method (Klein & Calderwood, 1989), the audits did provide very useful illustrations of task-specific expertise. What was also very useful here was that the participants instantly recognized how the representation of experience using the audit table could be beneficial to them when coaching or training new human resource personnel.

An example of the data that was illustrative of the simulation interview, Stage 3 of ACTA, is provided in Table 24.3. We found that collecting this data required a more detailed understanding of the task area. Probing the participants for further detail required two interviewers. It was interesting here that the SME's ability to innovate differed quite markedly between participants. The difficulty here was that the SMEs began to seek confirmation that they were giving the "right" answers. Although they were more than happy to suggest what they would do, at great length, they were more interested in other experts' solutions to the scenario. On reflection, because the scenario was seen by the SMEs to be of current critical importance, at times they were reluctant to be probed for further detail.

The use of the cognitive demands table (Table 24.4) assisted the analysis and sorting of the data. The practitioners were all very impressed by the output of this analysis, as it provided them with a clear framework from which to build new training awareness procedures for recruitment and selection. Reoccurring themes in the data were easily combined.

After completing the task analysis, each of the SMEs was asked to comment on the process and asked to provide a subjective evaluation. Each participant expressed that they had enjoyed the process of reflection on a crucial task and would feel comfortable using both the techniques and their outputs. We also found the techniques fairly practical and easy to use, with the exception of developing a suitable simulation scenario.

ACTA had great face validity with the human resource directors. The techniques were very relevant to their needs as practitioners and made efficient use of their time and therefore address some of the concerns about researching managerial cognition raised by Jenkins (1998).

The subjective evaluation reported here is limited; however, it does provide a real-world task from which more rigorous evaluations of ACTA may be completed. As Militello and Hutton (1998) stated, no well-established metrics to evaluate the reliability and validity of ACTA methods exist. They also suggested, and we agree, that much more research evaluating the

TABLE 24.4
Example of a Cognitive Demands Table

Difficult Cognitive Element	Why Difficult?	Common Errors	Cues and Strategies Used
Knowing where to find suitable candidates	Novices may not be trained to seek out alternative labor sources / Not everyone knows about the size or nature of the labor market	Novices would be likely to advertise positions where they had been placed previously	Start by reviewing the vacancy; consider the internal and external environment / Refer to external agencies for advice / Consider competitors, for example, other hotels, retail outlets

ACTA methods needs to be completed. In addition, given the findings reported here, it would be useful if further research commented on the difficulties or ease associated with the use of the techniques, in particular, the development of simulations. This small research exploration created a great deal of interest from the participants, and it is our aim to capitalize on this by exploring a range of tasks completed by human resource directors with the aid of ACTA.

REFERENCES

Brotherton, B. (Ed.). (1999). *The handbook of contemporary hospitality management research.* Chichester, England: Wiley.

Eden, C., & Spender, J. C. (1998). *Managerial and organizational cognition.* Newbury Park, CA: Sage.

Flanagan, J. C. (1954). The critical incident technique. *Psychological Bulletin, 51*(4), 327–359.

Hoffman, R. R., Crandall, B., & Shadbolt, N. (1998). Use of critical decision method to elicit expert knowledge: A case study in the methodology of cognitive task analysis. *Human Factors, 40,* 254–276.

Huff, A. S. (1994). *Mapping strategic thought.* Chichester, England: Wiley.

Jenkins, M. (1998). The theory and practice of comparing causal maps. In C. Eden & J. C. Spender (Eds.), *Managerial and organizational cognition* (pp. 231–245). Newbury Park, CA: Sage.

Klein, G. A., & Calderwood, R. (1989). Critical decision method for eliciting knowledge. *IEEE Transactions on Systems, Man and Cybernetics, 19,* 462–472.

Militello, L. G., & Hutton, R. J. B. (1998). Applied cognitive task analysis (ACTA): A practitioner's toolkit for understanding cognitive task demands. *Ergonomics, 41,* 1618–1641.

Militello, L. G., Hutton, R. J. B., & Miller, T. (1997). Applied cognitive task analysis [Computer software]. Fairborn, OH: Klein Associates.

Riley, M. (1991). An analysis of hotel labour markets. In C. Cooper (Ed.), *Progress in tourism, recreation and hospitality management* (Vol. 3, pp. 232–246). London: Belhaven Press.

Tolman, C. (1948). Cognitive maps in rats and men. *Psychological Review, 55,* 189–208.

25

Eliciting Knowledge From Military Ground Navigators

Barry Peterson
Jason L. Stine
Rudolph P. Darken
Naval Postgraduate School

Over the course of our study of tactical land navigators, we discovered the naturalistic decision making (NDM) framework and immediately valued its relevance to our project (Klein, Orasanu, Calderwood, & Zsambok, 1993; Orasanu & Connolly, 1993; Zsambok, 1997). Yet in the literature, we found only a handful of case studies (Brezovic, Klein, & Thordsen, 1990; Flin, Slaven, & Stewart, 1996) to guide us through the mechanics of applying the NDM framework to the conduct of a cognitive task analysis (CTA) of an actual domain. As pointed out by Hoffman, Crandall, and Shadbolt (1998), there is a lack of evaluative data in the literature, partially caused by a lack of publicly available methodological descriptions. This chapter marks our contribution to help remedy this shortfall.

We identify three areas most valuable to readers interested in conducting CTA to support training. First, we describe the knowledge elicitation method we used. Through this description, we document a case study to benefit others who plan to elicit knowledge in domains that share characteristics of the land navigation task domain. In this way, we contribute our experience to a small number of related infantry navigation studies (DuBois & Shalin, 1995; Klein & Wiggins, 1999). Second, we represent the product of our knowledge elicitation. This representation serves to both describe domain expertise and expose the chosen representational formats for review. Finally, the ultimate purpose of this project is to consider approaches to improve the ways navigators become proficient and expert (Dreyfus, 1997). We conclude with a discussion about how our current knowledge elicitation and representation efforts have affected our future plans regarding the design of virtual training applications.

METHODS

We were attracted to NDM because it respects the influence of the environment on performance and emphasizes study of performance out in the field in the environment in which it naturally takes place (Zsambok, 1997). Furthermore, the NDM framework emerged from inquiry into domains that were rather impervious to more traditional methods (Orasanu & Connolly, 1993; Zsambok, 1997). The profile of such domains defines properties of task and environment: The task is performed in dynamic, uncertain environments; the performer has to adapt to shifting, ill-defined, competing subgoals; the performer must continuously monitor and integrate environmental feedback from his or her actions; the performance conditions are high-tempo and high-stress; there are high stakes placed on performance quality; there are multiple players involved; and there are organizational goals and norms (Orasanu & Connolly, 1993; Zsambok, 1997). These properties matched our domain properties quite closely.

We selected the critical decision method (CDM) to guide our knowledge elicitation (Hoffman et al., 1998; Klein, Calderwood, & MacGregor, 1989). Our reasons for initially selecting the CDM are closely aligned with the reasons Klein, Calderwood, and MacGregor (1989) developed it. We wanted to study the skill in its natural setting but could not justify the intrusion of direct observation methods.

We interviewed each expert within an hour of his completion of the specific patrol incident. Our two-member team planned to conduct each individual interview within 75 min, but actual times varied between 45 and 60 min each. Altogether, we held eight interviews. The first six were conducted indoors, and the last two were done at night in the field. The planned elicitation protocol closely matched that protocol described by Hoffman et al. (1998):

1. The interviewer orients participant to the patrol just completed as the patrol of interest.
2. The expert recounts the entire patrol.
3. The interviewer retells the story back to the expert.
4. The interviewer and expert collaborate to create a mission sketch and a timeline.
5. The interviewer asks probe questions to deepen his or her understanding of the expert's mental processes.

Our actual application of the CDM deviates from this textbook version. Our critique of the method we used yields lessons learned regarding our planning, conduct, and analysis of the interviews.

Planning

Per the definition of the CDM, we want to identify decisions that truly are critical. We benefited basing our interviews on recently completed decisions, but we did not verify that for the interviewee this particular patrol "presented a unique level of challenge" (Klein et al., 1989, p. 466). Also, if we were able to schedule more time with each expert during the single or multiple sessions, we might have probed more deeply.

We were interested in gathering pertinent data across all phases of the patrol because we know that activities conducted during the patrol's planning phase have a strong influence on the later success or failure of the patrol. However, by asking them to draw a map during the session, we unintentionally focused the interview on the travel phase exclusively, thereby limiting the discussion of key points such as planning events and actions on the objective.

Procedure

We intended to generate two artifacts during each interview. The first was the participant's sketch of the patrol; the second was a timeline with key decision points indicated on it. After attempts to produce both artifacts, we dropped the timeline and focused effort on the patrol sketch. We found the sketch to be a good tool for investigating the travel phase but not for the planning phase. It seems that the sketch afforded a focal object for the discussion, whereas the timeline served to scatter the discussion too much. We may consider creating a new tool to facilitate discussions of planning issues. We created a timeline for each interview by ourselves after the interview, so we must figure out a way to incorporate its creation into the interview session itself.

We were fortunate. Most of the interviews were conducted indoors. Those conducted outdoors enjoyed clear, dry, and warm weather. We were not adequately prepared to collect data in unfavorable conditions such as extreme cold or wet weather. The field interviews were quite different from those done indoors. Indoors, we had two butcher pads with easels and lighting. In the field, we lay in the dirt shining a flashlight on the student's map while scribbling notes on a clipboard. This method was effective in that it focused and sped the interview; unfortunately, we did not get as much detail nor did we obtain patrol sketches.

Analysis

It is necessary to review the interviews and create the necessary descriptions of them as soon as possible. Notes alone would have been insufficient aids for this analysis; we needed sketches and other artifacts. We also learned to become aware that further questions will likely emerge from early analysis of the interviews. We yielded to the temptation to add on to the original protocol a few questions to probe into these emergences. Such tinkering changes the nature of the interviews so they are in essence two different creatures. By not following this advice, we possibly corrupted data from the last two experts we interviewed.

RESULTS

We represent the results of the interviews in three ways. First, we provide high-level general results. Second, within this description, we identify the cues on which expert navigators rely. Third, we supplement the general model with specific excerpts to show how expert navigators assess situations and make decisions.

General Results

Analysis of the interview data showed some commonality across experts, as each expert described the use of four key mental processes: they rely on high-fidelity mental maps, they blend multiple cues, they adjust and re-calibrate tools dynamically, and they visualize spatial information.

During mission planning, the expert spends a great deal of time and effort creating a highly detailed, three-dimensional mental map. This is much more than simple memorization of the paper map, as the expert visualizes the terrain as it will appear, including features hinted at but not included on the paper map. The mental map includes vegetation, relative terrain elevation, roads by type and quality, streams and lakes, and any man-made structures. During the execution phase, experts refer to the paper map only in extreme cases and trust the mental map completely. The mental map focuses on the planned route and builds from there. The route includes a general compass azimuth and measured distance, and indeed, this is exactly how the expert describes it to other patrol members. However, the expert views the route more as a corridor than as a line on the map, with key terrain features on each side serving as lateral boundaries. All experts include easily recognizable checkpoints to separate route legs and emphasize these to other patrol members. Routes always incorporate all key terrain features in visual range, including roads and streams, man-made objects, and changes in elevation.

While walking, experts process information from the environment and compare it to the mental map. They use selected cues from four categories and assign relative weights to each based primarily on environmental conditions. However, it seems that different experts may use significantly different relative weights, even in the same environment. Personal preferences, training, experience, or some other factors may cause these differences, which may be noted with further research. The major cues used are terrain features, compass azimuth, and pace count. The expert always monitors all three. However, his or her degree of reliance on each one is based on weather, vegetation, light, and visibility conditions. The minor cue categories are tactical and mission considerations. These affect the expert in more subtle ways but are still considered continuously. Adapted from Hoffman et al. (1998) we provide a cue inventory in Table 25.1, with cues grouped by category.

Experts can dynamically calibrate and correct navigation tools. They keep a pace count to estimate distance traveled and frequently check a magnetic compass heading. If either of these provides information in conflict with the mental map, the expert can approximate the error and recalibrate the tool on the fly. Experts frequently measure, over a known distance (usually 100 m), how many steps they take to cover that distance. However, pace counts may vary widely due to fatigue, visibility, and rough terrain. Experienced navigators can factor these into the pace count. Experts process pace count information, compare measured distance traveled with mental map distance, and adjust the pace count to correct any discrepancy. As the navigator walks, he or she cannot follow a straight azimuth but must move around trees, lakes, boulders, and other obstacles.

TABLE 25.1
Cue Inventory Grouped by Category

Category of Cue	Cue
Navigation tools	Compass azimuth
	Pacecount
	Paper map
	Mental map
Environmental conditions	Ground slope
	Vegetation
Mission conditions	Time
	Input from other patrol members
Terrain features	Road
	Body of water
	Topography
	Man-made feature

Moreover, unexpected enemy contact may require the patrol to deviate from the planned route. In these cases, experts can mentally compute new azimuth headings and implement them without stopping. Experts view a halt for a paper map check as an abject failure. Frequent stops lower the patrol's confidence in the navigator and thus overall morale. Dynamic adjustment allows experts to minimize these stops. Although the organizational expectation is minimal map checking, experts hold themselves to a higher standard—zero paper map checks.

From map study alone, experts can visualize three-dimensional terrain. They can also, while walking, visualize how real terrain would look on a two-dimensional paper map. These two related skills are vitally important characteristics of good navigators. The first, known as map-to-ground, is primarily important in the planning phase, as it allows the expert to create his or her detailed mental map. It enables the navigator to select the proper route and create a useful terrain model to explain the route to the rest of the patrol. Conversely, experts use the second skill, ground-to-map, during the execution phase. It allows comparison of real terrain to mental map and allows the expert to make necessary azimuth adjustments dynamically. This is a continuous process and is beyond the capabilities of novice navigators. Experts mention the development of spatial visualization as a key element in the development of expertise because without it, the other skills cannot develop fully.

Situation Assessment Records

As described by Hoffman et al. (1998), a situation assessment record highlights the points at which the expert made a decision based on a revised assessment of the situation. After examining the example presented there and comparing it to the elements of the recognition-primed decision (RPD) model as diagrammed by Klein (1998), we saw an opportunity to use the RPD pattern to describe expert navigation and describe the situation assessment record. There are three different variations of the flow through the RPD model, and each variation is related to the decision maker's recognition of the situation. Variation 1 describes episodes in which the expert recognizes a typical situation. The fact that the situation is typical means the expert takes action immediately, without thinking; the recognition of the situation primes the appropriate action. Variation 1 typifies the quick and accurate behavior frequently associated with expertise (Dreyfus, 1997).

Sometimes, even experts are faced with situations that are not immediately recognized as being typical. Here begins Variation 2. During these episodes, the expert directs mental effort to the process of recognizing the specific cues and patterns that comprise a situation. Once the situation is recognized, then the expert takes action as in Variation 1.

Variation 3 begins with the recognition of a situation. However, unlike Variations 1 and 2, in these cases, the expert does not immediately know what to do. Mental effort is expended not on situation recognition but response evaluation. In some ways, the expert behaves as a competent performer would (Dreyfus, 1997). She or he must figure out what to do.

The expert's recognition of the situation is the key. This recognition generates four "by-products" (Klein, 1998) or types of mental constructs useful to the expert's future performance: expectancies, relevant cues, plausible goals, and typical actions. For a given situation, there are associated expectancies about what will happen next. Sometimes, these expectancies are expressed in terms of relevant cues. The expert attends to the relevant cues to confirm or disconfirm the expectancies of the situation. The violation of an expectancy often triggers a new situation assessment. Also, the situation defines which goals are plausible. Decision-maker attention to relevant cues and input from the organization can cause the relative importance of these goals to shift, and sometimes these shifts will generate a new situational assessment. Finally, to achieve the goals, the expert has a set of typical actions associated with each situation. In Variations 1 and 2, the expert implements one action from this set without evaluating each possibility (Klein, 1998).

We represent our characterization of expert land navigation performance using this model. Drawn from our interview data, we present one story, in situation assessment record format, to illustrate each of the three variations. In these stories, the relevant cues, plausible goals, and typical actions are all drawn from the standard sets, as listed in Table 25.2, using the structure described in Klein (1998).

All of our participants operated under Variation 1 conditions most of the time. They recognized the navigation situation as being typical, and they just continued to navigate—walking and scanning the environment. The record is shown in Table 25.3.

As shown in Table 25.4, our example of Variation 2 comes from a participant who was able to recognize his own error and correct it dynamically on the move without disrupting the flow of the patrol's movement. It is likely that the other patrol members were not even aware that the error occurred.

Our example of Variation 3 illustrates the navigator's response to the identification of an anomaly in expectations. It is presented in Table 24.5.

DISCUSSION AND FUTURE WORK

Within each organization, there can be a variety of motivations for expending the resources to conduct a CTA. Our motivation was to represent land navigation expertise in a way that would permit inexperienced navigators to learn from expert navigators. In essence, we saw the CTA as a tool that

TABLE 25.2

The Specific By-Products of Land Navigation Situation Assessment

---

The Standard By-Products of the Expert Navigator's Situation Assessment

---

Expectancies	*Relevant Cues*
Generated by evaluation of the situation with regard to the mental map	Selected from the cue inventory categories • Navigation tools • Environmental conditions • Mission conditions • Terrain features

---

Plausible Goals	*Typical Actions*

Selected from the list of standard goals
• Maximize speed
• Maximize stealth
• Minimize exertion
• Maintain orientation
	The standard typical action is one of three methods
	• Arrive at checkpoint method
	  1. Confirm checkpoint if needed
	  2. Reset pacecount
	  3. Change azimuth if needed
	• Confirm route method
	  1. Maintain pacecount
	  2. Maintain azimuth
	• Error recovery method
	  1. Confirm checkpoint if needed
	  2. Reset tools if needed
	  3. Map check if needed

TABLE 25.3

Situation Assessment Record for Recognition-Primed Decision Variation 1

Example of Variation 1: "I know the situation, therefore I know the course of action."
"Continue Mission"

Situation 1: On course between start point and Checkpoint 1
  Relevant cues: Standard
  Plausible goals: Standard
  Typical actions: Standard
  Expectations: Expect to begin by moving uphill, then we will cross a road
  Course of action: Arrive at checkpoint method
Situation 2: On course between Checkpoint 1 and Checkpoint 2
  Relevant cues: Standard
  Plausible goals: Standard
  Typical actions: Standard
  Expectations: After crossing the road, we will hit a draw; we will box around the draw;
    then we should cross a road
  Course of action: Arrive at checkpoint method

TABLE 25.4

Situation Assessment Record for Recognition-Primed Decision Variation 2

Example of Variation 2: "What is the situation?"
"I'll just do a quick dynamic pacecount recalculation . . ."

Situation 1: On course between Checkpoint 2 and Checkpoint 3
  Relevant cues: Standard
  Plausible goals: Standard
  Typical actions: Standard
  Expectations: We will cross another road; the pacecount here should be 300 m; we will
    next be able to see a hill; next, we should cross a major road
  Course of action: Arrive at checkpoint method
Situation 2: On course between Checkpoint 3 and Checkpoint 4
  Relevant cues: Standard
  Plausible goals: Standard
  Typical actions: Standard
  Expectations: We will identify a bend in the road at 450 m
  Course of action: Arrive at checkpoint method
Situation 3: On course between Checkpoint 4 and Checkpoint 5
  Relevant cues: Standard
  Plausible goals: Standard
  Typical actions: Standard
  Expectations: We will cross a major road at 1000 m
  Course of action: Arrive at checkpoint method
Situation 4: On course between Checkpoint 5 and Checkpoint 6
  Relevant cues: Standard
  Plausible goals: Standard
  Typical actions: Standard
  Expectations: We will cross two draws; then we will cross an improved road
  Anomaly: We crossed a secondary road at 100 m
  Diagnose:
  Feature matching: Traveling on the right compass heading; pacecount is 100 m; crossed
    road; have not crossed draws (matches these features to the features of his mental
    map)
  Story: On our last leg, we must have stopped too far north; then on this leg, we crossed
    the major road too far to the north/east
Situation 5: Off course between Checkpoint 5 and Checkpoint 6
  Goal: Reorient and compensate for error
  Course of action: Error recovery method; recalculate pacecount, mentally change the
    route
Situation 6: On course between Checkpoint 5 and Checkpoint 6
  Relevant cues: Standard
  Plausible goals: Standard
  Typical actions: Standard
  Expectations: We will cross two draws; then we will cross an improved road
  Course of action: Confirm route method

TABLE 25.5
Situation Assessment Record for Recognition-Primed Decision Variation 3

Example of Variation 3: "I know the situation ... what do I do about it?"
"Hey, we're too far from the checkpoint ..."

Situation 1: On course between Checkpoint 2 and Checkpoint 3
  Relevant cues: Standard
  Plausible goals: Standard
  Typical actions: Standard
  Expectations: Should enter triangular open area between roads; should then cross major road
 Course of action: Confirm route method
Situation 2: On course between Checkpoint 4 and Checkpoint 5
  Relevant cues: Standard
  Plausible goals: Standard
  Typical actions: Standard
  Expectations: Checkpoint 5 should be 400 m away on set azimuth
  Anomaly: At 400 m, did not hit Checkpoint 5
  Diagnose:
  Feature matching: Two hills, one to our left and one to our right; our compass heading is
    correct; pacecount is 400 m; estimate that Checkpoint 5 is still 300 m away distance
  Story: We must have misplaced Checkpoint 4; on our last leg, we did not go far enough
    east; that means we are 400 m west of our planned location
Situation 3: Erroneous Checkpoint 4
  Goal: Maximize stealth; minimize patrol confusion
  Typical actions: None
  Evaluate actions: Go back and move patrol's defensive position (mental simulation)
  Will it work: No ... patrol is preparing to engage the enemy
  Evaluate actions: Change route from Checkpoint 4 to Checkpoint 5 (mental simulation)
  Will it work: Yes, but ... patrol leader must be informed
  Evaluate actions: Change route and inform patrol leader (mental simulation)
  Will it work: Yes
  Course of action: Change route and inform patrol leader
Situation 4: On course between Checkpoint 4 and Checkpoint 5
  Relevant cues: Standard
  Plausible goals: Standard
  Typical actions: Standard
  Expectations: Checkpoint 5 should be 700 m away on new azimuth
  Course of action: Confirm route method

would allow us to "get inside" the heads of experts and to describe the environmental objects of their attention and their decision-making processes. In hindsight, we can identify three core themes that characterize our approach: (a) we wanted to minimize our intrusion into the participant's world, (b) we wanted to work only with experts, and (c) we assumed that we could adequately represent the knowledge that we elicited.

As mentioned earlier, one reason we selected the CDM method was that we felt an interview would be less intrusive than an observation-based method. In addition, we were able to conduct the elicitations relatively quickly. However, there were also a few disadvantages of following this route. Because the interview was conducted in a setting that was similar but not identical to the performance environment, it was not possible to capture the richness of the environment. We had no way to preserve the visual, auditory, or kinesthetic sensations that the expert actually experienced; instead, we had to settle for a verbal description of these naturally rich experiences. In addition, we met with each participant only once and we never observed his or her actual performance. Therefore, we were completely reliant on the self-descriptions; we had no way to validate these descriptions. Again, in the name of maximizing efficiency and minimizing intrusion, we attempted to simplify and streamline our access to the participants. We asked the unit to supply us with expert navigators, but we were unable to either control or evaluate directly the skill level of the navigators sent to us.

There were many reasons that we wanted to work with experts. Primarily, most of these reasons were based on the belief that experts have the most valuable knowledge, so the value of the knowledge we elicited would be highest if we selected them. In addition, this decision favorably constrained our method selection because the CDM works best with participants who are experienced enough to have a rich set of domain cases in their background. However, we now recognize two potential dangers underlying the decision to work with experts. The first is that experts may not necessarily be the best people to model. There are many factors to consider when profiling the study participants, and one of these is the individual student's skill level. There are cases in which a beginner student might be overwhelmed when attempting to learn from an expert; indeed, it might be more appropriate for a beginner to learn from an advanced beginner or an intermediate performer. Even if we were able to concretely identify the optimal skill level of our participants based on our student skill levels, it does not mean that we can easily, quickly, or accurately categorize his or her skill level. Both of these dangers dictate that before we direct our CTA to specific skill levels, we must build a model of skill development in the domain. Construction of this model is a viable goal for CTA efforts, but its completion should be considered a prerequisite to the actual CTA that informs the design of the training system.

Finally, we assumed that the knowledge we elicited could be adequately described using the representation methods available. Although the situation awareness records did provide a useful representation, there were some weaknesses associated with them. The most glaring issue involves the nature of the knowledge that expert navigators rely on; experts use their perceptual knowledge to quickly assess the situation. A text-based representation cannot capture the richness of the environmental cues that the expert uses. The knowledge representation must provide features that integrate visual images with the textual descriptions.

In summary, the CTA method we used allowed us to adequately describe a portion of the knowledge that an expert navigator relies on with minimal intrusion. However, the bottom line is that the results of this initial CTA effort are not substantial enough on which to base the design of a training system. Rather, this CTA is a point from which we can launch more detailed and exhaustive efforts. To increase the odds of success, it is important to spend ample time in the performance environment to gain a thorough understanding of the task domain. In addition, the goal of the CTA should be as clear and detailed as possible. The results of our elicitation and representation efforts led us to reconsider two elements of future work. First, we discuss the methods we plan to use for future knowledge elicitation. Next, we consider how to best use this knowledge to improve current training practices.

Knowledge Elicitation

The CDM was an appropriate method for us, but in our future work, we want to probe more deeply. Therefore, we are considering combining the CDM with direct observation methods, reminiscent of previous work done by Klein Associates in the study of tank platoon leaders (Brezovic et al., 1990). At an early stage of the project, we rejected direct observation primarily because we deemed it to be too intrusive and felt that we would not capture the underlying cognitive processes that control overt behavior. Also, we felt that interviewing would be more efficient in that we could mentally direct the expert to decisions that were critical rather than potentially spending hours in the field observing an expert perform relatively uncritical tasks that never fully demanded his expertise.

Training Navigators

One component of the envisioned training system was an apprentice–master relationship between the student and virtual coach. We still value this virtual apprenticeship model, but we now believe that students would benefit from applying that approach to traditional physical practice as well as virtual practice. Short of the whole-task virtual apprenticeship, we are considering two computer-mediated, part-task approaches to train the critical visualization skills. A system that focuses on the route planning phase, hence training map-to-ground visualization, could be driven by an executable representation that makes decisions based on the perception of two-dimensional map features. Stine (2000) built one route-planning prototype based on an adaptive agent representation. To train ground-to-map visualization, we envision a system that does not require an executable representation. The route execution system would provide virtual practice for the student as he or she virtually travels from checkpoint to checkpoint along a prescribed route. The computer could provide direct feedback by displaying geocentric map perspective views and egocentric navigator perspectives under the student's control. McLean (1999) already successfully prototyped a similar system for helicopter pilots.

As it should, this discussion circles back to the practical limitations on military training resources. If we had plenty of experts who were superb instructors and we provided them with the requisite time for training the apprentice, we probably would have no need to consider other alternatives. However, the reality of the situation is that there are few experts, and of them, even fewer are also expert instructors. The units who do have these experts certainly are unable to maximize the contact time between them and the students. Cast under the light of these practical constraints, virtual apprenticeships make more sense. The path to that place requires intensive and efficient study of expert navigators coupled with the appropriate representational forms. Perhaps a by-product of our studies will be our own successful navigation toward that objective.

ACKNOWLEDGMENTS

We acknowledge the professional soldiers whose cooperation and expertise made this research possible. We thank Commander of F Company, 1–1 SFTG (A), LTC William Banker, his chain of command, the instructors and students, Special Warfare Instruction Center, Fort Bragg, NC, for hosting the individual interviews. We thank Commander of the 4th Ranger Training Battalion, LTC Chinn, his chain of command and the instructors, U.S. Army Ranger School, Fort Benning, GA, for hosting the initial focus group that directed us to Fort Bragg. This research has been sponsored by the U.S. Office of Naval Research, Cognitive and Neural Science and Technology Division.

REFERENCES

Brezovic, C., Klein, G., & Thordsen, M. (1990). *Decision making in armored platoon command* (ARI Tech. Rep. No. KA–TR–858 (B)–05F). Alexandria, VA: U.S. Army Research Institute for the Behavioral and Social Sciences.

Dreyfus, H. L. (1997). Intuitive, deliberative, and calculative models of expert performance. In C. E. Zsambok & G. Klein (Eds.), *Naturalistic decision making* (pp. 17–28). Mahwah, NJ: Lawrence Erlbaum Associates.

DuBois, D., & Shalin, V. L. (1995). Adapting cognitive methods to real-world objectives: An application to job knowledge testing. In P. D. Nichols, S. F. Chipman, & R. L. Brennan (Eds.), *Cognitively diagnostic assessment* (pp. 189–220). Hillsdale, NJ: Lawrence Erlbaum Associates.

Flin, R., Slaven, G., & Stewart, K. (1996). Emergency decision making in the offshore oil and gas industry. *Human Factors, 38*, 262–277.

Hoffman, R. R., Crandall, B., & Shadbolt, N. (1998). Use of the critical decision method to elicit expert knowledge: A case study in the methodology of cognitive task analysis. *Human Factors, 40*, 254–276.

Klein, G. (1998). *Sources of power: How people make decisions.* Cambridge, MA: MIT Press.

Klein, G. A., Calderwood, R., & MacGregor, D. (1989). Critical decision method for eliciting knowledge. *IEEE Transactions on Systems, Man and Cybernetics, 19*, 462–472.

Klein, G. A., Orasanu, J., Calderwood, R., & Zsambok, C. E. (Eds.). (1993). *Decision making in action: Models and methods.* Norwood, NJ: Ablex.

Klein, G., & Wiggins, S. (1999, September). *Cognitive task analysis.* Paper presented at Workshop 6 of the 43rd Annual Meeting of the Human Factors and Ergonomics Society, Houston, TX.

McLean, T. D. (1999). *An interactive virtual environment for training map-reading skill in helicopter pilots.* Unpublished master's thesis, Naval Postgraduate School, Monterey, CA.

Orasanu, J., & Connolly, T. (1993). The reinvention of decision making. In G. A. Klein, J. Orasanu, R. Calderwood, & C. E. Zsambok (Eds.), *Decision making in action: Models and methods* (pp. 3–20). Norwood, NJ: Ablex.

Stine, J. L. (2000). *Representing expertise in tactical land navigation.* Unpublished master's thesis, Naval Postgraduate School, Monterey, CA.

Zsambok, C. E. (1997). Naturalistic decision making: Where are we now? In C. E. Zsambok & G. Klein (Eds.), *Naturalistic decision making* (pp. 3–16). Mahwah, NJ: Lawrence Erlbaum Associates.

There Is More to Seeing Than Meets the Eyeball: The Art and Science of Observation

Raanan Lipshitz
University of Haifa

Observational methodology contrasts with experimental and quasi-experimental methodologies inasmuch as it does not use the design and statistical controls that they employ to warrant the validity of research findings. Lack of controls is both a blessing and a handicap. On one hand, it makes observation the quintessential naturalistic methodology. On the other hand, it creates a dilemma: How can observational studies be performed rigorously without controls over the research setting? In this chapter, I address this question by drawing a distinction between two conceptions of human observers: *the observer as recorder*, who is required to emulate mechanical data collection devices by refraining as much as possible from bringing in his or her own interpretation into the observation process, and *the observer as interpreter*, who is encouraged to convey his or her interpretations provided he or she makes the process of moving from data to conceptualization transparent to others. The two conceptualizations entail different notions of "good data," validation of findings, and rigorous observation. Following a critique of the first conception in terms of arguments inspired by the second conception, I discuss how the latter can be practiced by decision researchers compatibly with scientific rigor. I begin with a vignette from a review by Thorngate (1978) of an edited volume of group decision making that has stuck in my mind ever since I came across it quite a few years ago:

A few years ago I had the good fortune to observe a respected zoologist ob-serving the social behavior of a colony of monkeys in the hills behind Berke-ley. She had been collecting notes on the colony for about five years. In a vain attempt to engage in interdisciplinary dialogue, I asked her if she was plan-ning any experimental manipulations of her subjects. She looked at me in stunned disbelief: "Why should I?" she said, "There is so much more to learn by watching!"

Thelma Rowell's epithet often haunts me, especially after reading such books as *The Dynamics of Group Decisions*. . . . Despite the inventiveness of the experiments reported in [this book] I strongly suspect that the laboratory methodology they employ has reached [its] limit. . . . Dynamics are studied with scant regard for group structure, contents, functions, or history. Deci-sions are studied with scant regard for realism, or for their relationships to previous or forthcoming decisions. Research ideas appear to be thralls of lab-oratory methods. Statistical tails appear to be wagging theoretical dogs.

I have no doubt social psychologists who study group dynamics by tradi-tional experimental and statistical methods can, and will, continue to do so for years to come. I shudder at the prospect. It would seem much more saga-cious to follow the lead of Thelma Rowell, to search before we research, to trust our senses as much as our designs. I think we are far more likely to gain understanding about group dynamics by watching how real groups handle their own problems in their own historical and normative context than by watching what ersatz do in the context of our laboratories. In short, I think so-cial psychology is in dire need of naturalists. . . . Experimentation often viti-ates understanding. (Thorngate, 1978, p. 457)

What made this episode into a sort of epiphany for me was on one hand Rowell's conviction that "there is so much more to learn by watching," to which I, as an admirer of Darwin, could easily relate, and the ready accep-tance by Thorngate (who is, I assume, an experimenter by training like myself) of this conviction, nicely put as the dictum "Search before [you] research, [and] trust [y]our senses as much as [y]our designs." Still, ow-ing to my training (or possibly to some character flaw), some doubts persistently clouded my delight in Thorngate's (1978) call for naturalists to step forward in psychological research. Does "trust[ing] our senses" mean that everything goes? Put differently, how can we make warranted asser-tions based on observational studies? What I wish to do in this presenta-tion is share with you some tentative conclusions to which my preoccupa-tion with this question has led me to date. These conclusions assume that rigorous research, irrespective of methodology, satisfies two principles—"collect good data" and "draw sound conclusions"—and basically unpack the implications of these principles for observational methodology. To be-gin with, I consider some general observations on the problem of rigor in this methodology.

RIGOR IN OBSERVATIONAL METHODOLOGY

According to Weick (1968), "The term 'observational methods' is often used to refer to hypothesis-free inquiry, looking at events in natural surroundings, nonintervention by the researcher, unselective recording, and avoidance of manipulation in the independent variable" (p. 362). Lack of controls is both a blessing and a handicap. On one hand, it makes observation the quintessential naturalistic methodology. On the other hand, it creates a dilemma: How can observational studies be performed rigorously without imposing controls that vitiate its basic nature?

The key to this question is the distinction between the observer as recorder and the observer as interpreter. Observation inevitably involves interpretation, and observation rests more on rigorous interpretation than on faithful collection of objective (i.e., uninterpreted) facts. These are not new ideas, as the numerous sources on which I rely show. My modest contribution may lie in pulling them together and working out their implications for naturalistic decision making (NDM) methodology, thereby calling the attention of researchers within this framework to the advantages of rigorous naturalistic observation. My presentation consists of four sections: the observer as a recorder, the observer as an interpreter, a short case analysis of an observational study of decision making, and a conclusion. In the first section, the observer as recorder, I describe the approach to observational methodology adopted (I think) by most NDM observational studies.

THE OBSERVER AS A RECORDER

A distinctive feature of observational methodology is that the researcher is the instrument of data collection. This aspect of observation, which I term the observer as a recorder, accentuates the weaknesses of systematic observation as a means for fulfilling the first principle of good research, namely, Collect good data. To quote Weick (1968) once again, "Observational methods are more vulnerable to the fallibilities of human perceivers than almost any other method" (p. 428). "All products of observation involve simplification, editing, imposed meaning, and omission. Because of this selectivity, all observers essentially gloss what they observe" (Weick, 1985, p. 569).

For the reasons noted by Weick above, unaided observation (and its close kin, the unstructured interview) are considered weak (or soft) methodologies. Solutions for this problem typically attempt to minimize the observer's contribution to the recorded datum. "In any given study, between the record and the behavior it is supposed to represent should be interposed only the most primitive act of judgment or discrimination possible—the one needed to perceive whether the behavior has occurred or not" (Medley & Mitzel, 1963, p. 263). Following this line of reasoning, Weick (1968) advocated the use of "simple, unequivocal behavioral indices [i.e., checklists] ... [which] require only a judgment of present-absent, ... are easy to grasp [i.e., easy to train to a high degree of interjudge agreement], and do not impose excessive demands on the observer" (p. 433). Alternatively, Weick (p. 432) suggested dividing the task of observation among multiple observers, each watching a part of the phenomenon of interest. Thus, "Sherif and Sherif (1964) and Haeberle (1959) reduced the memory loss by having observers focus on one small portion of the problem at a time.... The general rule of thumb concerning reliability of categories is summarized by Gellert: The fewer the categories, the more precise their definition, and the less inference required in making classifications, the greater will be reliability of the data (Gellert, 1955, p. 194)" (Weick, 1968, p. 403).

It is fair to say that the ideal human observer-cum-recorder mimics mechanical data collection instruments that provide identical readings under identical situations. Employing interjudge reliability to gauge the credibility of an observation system basically extends Turing's test to the realm of observation: an observation system consisting of several human and nonhuman observers is perfectly calibrated if it is impossible to determine, solely from the system's output, which observer is the source of which observation.

The premise underlying the notion of the observer as a recorder is that human observers are valuable research tools to the extent that they faithfully collect and report "raw data," namely uninterpreted objective givens about the phenomenon under investigation. The limitations of this premise, which is particularly suitable for quantitative, hypothesis-driven research, are revealed by the alternative notion of the observer as an interpreter.

THE OBSERVER AS AN INTERPRETER

The basic premise underlying the observer as an interpreter is that observation and interpretation (and more generally, data and theory) are inextricable:

> I contend that observation and interpretation are inseparable—not just in that they never do occur independently, but rather in that it is inconceivable that either could obtain in total isolation from one another.... The Neo-positivistic model of observation—wherein our sensational data-registration and our intellectual constructions thereupon are cleft atwain—is an analytical stroke tantamount to logical butchery. (Hanson, 1964, p. 1)

"To observe X is to observe X as something or other" (Hanson, 1964, p. 2):

> Enter a laboratory; approach the table crowded with an assortment of apparatus, an electric cell, silk-covered copper wire, small cups of mercury, a mirror mounted on an iron bar; the experimenter is inserting into small openings the metal ends of ebony-headed pins; the iron oscillates and the mirror attached to it throws a luminous band upon a celluloid scale; the forward-backward motion of this spot enables the physicist to observe the minute oscillations of the iron bar. But ask him what he is doing. Will he answer "I am studying the oscillations of an iron bar which carries a mirror?" No, he will say that he is measuring the electrical resistance of the spools. If you are astonished, if you ask him what his words mean, what relation they have with the phenomenon he has been observing and which you have noted at the same time as he he will answer that your question requires a long explanation and that you should take a course in electricity. (Pierre Duhem, as cited in Hanson, 1964, p. 6)

Duhem (Hanson, 1964) contrasted two factual (i.e., minimally interpreted) reports of what goes on in a particular laboratory. One report is provided by a naive (i.e., nonphysicist) observer, the other by a knowledgeable observer (a professional physicist). The naive report is more factual inasmuch as making such reports or making sense of such reports requires only a knowledge of the English language. Making (and particularly making sense of) the physicist's report requires, in addition, a rudimentary knowledge of physics. Yet it is the less factual report that is more informative or, indeed, that is worthy of the label *meaningful*. Conceptualizing human observers uniquely as recorders and applying Turing's test to assess their value is misguided. In the final analysis, expert human observers are valued not for mimicking mechanical devices but for their ability to (a) note details (raw data) that others cannot spot and (b) their ability to "see the invisible" (Klein & Hoffman, 1993), that is, their ability to perceive meanings and offer interpretations that others cannot. The fact that unlike mechanical data collection instruments human observers inevitably make sense of the data that they collect and that there lies their principal contribution has been pointed out by Bloor (1978) in connection with an observational study of decision making:

> The initial research task was the accomplishment of descriptions of each specialist's assessment practices. These ethnographies cannot be compendia of specialists' actions and utterances in the clinic setting. I say "cannot" both because such compendia would be very deficient since all observation is, by its very nature, selective, and because such compendia—devoid of any attribution of meaning to those actions and utterances—would be a mere babble, a senseless jumble. What is required is a description of specialists' assessment

practices informed by a knowledge of specialists' definitions of the situation and of their practical purposes at hand. (p. 546)

So far, I have argued that pure factual reports devoid of conceptualization or interpretation are either impossible or not meaningful. According to Kirk and Miller (1986), under certain conditions they can actually be misleading:

> Human beings do not simply perceive, then interpret, but rather go through a process called cognition. The normal adult human is not ordinarily fooled by his or her visual perspective into thinking people going toward him or her are growing taller, or that a disc seen from an angle is elliptical. Prior to interpreting cognitive experience, people match visual and other input with stored percepts in particular ways. This is to say they actually require a theory (e.g., of stimulus constancy) in order to be able to see an object as approaching rather than growing. . . . Reliability, then—like validity—is meaningful only by reference to some theory. The implicit theory that requires all observations be identical is . . . appropriate . . . only in artificial experimental situations. (p. 50)

In conclusion, obtaining good data by constraining observers to a simple yes or no with respect to predetermined features of the phenomenon under observation creates several problems.

1. Rather than eliminating interpretation, it imposes on the situation preconceived interpretations embedded in the predetermined categories supplied to the observer.

2. It ignores the fact that the forte of human observers is the ability to take advantage of serendipity, which lies at the heart of all true discoveries. Because they make sense online of the data that they collect, human observers can shift their focus from data that they set out to collect or subjects that they set out to explore to vicariously noted data that they should have planned to collect or subjects that they should have set out to explore.

3. Yes–no categories (and frequency counts based on such categories) cause serious difficulties in understanding their meanings or evaluating their interpretation. Proper evaluation of the credibility of assertions based on observational data requires independent observers or readers of the research report to understand precisely what the observer saw, that is, what they consider to be the "bare facts" from which they construe their interpretations. In this way the audience can (a) evaluate the researcher's move from facts to explicit interpretations and the scope of these interpretations beyond the context in which they were made, (b) derive and evaluate the implicit interpretations embedded in the researcher's factual accounts, and (c) draw alternative conclusions and interpretations.

In conclusion, the methods of obtaining and the nature of what constitute good data according to the observer as a recorder and the observer as an interpreter are radically different. The former constrain the observer to producing thin, information-poor, categorical reports that are amenable to quantitative analysis and statistical hypothesis testing. The latter require the observer to provide "thick," namely, detailed and concrete descriptions, that allow the separation of low-level factual statements from high-level conclusions and interpretations. One way of obtaining such good observational data is to use video recording, as Bloor (1978) did in the passage quoted earlier. Another, less trustworthy but more flexible method is to record data in the form of anecdotes, specimen records, or observational notes. "Anecdotes are detailed, objective statements of settings and actions that can be classified, quantified, and organized to test hypotheses.... Specimen records are 'sequential, unselective, plain narrative description of behavior with some of its conditions' " (Weick, 1985, p. 593).

Observational notes are the "hard evidence," containing as little interpretation as possible, specifying who said or did what under stated circumstances. "Writing in concrete language is difficult because most people have had years of training to condense, summarize, abbreviate, and generalize. ... In writing up field notes we must reverse this deeply ingrained habit of generalization and *expand, fill out, enlarge*, and give as much specific detail, as possible" (Spradley, 1980, p. 68, italics added).

So far, I have discussed the different implications of the notion of the observer as a recorder and the observer as an interpreter for the first principle of rigorous research, Collect good data. I turn next to the application of the second principle, Draw sound conclusions, to observational research, reframing it slightly to "Base inference on transparent, and plausible assumptions." Weick (1985) explicated this principle as follows:

> To deal methodically with validity is to make explicit one's status on those dimensions that are central to the assessment of credibility as defined by different audiences. In the case of an audience of scientists, observers provide evidence that displays their reasoning process including summary statements, supporting and non-supporting anecdotes, and information on the range of variation in the phenomena observed. (Agar, 1980, as cited in Weick, 1985, p. 593)

To the best of my knowledge, there is no established theory or methodology for evaluating the plausibility of assumptions or the adequacy of Weick's earlier recipe for demonstrating validity. In the absence of such methodology, it may be helpful to examine a particular example of how a researcher presents and establishes the plausibility of his interpretation. I chose Mehan's (1983) study of the decision process of placing handicapped students because it provides a particularly lucid demonstration of the two building blocks of rigorous interpretation: paraphrasing and theorizing. Mehan did not work within NDM (I am not familiar with an NDM study that applies them). However, his basic conclusion, that exploring social setting is sometimes more pertinent to understanding real-world decision processes than investigating processes of choice, is certainly compatible with it.

BUILDING BLOCKS OF RIGOROUS INTERPRETATION: PARAPHRASING AND THEORIZING

Mehan (1983) collected data from various sources including official records, interviews with teachers and school officials, and observations of various meetings. The bulk of his analysis refers to transcribed video recording of committee discussions regarding referral decisions of potentially handicapped students to various remedial programs. The data are the verbatim statements of committee members, including notations designating speech flow, intonation, and accentuation. These data are interpreted by a two-stage procedure consisting of paraphrasing and theorizing.

Paraphrasing

Paraphrasing is a process of pattern identification whereby significant aspects of the data are abstracted while being "disciplined by continual reference to concrete qualitative data" (Weick, 1985, p. 579). In other words, this is a process of interpretation (mapping raw data onto conceptual abstract categories) and allowing readers to judge the plausibility of this move for themselves. Mehan (1983) paraphrased the transcripts presented in his paper (but not here) as follows:

> In sum, the mother's and teacher's reports have the following features in common:
>
> 1. They were elicited.
> 2. They were made available by people who occupy either low status or temporary positions (in terms of institutional stratification and distribution of technical knowledge).
> 3. Their claims to truth were based on commonsense knowledge.
> 4. Their reports were based on direct albeit unguided or unstructured observations.
> 5. They offered contingent assessments of student performance.
> 6. They resulted in a context bound view of student disability.

By contrast, the psychologist's and the nurse's reports had the following features in common:

1. They were presented, not elicited.
2. They were presented by people who occupy high status and permanent positions.
3. Their claims were based on technical knowledge and expertise.
4. They were based on indirect albeit guided or structured observations.
5. They offered categorical assessments of student performance.
6. They resulted in a context free view of student disability. (p. 203)

The validity of Mehan's (1983) paraphrasing can be established in a straightforward fashion. As each of the six contrasts between the two subgroups demonstrated by Mehan, the reader can (a) understand their relation to the data unambiguously and (b) if the reader wishes, compare their adequacy with alternative paraphrasing. The function of paraphrasing is also twofold: (a) to point to what is puzzling, interesting, or requiring explanation in the phenomenon under investigation, and (b) to prepare the ground for the explanation to be presented in the next stage, namely theorizing.

Theorizing

Theorizing is the stage at which the various strands identified in the process of paraphrasing are integrated into broader conceptual frameworks such as new or existing models or theories. Mehan's (1983) theorizing is illustrated as follows. "Both the professionals and the lay members make claims for the authority of their reports. The differences in the authority of the professionals' and lay members' recommendations are found in the differences in the structure of language used in assembling these two kinds of reports" (Mehan, 1983, p. 205).

In his theorizing, Mehan (1983) asserted that it is the linguistic aspect of the decision-making process rather than aspects traditionally considered by students of decision making that explains how the decisions that he studied were made:

There is a certain mystique in the use of technical vocabulary, as evidenced by the special language of doctors, lawyers, and businessmen in our society (Shuy, 1973; Philips; 1977; Shuy & Larkin, 1978). Technical language can be mystifying (Marcuse, 1964; Laing, 1967; Habermas, 1970). The use of technical language indicates a superior status and a special knowledge based on specialized qualifications.

When technical language is used and embedded in the institutional trappings of the proceedings of a meeting, the grounds of negotiating meaning are removed from under the conversation. . . . To interrupt, to question, to request a clarification of the psychologist, then, is to challenge the authority of the official position of the district and its representative concerning the child. The hearers are placed in the position of assuming the speaker is speaking knowledgeably, and they do not have the competence to understand.

We now return to the question that was raised at the outset of this paper: How is it arranged that committees of educators meet and make decisions without seeming to do so? ... The ambiguity of professional reports is not challenged because the obscurity of professional language removes the grounds for ... [deliberation]. ... The professional report gains its status and authority by virtue of the fact that it is obscure, difficult to understand, and is embedded in the institutional trappings of the formal proceedings of the committee meeting. (Mehan, 1983, pp. 207–209)

The frequent references to published research in Mehan's (1983) theorizing reveal that whereas validity checking at the stage of paraphrasing relies on the transparency and plausibility of the mapping of data to concepts, validity checking at the stage of theorizing relies on the transparency and accuracy of relating the emerging interpretation to existing empirical findings, theories, and conceptual frameworks. The validity of observation relies to a great extent on the validity of the conceptual lenses through which the observation is made and on the observer's skill in applying these lenses to the concrete situation that he or she is observing.

Mehan's (1983) example shows that because observation and interpretation are inextricable (as Hanson, 1964, pointed out) and because the distinction between factual data and theoretical interpretation is essentially an analytic device based on convention (as Weick, 1985, implied), rigor in observation is not to be obtained by curtailing observers' interpretation but by making their operation as explicit as possible. To this end, observers should allow their audience to understand (a) what they consider as data, (b) the method by which and the context within which data were collected, (c) the interpretations (i.e., conceptual conclusions or attributions of meanings) that they give to the data, and (d) the method of transition from data to theory. The two-stage sequence of paraphrasing followed by theorizing is one method for satisfying requirements (c) and (d).

The assertion that the four preceding requirements are basic for the rigor of observational studies is grounded in a general approach to the validation of qualitative studies whereby the consumer (or reader) of a research report, rather than the researcher, is responsible for establishing the validity of research outcomes (Firestone, 1993; Marshall, 1990; Mishler, 1990). Dyer and Wilkins (1991) explained the vital role of good (i.e., detailed and concrete) data in establishing validity this way:

The classic case study approach has been extremely powerful because it described general phenomena so well that others have little difficulty seeing the same phenomena in their own experience and research.... For example, it is the demonstration of informal status in the bowling matches and the personal feelings described by Whyte's (1943) [classic study *Street Corner Society*] as much as the theoretical statement of informal status that creates the impact.... More than once we had an "aha" experience when reading such studies because the rich descriptions have unveiled the dynamics of the phenomena and helped us identify similar dynamics in our own research or our daily lives. (p. 617)

In conclusion, obtaining good data, paraphrasing, and theorizing as explicated previously allow examination of the following questions. Are the data trustworthy and pertinent to the research question? Should additional or different data have been collected? Is the transition from data to conclusions and conceptualization credible? Are the researcher's conclusions and interpretations more plausible than alternative conclusions and interpretations? Are the assumptions underlying the whole enterprise plausible? A study is rigorous to the extent that it permits readers to answer these questions. Its conclusions are valid to the extent that the answer to these questions is affirmative. Needless to say, researchers who include persuasive affirmative answers to these questions in their reports do their research a good service.

Although this chapter is concerned with observational studies, it is fair to conclude that the process of criticism outlined previously is generally applicable to all research efforts:

When Kenneth Arrow was asked by George Feiwel what criteria he uses to judge competing theories in economics he answered "Persuasiveness." Does it correspond to our understanding of the economic world? I think it is foolish to say that we rely on hard empirical evidence completely. A very important part is just our perception of the economic world. If you find a new concept the question is, does it illuminate our perception? Do you feel it suits what is going on in everyday life? Of course, whether it fits empirical and other tests is also important. (McCloskey, 1985)

CONCLUSIONS

Observation is an ideal methodology for NDM inasmuch as it allows researchers to study their subject matter in situ. According to standard notions of rigor, observational methodology lacks rigor because it minimizes control by means of experimental design or statistical methods. In my presentation, I attempted to show that the notion that observational methodology is soft is misconceived. Observational studies should and can be held accountable to the same basic principles of rigor as studies using experimental or quasi-experimental designs. This, however, does not entail that these principles should be operationalized in the same fashion for the two methodologies. In fact, as I attempted to show, applying experimental or quasi-experimental standards of rigor to observational study leads to a narrow conceptualization of human observers as recorders, thereby depriving them of their principal advantage, namely the ability to interpret or literally "go beyond the data given."

What standards of rigor are appropriate for observational studies of decision making? A complete answer to this question exceeds the scope of this presentation (and, to be frank, my current state of knowledge). To start with, I venture some leads which hopefully are both plausible and useful:

• The standards should pertain to rigor in interpretation or argumentation. The notion of good data as rich data and the sequence of collecting good data, paraphrasing, and theorizing can serve as useful building blocks for conceptualizing rigor in observational studies.

• The standards appropriate for observation may turn out to be very different from traditional (i.e., positivistic) standards of rigor. A particularly intriguing difference pertains to the difference between discrepant observations and discrepant interpretations. Whereas the former are traditionally regarded as outliers, noise, or unreliable observations that reduce the validity of findings, the latter are regarded as marks of originality and perceptiveness that set the expert apart from the journeyman.

• The literature on naturalistic, qualitative, or inquiry-guided research is potentially a useful source for developing standards appropriate for NDM in general and observational studies of decision making in particular.

• An alternative method for developing standards of rigor for observational studies of decision making is fine-grained analysis of the differences between observational studies that seem to produce worthwhile, credible, findings and those that justify the suspicion that observation is an inferior methodology. In essence, this amounts to asking NDM researchers to apply to themselves the sine qua non of their own discipline, namely, learning how decisions are and ought to be made by studying how competent practitioners actually make decisions.

• Finally, because real-world decision makers typically develop (or fail to develop) accurate situation awareness by interpreting observational data, NDM studies of how real-world decision makers perform well and the models that they produce (e.g., Cohen's R/M model and its associated STEP procedure; Cohen & Freeman, 1997) can also serve as a source for standards of rigorous observation. This suggestion is not as odd as it may sound. As Phillips (1990) noted, "Recent work has shown that scientists, like workers in other areas, are in the business of providing reasonable justifications for their assertions, but nothing they do can make these assertions absolutely safe from criticism and potential overthrow" (p. 31).

REFERENCES

Agar, M. (1980). Getting better quality stuff: Methodological competition in an interdisciplinary niche. *Urban Life, 9*, 34–50.

Bloor, M. (1978). On the analysis of observational data: A discussion of the worth and uses of inductive techniques and respondent validation. *Sociology, 12*, 545–552.

Cohen, M. S., & Freeman, J. T. (1997). Understanding and enhancing critical thinking in recognition-based decision-making. In R. Flin, E. Salas, M. Strub, & L. Martin (Eds.), *Decision making under stress: Emerging themes and applications* (pp. 161–169). Aldershot, England: Ashgate.

Dyer, W. G., & Wilkins, A. L. (1991). Better stories, not better constructs: A rejoinder to Eisenhardt. *Academy of Management Review, 16*, 613–619.

Firestone, W. A. (1993). Alternative arguments for generalizing from data as applied to qualitative research. *Educational Researcher, 22*, 16–23.

Gellert, E. (1955). Systematic observation: A method in child study. *Harvard Educational Review, 25*, 179–195.

Habermas, J. (1970). Toward a theory of communicative competence. In H. P. Dreitzel (Ed.), *Recent Sociology #2 Patterns of communicative behavior*. New York: Macmillan.

Haeberle, A. W. (1959). Quantification of observational data in various stages of the research. *American Journal of Orthopsychiatry, 28*, 169–191.

Hanson, N. R. (1964). Observation and interpretation. *The voice of America forum lectures. Philosophy series, 9.*

Kirk, J., & Miller, M. L. (1986). *Reliability and validity in qualitative research*. Beverly Hills: Sage.

Klein, G. A., & Hoffman, R. R. (1993). Seeing the invisible: Perceptual-cognitive aspects of expertise. In M. Rabinowitz (Ed.), *Cognitive science foundations of instruction* (pp. 203–226). Hillsdale, NJ: Lawrence Erlbaum Associates.

Laing, R. D. (1967). *The politics of experience*. New York: Pantheon.

Marcuse, H. (1964). *The one dimensional man*. Boston: Beacon.

Marshall, C. (1990). Goodness criteria: Are they objective or judgment calls? In E. G. Guba (Ed.), *The paradigm dialog* (pp. 188–201). Newbury Park, CA: Sage.

McCloskey, D. (1985). *The rhetoric of economics*. Madison: University of Wisconsin Press.

Medley, D. M., & Mitzel, H. E. (1963). Measuring classroom behavior by systematic observation. In N. L. Gage (Ed.), *Handbook of research on teaching* (pp. 247–328). Chicago: Rand.

Mehan, H. (1983). The role of language and the language of role in institutional decision making. *Language and Society, 12*, 187–211.

Mishler, E. G. (1990). Validation in inquiry-guided research: The role of exemplars in narrative studies. *Harvard Educational Review, 60*, 415–441.

Phillips, D. C. (1990). Postpositivistic science: Myths and realities. In E. Guba (Ed.), *The paradigm dialog* (pp. 31–45). London: Sage.

Sherif, M., & Sherif, C. W. (1964). *Reference groups*. New York: Harper.

Shuy, R. (1973, August). *Problems of communication in the cross-cultural medical interview*. Paper presented at the American Sociological Association meeting, New York.

Shuy, R., & Larkin, D. L. (1978). Linguistic considerations in the simplification/clarification of insurance policy language. *Discourse Processes, 1*, 305–321.

Spradley, J. P. (1980). *Participant observation*. New York: Holt.

Thorngate, W. (1978). Will group dynamics research ever change? [Review of the book *Dynamics of group decisions*]. *Contemporary Psychology, 25*, 457.

Weick, K. E. (1968). Systematic observational methods. In G. Lindzey & E. Aronson (Eds.), *The handbook of social psychology* (Vol. 2, 2nd ed., pp. 357–451). Reading, MA: Addison Wesley.

Weick, K. E. (1985). Systematic observational methods. In G. Lindzey & E. Aronson (Eds.), *The handbook of social psychology* (Vol. 2, 3rd ed., pp. 567–644). Reading, MA: Addison Wesley.

Whyte, W. F. (1943). *Street corner society.* Chicago: Chicago University Press.

27

Using Observational Study as a Tool for Discovery: Uncovering Cognitive and Collaborative Demands and Adaptive Strategies

Emilie M. Roth
Roth Cognitive Engineering

Emily S. Patterson
Ohio State University

One of the primary strengths of naturalistic observations is that they support a discovery process (Mumaw, Roth, Vicente, & Burns, 2000; Woods, 1993). They serve to draw attention to significant phenomena and suggest new ideas whose validity and generality can then be evaluated through additional studies.

Field observations afford the opportunity to gain a realistic view of the full complexity of the work environment and empirically grounded hypotheses for how interventions could impact the nature of work in that setting. They enable researchers to uncover and document cognitive and collaborative demands imposed by a domain, the strategies that practitioners have developed in response to those demands, and the role that existing artifacts play in meeting domain demands. The results can be used to point to and guide the development of new types of support systems.

In this chapter, we use two studies to illustrate this approach. In the first case (Roth, Malsch, & Multer, 2001), a series of field observations and structured interviews were conducted at train dispatching centers to inform the design of a "data-link" technology intended to improve performance by reducing communications on an overloaded audio channel. In the second case (Roth & O'Hara, 1999, 2002), observations of the use of advanced human–system interfaces (HSIs) in a nuclear power plant simulator were conducted prior to implementation in the plant to uncover and document unanticipated changes in cognitive and collaborative demands as a result of the introduction of the new technology.

The studies illustrate the methods used in conducting and analyzing the results of observational studies, as well as the kinds of insights that can be gained from observational studies.

OVERVIEW OF METHODOLOGY

Naturalistic observation studies employ a methodology similar in approach to other ethnographically derived methods (e.g., Jordan & Henderson, 1995; Nardi, 1997) and the European field study tradition (De Keyser, 1990; Heath & Luff, 2000). Observers are placed in the actual work setting to observe and interview domain practitioners as opportunities arise. Particular attention is placed on detailed capture of illustrative incidents that provide concrete examples of the kinds of complexities that can arise in the environment, the kinds of cognitive and collaborative strategies and facilitating activities that domain practitioners use to handle these situations, and how existing artifacts are tailored to meet situation demands. These illustrative incidents may be examples of practitioner performance in routine situations that arise often, or they may represent a response to a relatively rare occurrence (e.g., equipment malfunction, accident) that arises during the observational study.

Exploratory observational studies contrast to other scientific methods in that the focus of the observations and analysis is on discovery rather than hypothesis testing. Different analysts looking at the same domain might very well focus on different aspects and uncover different insights if they draw on very different conceptual frameworks in selecting what is "interesting" to capture. In this type of research, what matters is not reliability, that is, would different analysts working independently have focused on the same observations, but rather how generative the work is—Are the results insightful and productive with respect to pointing to sources of performance problems and opportunities for improvement?

Figure 27.1 provides a graphic representation of the data analysis and abstraction process used to derive generalizations from the specific observations (cf. Hollnagel, Pederson, & Rasmussen, 1981; Patterson & Woods, 2001). Observations and analyses are guided by (a) the questions that the study is intended to address, (b) the sample of practitioners and activities observed, and (c) the conceptual frameworks that the observers bring to bear.

Conceptual frameworks play an important role in guiding observations and analyses (cf. Lipshitz, chap. 26, this volume). In the studies described below observations and analyses are informed by bodies of knowledge

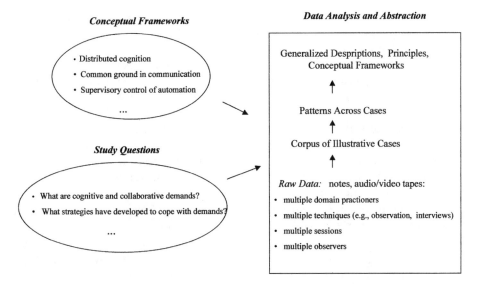

FIG. 27.1. Conceptual frameworks and study questions that define what is "interesting" and thus guide observations and the data analysis and abstraction process.

about the interaction of humans and supporting artifacts in complex, sociotechnical systems from the field of cognitive engineering and related behavioral and social sciences (e.g., theory of distributed cognition, role of common ground in multiagent communication, and principles of human automation interaction and the consequences of clumsy automation).

These conceptual frameworks guide the identification of interesting cases to capture. They enable tractability in data collection, both in terms of the amount of time spent observing (on the order of days instead months) and the level of detail of what is recorded and analyzed (sequences of events rather than second-by-second behavioral and verbal interactions).

Conceptual frameworks also support the analysis process used to identify common patterns across a corpus of illustrative cases and to draw generalizations that have applicability beyond the particular cases examined. These analysis stages correspond to the levels of analysis that Lipshitz (chap. 26, this volume) refers to as "paraphrasing" and "theorizing."

We point out that conceptual frameworks not only serve as a starting point for observation and analysis—a framework by which to interpret and aggregate findings—they are also an output of the analysis process. The general descriptions, principles, and conceptual frameworks that emerge from the analysis of an observational study are expressed at a level of abstraction that allow across-domain comparison and application. The results of observational studies can be used to support, expand, refine, or refute existing conceptual frameworks. They can also be used to generate new conceptual frameworks and theories.

The differences in goals of exploratory observational studies as contrasted with studies designed to test a specific hypothesis leads to different study design considerations. For example, an important consideration in a study designed for hypothesis testing is to control the conditions of observation and minimize variability (both with respect to the range of situations observed and with respect to what observers record). In contrast, because the focus of exploratory observational studies is on discovery, the objective is to broaden the set of observations and conceptual frameworks brought to bear by observers to maximize the opportunity for uncovering interesting findings and drawing productive insights.

Several techniques are used to broaden the set of observations and conceptual frameworks brought to bear. These include

- Broadly sampling domain practice (e.g., multiple shifts, multiple practitioners at multiple levels of experience, multiple sites).
- Use of multiple converging techniques (e.g., field observations, structured interviews, questionnaires).
- Use of multiple observers who are likely to bring different conceptual frameworks.

Another important concern in performing observational studies is the accuracy of the observations and their interpretation. Several strategies are used to guard against errors and biases. First, observers take advantage of opportunities to continuously "bootstrap" their understanding of the domain complexities, iteratively refining their notion of what is interesting to collect and analyze (Lipshitz, chap. 26, this volume). Each observation provides a potential opportunity to generate new conjectures as well as test conjectures generated in prior observations. This approach allows the discovery of critical factors not predicted in advance as well as the opportunity to discard early conjectures that are not supported by later observations. The ability to sample the domain of practice broadly (multiple practitioners, multiple levels of domain expertise, multiple session, multiple sites) provides an opportunity to look for commonalities across cases as well as divergences (contrasting cases) that reveal interesting insights.

A second technique to promote accurate interpretation is to compare the insights and perspective of multiple observers. It is common to hold meetings immediately following an observation period in which the multiple observers share their observations and interpretations. This reduces the risk of forgetting critical details that are "in the head" of the observer but not recorded in the "raw data" and improves data reliability by allowing multiple observers to contrast their interpretations of events while their memories are still fresh and it is still possible to pursue additional data to resolve ambiguities or differences in interpretation (e.g., by conducting follow-up interviews with the domain practitioners who were the subject of the observations).

The ultimate criterion in evaluating the results of an observational study is that once the insights are made and pointed out, other analysts (or more relevantly, the domain practitioners themselves) would agree with the findings and interpretation. A common practice is to present the results of the study to domain practitioners (either in the form of a report or a presentation) and solicit feedback on the accuracy of the base observations and their interpretations.

The two studies summarized below provide concrete illustrations of this methodological approach.

STUDY I: INFORMING THE DESIGN OF NEW TECHNOLOGY

New technologies often fail to have the desired effects on human performance when introduced into actual complex, sociotechnical settings. This first study (Roth et al., 2001) illustrates how a series of observations and interviews uncovered cognitive and collaborative demands and adaptive strategies with current communication technology that had implications for the design of a new data-link communication technology for train dispatching (Roth et al., 2001; Roth, Malsch, Multer, & Coplen, 1999).

Currently, voice radio is the primary means of communication between railroad dispatchers and the railway workers they interact with (e.g., locomotive engineers; maintenance of way workers). The radio channels are overloaded, however, creating a data overload situation for train dispatchers. The railroad industry has been examining the use of data-link technology to communicate in place of or in addition to voice radio communication. The guiding questions for this study were (a) What activities could be supported more effectively with data-link digital communication systems? and (b) What features of the existing technology are important to effective dispatcher performance and therefore need to be considered when deploying new technology?

The study combined field observations at dispatch centers with structured interviews with experienced dispatchers. In the first phase, railroad dispatchers were observed as they went about their job in a railroad dispatch center that primarily handled passenger trains. Two observers participated. Each observer sat next to a different railroad dispatcher and observed the communications he or she engaged in and the train routing and track management decisions that were made. The observer asked the dispatcher questions during low work load periods. Questions were guided by a checklist of predefined topics and by the observed behavior.

A total of eight dispatchers were observed across two shifts. Observations included high work load early morning rush-hour periods, lower work load midday periods, and shift turnovers. Phase 2 consisted of structured interviews with experienced railroad dispatchers and related personnel from the same railroad dispatch. Phase 3 involved field observations at a second dispatch center that primarily handled freight trains. This was to assess the generality of the results obtained at the first dispatch center. Phase 4 involved a second set of field observations at the same dispatch center observed during Phase 1. This was to verify and expand on the results obtained in the previous three phases. In general, the results from each phase confirmed and extended the results from the previous phase.

Uncovering the Role of Radio "Party Line" in Facilitating Railroad Dispatching

Railroad dispatching involves extensive communication and coordination among individuals distributed in time and space. In a typical railroad dispatch center, there are multiple dispatchers working in parallel, each responsible for different territories, who must coordinate with others to manage track usage efficiently and minimize train delays. Observations were therefore guided by concepts from the distributed cognition, common ground, and distributed planning literatures. The observations revealed the cognitive and collaborative demands and the cooperative planning and error detection strategies that dispatchers have developed with the current technology.

What makes railroad dispatching cognitively difficult is the need to deal with unplanned demands on track usage (e.g., the need to accommodate unscheduled trains and requests for time on the track for maintenance work) and the need for dynamic replanning in response to unanticipated events (e.g., train delays, track outages). The observational study revealed that successful performance depends on the ability of dispatchers to monitor train movement beyond their territory, anticipate delays, balance multiple demands placed on track usage, and make rapid decisions. This requires keeping track of where trains are, whether they will reach destination points (meets, stations) on time or will be delayed, and how long the delays will be.

To meet these demands, dispatchers have developed information-gathering strategies that allow them to anticipate requirements for changes to schedules and planned meets early so as to have time to take compensatory action. Many of these strategies depend on communication and coordination among individuals distributed across time and space. This includes coordination among dispatchers managing abutting territories within a dispatch center as well as coordination among the various crafts within a railroad (e.g., locomotive engineers, train masters, dispatchers, and roadway personnel).

One of the most salient findings was that railroad dispatchers took advantage of the broadcast or party-line feature of radio to anticipate and plan ahead. The ability to listen in on communications directed at others that have a bearing on achievement of your own goals and to recognize when information in your possession is of relevance to others and broadcast it were found to be important contributors to efficient management of track use (cf. the use of voice loops technology in space shuttle mission control; Patterson, Watts-Petrotti, & Woods, 1999).

Dispatchers routinely listen for information on the radio channel that is not directly addressed to them but provides important clues to potential delays, problems, or need for assistance. As one dispatcher put it, "after a while you kind of fine tune your ear to pick up certain key things." Examples include the following:

- Identifying when a train has left a station. A train conductor will generally tell the locomotive engineer "OK out of New London." By comparing the actual departure time to the scheduled departure, a dispatcher can calculate train delays.
- Identifying equipment problems. By overhearing conversation between a locomotive engineer and the mechanical department, the dispatcher gets early notice of malfunctioning train engines that will need to be replaced.
- Listening for and/or heading off potential interactions and conflicts. Dispatchers listen for commitments made by others that may impact activity in their territory. The ability to listen ahead allows dispatchers to nip potential conflicts before they arise.
- Listening for mistakes. An experienced train dispatcher will pick up key information that may signal a misunderstanding, confusion, or error.

Implications for the Design of Data-Link Technology

There are several implications for the design of data-link technology from this study. First, there was clear evidence from several observed incidents that the radio channel is now overloaded and that there is a need to off-load some of the communication onto other media. Data-link technology provides a vehicle for taking information that is now communicated orally and instead presenting it visually on a computer display. This has clear benefits for certain types of information. For example, having dispatchers read aloud and train crews repeat back complicated movement authorization forms is time consuming and error prone. Transmitting the information as a visual text or graphical display should reduce radio congestion and may reduce the number of read-back errors and other errors of confusion and misunderstanding that sometimes occur during verbal radio transmissions.

At the same time, the results of the observational study revealed the importance of the broadcast or party-line aspect of radio communication that provides a shared frame of reference and allows dispatchers and others working on the railroad to anticipate situations and act proactively. The study identified the need to preserve the broadcast or party-line aspect of radio communication when shifting to data-link technology.

Although data-link technology is often implemented as a private communication channel where only the specified receiver has access to the information transmitted, this is not an inherent characteristic of the technology. It is possible to envision broadcast versions of data-link technology in which multiple individuals can access a transmitted message or view common graphical displays regarding real-time status of track and train information.

To explore this hypothesis under more controlled conditions, a follow-on laboratory study was conducted. Malsch (1999) implemented two data-link systems: a directed system with no broadcasting capacity and a broadcast system. The systems were compared for their effectiveness in a simulated railroad dispatching task with scenario elements abstracted from the observed incidents. Although both versions of data link resulted in more efficient communication as compared to radio transmission, the broadcast version of data-link produced better dispatcher performance than the directed data-link system on several measures such as train safety.

In summary, one of the most significant contributions of the study was that it revealed the important role that the broadcast or party-line feature of the radio communication media played in facilitating safe and efficient dispatch operation in the current environment. Observed illustrative incidents suggested that changing the design of the new data-link technology to preserve this broadcast aspect of dispatcher communication would improve performance, which was then confirmed in a follow-on controlled laboratory study.

STUDY 2: MAKING THE INTRODUCTION OF NEW TECHNOLOGY SAFER

The first study illustrated the use of an observational study to improve the design of a new technology prior to implementation. Observational studies can also be used to reduce unintended effects on performance from the introduction of new technology into a field of practice by identifying training and operational changes that should accompany the implementation. The next study (Roth & O'Hara, 1999, 2002) illustrates how observation of operators in a high-fidelity simulator was used to identify new training and operational needs prior to the implementation of advanced HSIs in a nuclear power plant.

Introduction of new technology inevitably changes the nature of cognitive and collaborative work. Some of these changes are explicitly engineered with the goal of improving performance. However, there can also be unanticipated effects. It is easy to find examples in which the introduction of new systems have had unanticipated negative effects, creating new burdens for practitioners often at the busiest or most critical times (Mackay, 1999; Roth, Malin, & Schreckenghost, 1997; Vicente, Roth, & Mumaw, 2001; Woods, Johannesen, Cook, & Sarter, 1994).

Woods (1998) and his colleagues (Potter, Roth, Woods, & Elm, 2000) have argued that new support technologies should be regarded as hypotheses about what constitutes effective support and how technological change is expected to shape cognition and collaboration in the "envisioned world" that contains the new technology (Dekker & Woods, 1999). Observational studies provide a powerful tool for exploring the envisioned world both to evaluate the validity of designer assumptions and to drive further discovery and innovation. In this study, advanced HSIs, including a computer-based procedure system, an advanced alarm system, and a graphic-based plant information display system, were in the final phases prior to implementation in a conventional nuclear power plant control room. Operators were undergoing training on the use of the new interfaces on a high-fidelity, full-scope simulator, which provided an opportunity to observe the use of the technology by experienced operators while handling plant disturbances and interview the operators immediately following the simulation. The guiding questions for this study were (a) What aspects of the new HSIs were clear improvements over traditional control boards? and (b) Were there any new unanticipated challenges or issues that emerged with the introduction of the new HSIs?

The cognitive engineering literature on teamwork, the importance of shared representations for supporting communication and coordination among team members, and the potential for new technologies to create private "keyholes" that can disrupt individual and team situation awareness are examples of conceptual frameworks that were relevant and served to guide observations and inquiries.

Approach

Five professional operating crews were observed and interviewed during a week of training in a full-scope, dynamic plant simulator. Each crew was unobtrusively observed during four simulated emergency scenarios by two observers placed in an observation deck (instructor's area.)

At the end of the 2 days of observation, the operators were interviewed in crews. The primary purpose of the interviews was to obtain the operators' perspective on how they used the new HSI systems and how the new systems affected their performance as individuals and as a team. Questions probed the perceived impact of the new systems on operator work load, situation awareness, distribution of tasks and responsibilities among team members, and communication and coordination among the team members.

Controlling a nuclear power plant involves dynamic, real-time communication and coordination among individuals with dedicated roles and responsibilities. A control room crew is typically made up of three individuals: a shift supervisor and two board operators, although others augment the crew during emergencies. When there is an emergency that causes the plant to shut down (i.e., a plant trip) in the current environment, the shift supervisor reads aloud paper-based procedures, called "Emergency Operating Procedures," that guide the crew step-by-step through the emergency response. The board operators' job is to read plant parameter values from the board for the shift supervisor and take control actions as directed by the procedures that the shift supervisor reads aloud. With the new HSI design, the parameters are automatically provided to the shift supervisor as part of the computer-based procedure system.

Findings About Individual and Team Situation Awareness

We identified several aspects of the new HSIs that gave clear improvements as well as had unanticipated impacts on individual and team situation awareness. One of the most interesting findings of the study was the impact of the HSI systems on the structure and dynamics of the crew. The introduction of the new HSI affected the scope of responsibility of the different crew members, the communication pattern among crew members, and the situation awareness of the different crew members.

The new HSIs removed the need for detailed communication between the shift supervisor and the board operator because the computer-based procedure automatically provided the shift supervisor with the plant parameter data required for him or her to work through the procedures. The shift supervisor and board operators were able to work more in parallel. The shift supervisor concentrated on working through the procedures, and the board operators concentrated on monitoring the advanced alarm system, graphics display, and control board HSIs. As a result, the shift supervisor and the board operators individually reported improved situation awareness and greater confidence in the accuracy and speed of their performance within their own locus of responsibility.

There was an unanticipated effect, however. Operators reported that more conscious effort was required to maintain awareness of each other's situation assessment and activities than with the older, hard-wired control board technology. Although the computerized procedure reduced the shift supervisor's overall work load, it also introduced a new demand—the need to keep the crew informed of his or her assessment of the situation and the status and direction of the procedural path as he or she worked through the procedure. Shift supervisors reported a need to consciously remember to inform the crew of their status through the procedure and to consciously formulate what to communicate. The new communication requirement is a substantial cognitive task that appeared to improve with training and experience.

Findings on the Ability to Monitor Effectiveness of Procedures

Another question of interest was the impact of the new HSIs on the ability of crews to monitor the effectiveness of the procedures in handling emergency scenarios. This included the ability of the crews to detect and respond to cases in which the actions specified in the procedures were not fully appropriate to the specific situation. Several studies examining both actual and simulated incidents have shown that conditions sometimes arise in which response guidance in the procedures is not fully appropriate to the situation (Kauffman, Lanik, Trager, & Spence, 1992; Roth, Mumaw, & Lewis, 1994). In those cases, the ability of the crews to recognize that the actions specified in the procedures are not fully appropriate to the specific plant conditions and to take corrective action are important cognitive activities. As a consequence, one of the points of focus in the Roth and O'Hara (2002) study was on how the computer-based procedures affected the operators' ability to monitor the effectiveness of the procedures and detect and respond to situations in which the actions specified by a procedural step were not fully appropriate to the situation.

In the study, three instances arose in which the computer-based procedure provided misleading information or directed the operators down the wrong procedural path. These instances constitute an "existence proof" of the fact that situations can arise in which the procedural path taken is not appropriate to the situation.

Given that situations can arise in which the decision aid is off track, important questions are, Can operators detect when the decision aid is off track?, and Are they able to redirect the decision aid and get back on track? In all three cases observed in the study, the operators were able to correctly detect that the computer-based procedure direction was inappropriate to the situation and overrode it. The examples illustrated important positive features of the computer-based procedure and raised questions about the conditions that are necessary to foster the ability of crews to detect that a computer-based procedure is off track and redirect it.

Implications for Training With the Introduction of the HSIs

The Roth and O'Hara (2002) study suggested ways to make the introduction of the new HSIs safer through training and operational changes. First, the new demand of supporting team situation awareness given the elimination of low-level communication about parameter values between the board operator and shift supervisor can be addressed by explicit training and changes to communication protocols to include periodic updates from the shift supervisor to the team about his or her assessment of the situation and the location in a procedure.

Second, the three observed instances in which the computer-based procedure was not appropriate constituted an existence proof that instances in which the computer-based procedure is off track can occur and consequently, that the task of detecting and redirecting the computer-based procedure needs to be supported. The findings suggest the importance of having (a) multiple diverse sources of information available to operators in the control room and (b) effective communication among the operators to detect and correct cases in which the computer-based procedure is off track.

The ability of the operators to recognize that the actions specified were inappropriate seemed to depend on three factors that have implications for training and operational changes:

1. Accurate understanding of current plant state.
2. Solid knowledge of the goals and assumptions of the procedures and the consequences of the actions indicated by the procedure.
3. Strong communication between the shift supervisor and the board operators that allowed the board operators to keep track of the procedural path that the shift supervisor was following.

Although the Roth and O'Hara (2002) study provided some suggestive evidence of the kinds of factors that contribute to the ability of crews to detect if a computer-based procedure was off track, clearly, more research is required to fully address this issue. First, only three instances were observed and analyzed, and these three instances might not be a representative sampling of the ways in which the computer-based procedure could be misleading. Second, the observation that the control room crew easily detected that the computer-based procedure was inappropriate for the situation may not generalize to other individuals, teams, or situations, particularly because only one instance of each situation was observed; therefore, there was no way to measure response variability. Further research is needed to generate detailed recommendations for change and to verify that the recommendations would have the desired effects on performance.

In summary, this study illustrates three important roles of observational studies:

1. Uncovering new cognitive and collaborative demands that were previously unanticipated and could be addressed with training before the implementation of a system in an actual, high-consequence work setting.
2. Documenting illustrative cases that provide an existence proof that certain situations can arise that need to be explicitly considered by system designers, trainers, evaluators, and managers.
3. Providing suggestive evidence that inform hypotheses for improving performance by changes to training and operational procedures that can then be explored under more controlled conditions.

DISCUSSION

In this chapter, we used two recent studies to illustrate the ways in which observational studies can contribute to the growth of knowledge on human decision making in complex domains. In the first study, an important function of the current communication technology in railroad dispatching was uncovered that had significant implications for the design of a new data-link digital communication technology. This function was hidden in the sense that it was an adaptation that was not officially supported by the current technology and unlikely to have been reported by the operators to be important. In the second study, a new demand for shift supervisors to explicitly communicate situation assessments to a team using new advanced displays in a nuclear power plant control room was uncovered. Because the observations were conducted in a high-fidelity simulator prior to implementation, this new demand could be included in training and operational changes that could be implemented at the same time as the new system, therefore making the transition period safer. In addition, three instances in which the computer-based procedure was inappropriate to the situation were uncovered and documented. They provided an existence proof that the situations could arise and therefore that provisions to support these situations and other similar situations needed to be made.

There are two phases that are important to the advancement of science. One is the controlled experiment phase that is used to confirm a hypothesis by controlling for, and thus eliminating, all other possible explanations for a given phenomenon. This controlled experiment phase is generally associated with the concept of science. However, there is also another element of the scientific process that is less widely discussed and that is the discovery phase. This is the phase during which fruitful conjectures are generated that can then be tested under more controlled conditions.

Naturalistic observation studies are one of the tools that support this discovery phase of the scientific process by increasing the empirical grounding of hypotheses about how tools will affect work in complex settings. They serve to draw attention to significant phenomena and relations that might otherwise have been missed, which can then be further explored in more controlled investigations.

REFERENCES

Dekker, S., & Woods, D. D. (1999). Extracting data from the future: Assessment and certification of envisioned systems. In S. Dekker & E. Hollnagel (Eds.), *Coping with computers in the cockpit* (pp. 7–27). Aldershot, England: Ashgate.

De Keyser, V. (1990). Why field studies? In M. Helander (Ed.), *Human factors in design for manufacturability and process planning* (pp. 305–316). Geneva: International Ergonomics Association.

Heath, C., & Luff, P. (2000). *Technology in action*. Cambridge, England: Cambridge University Press.

Hollnagel, E., Pederson, O., & Rasmussen, J. (1981). *Notes on human performance analysis* (Tech. Rep. No. Riso–M–2285). Roskilde, Denmark: Riso National Laboratory.

Jordan, B., & Henderson, A. (1995). Interaction analysis: Foundations and practice. *Journal of the Learning Sciences, 4,* 39–103.

Kauffman, J. V., Lanik, G. F., Trager, A., & Spence, R. A. (1992). *Operating experience feedback report—Human performance in operating events* (NUREG–1275). Washington, DC: Office for Analysis and Evaluation of Operational Data, U.S. Nuclear Regulatory Commission.

Mackay, W. (1999). Is paper safer? The role of paper flight strips in air traffic control. *ACM Transactions on Computer-Human Interaction, 6,* 311–340.

Malsch, N. F. (1999). *Design and testing of a railroad dispatching simulator using data-link technology.* Unpublished master's thesis, Department of Aeronautics and Astronautics, Massachusetts Institute of Technology.

Mumaw, R. J., Roth, E. M., Vicente, K. J., & Burns, C. M. (2000). There is more to monitoring a nuclear power plant than meets the eye. *Human Factors, 42,* 36–55.

Nardi, B. A. (1997). The use of ethnographic methods in design and evaluation. In M. Helander, T. K. Landauer, & P. Prabhu (Eds.), *Handbook of human-computer interaction* (2nd ed., pp. 361–366). Amsterdam: North-Holland.

Patterson, E. S., Watts-Perotti, J., & Woods, D. D. (1999). Voice loops as coordination aids in space shuttle mission control. *Computer Supported Cooperative Work: The Journal of Collaborative Computing, 8,* 353–371.

Patterson, E. S., & Woods, D. D. (2001). Shift changes, updates, and the on-call model in space shuttle mission control. *Computer Supported Cooperative Work: The Journal of Collaborative Computing, 10,* 317–346.

Potter, S. S., Roth, E. M., Woods, D. D., & Elm, W. (2000). Bootstrapping multiple converging cognitive task analysis techniques for system design. In J. M. Schraagen, S. F. Chipman, & V. L.

Shalin (Eds.), *Cognitive task analysis* (pp. 317–340). Mahwah, NJ: Lawrence Erlbaum Associates.

Roth, E. M., Malin, J. T., & Schreckenghost, D. L. (1997). Paradigms for intelligent interface design. In M. Helander, T. Landauer, & P. Prabhu (Eds.), *Handbook of human-computer interaction* (2nd ed., pp. 1177–1201). Amsterdam: North-Holland.

Roth, E. M., Malsch, N., & Multer, J. (2001). *Understanding how train dispatchers manage and control trains: Results of a cognitive task analysis* (DOT/FRA/ORD–01/02). Washington, DC: U.S. Department of Transportation/Federal Railroad Administration. Retrieved http://www.fra.dot.gov/pdf/cta.pdf

Roth, E. M., Malsch, N., Multer, J., & Coplen, M. (1999). Understanding how train dispatchers manage and control trains: A cognitive task analysis of a distributed planning task. In *Proceedings of the Human Factors and Ergonomics Society 43rd Annual Meeting* (pp. 218–222). Santa Monica, CA: Human Factors and Ergonomics Society.

Roth, E. M., Mumaw, R. J., & Lewis, P. M. (1994). *An empirical investigation of operator performance in cognitively demanding simulated emergencies* (NUREG/CR–6208). Washington, DC: U.S. Nuclear Regulatory Commission.

Roth, E. M., & O'Hara, J. (1999). Exploring the impact of advanced alarms, displays, and computerized procedures on teams. In *Proceedings of the Human Factors and Ergonomics Society 43rd Annual Meeting* (pp. 158–162). Santa Monica, CA: Human Factors and Ergonomics Society.

Roth, E. M., & O'Hara, J. (2002). *Integrating digital and conventional human system interfaces; Lessons learned from a control room modernization program.* Washington, DC: U.S. Nuclear Regulatory Commission. (NUREG/CR–6749 also BNL–NUREG–52638). Retrieved http://www.nrc.gov/ reading-rm/doc-collections/nuregs/contract/cr6749/6749-021104.pdf

Vicente, K. J., Roth, E. M., & Mumaw, R. J. (2001). How do operators monitor a complex, dynamic work domain? The impact of control room technology. *International Journal of Human–Computer Studies, 54,* 831–856. Retrieved http://www.idealibrary.com

Woods, D. D. (1993). Process-tracing methods for the study of cognition outside the experimental psychology laboratory. In G. A. Klein, J. Orasanu, R. Calderwood, & C. E. Zsambok (Eds.), *Decision making in action: Models and methods* (pp. 228–251). Norwood, NJ: Ablex.

Woods, D. D. (1998). Designs are hypotheses about how artifacts shape cognition and collaboration. *Ergonomics, 41,* 168–173.

Woods, D. D., Johannesen, L. J., Cook, R. I., & Sarter, N. B. (1994). *Behind human error: Cognitive systems, computers, and hindsight.* Crew Systems Ergonomic Information and Analysis Center (CSERIAC), Dayton, OH (State of the Art Report).

28

Naturalistic Analysis of Anesthetists' Clinical Practice

Leena Norros
Technical Research Centre of Finland

Ulla-Maija Klemola
Helsinki University Central Hospital

Anesthesia is characterized by complex interactions between physiological systems with largely unknown cause–effect relations and extremely tight time pressure. Information available during anesthesia concerning the changes in patient's state is inadequate and often contradictory (Severinghaus & Kelleher, 1992), increasing the uncertainty owing to complexity (Gaba, Maxwell, & DeAnda, 1987). Application of formal knowledge based on statistical inferences in obscure practical situations leaves the particular case open to alternative interpretations. Interpretation and judgment indeed play a central role in this activity.

Cognitive analyses of anesthesia practice have their background in the human-factors analyses of comparable process-control work that requires high reliability. These studies usually deal with domains in which the objects of control are man-made systems involving natural processes that are kept under control through extensive technical and information technological means. Due to this, objects become mediated and complex. Coping with this complexity seems to be the major demand on the human operator in the mastery of the dynamic courses of the processes, whereas the uncertainties in the process are interpreted as disturbances in the proper functioning of a technical system. Uncertainty, thus, is equated with an unanticipated event or disturbance.

In contrast to the preceding, in anesthesia, the uncertainties of the natural phenomena (patient's condition) must be mastered through the continuous practice of the anesthetists without any massive technical built-in defenses, while at the same time information of the physiological states of the patient is mediated through technical monitoring devices. Uncertainty regarding the patient's condition and the mediated nature of the information regarding this condition make anesthesia a challenging topic for study. An adequate methodology for such study requires a participant-centered approach that takes into account the physician's construction of the situation that controls his or her actions and acknowledges the continuity of this interaction with an uncertain world. Hereby, we pick up the gauntlet of studying action as a dynamic nonevent as thrown by Reason (2000).

METHODOLOGICAL PRINCIPLES

Our method is based on the cultural-historical theory of activity founded by Vygotsky, Leontjev, and Luria. The particular value of this theory is that it provides a framework for interpreting situated action in a perspective that takes into account the development of activities. According to the theory (Vygotsky, 1978), human activity is, first and foremost, defined through its object, namely, those parts of the environment that potentially fulfill the actor's needs and intentions (Leontjev, 1978). The object thus connects individual actions in a broader activity system (Engeström, 1987), thereby endowing individual actions with meaning. One of the principles of this approach is the historicity of human activity (Luria, 1976). The historical point of view implies that explaining activities requires understanding of their sociohistorical development. The subjective framing of the object forms a personal sense of actions and this relation serves as means to control practical actions.

In addition to the sources listed previously, our method draws on the pragmatist habit-centered conception of action in general and the concept of habit in particular. According to Peirce (1903/1998) and Dewey (1929/1999), habits are acquired dispositions to act that convey societal meaning. The purpose of our methodology is to identify habitual relationships in the texture of practical interactions between the anesthetist and the anesthetized patient, and to reveal the personal sense and societal meaning of observed courses of meaning.

Taking a participant-centered approach, we initiated our study of anesthetists' practices with an inquiry into their framing of their object of activity, that is, the patient under anesthesia. We refer to this framing as the anesthetists' orientation (Norros, 1995).

The empirical analysis of actions is based on conceptual inferences from qualitative data acquired through theme interviews and observations of actual behavior backed up with video recordings and completed with process-tracing interviews that make the actions intelligible (Klemola & Norros, 1997;

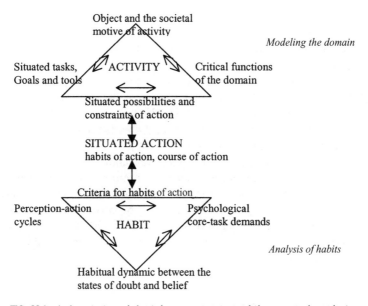

FIG. 28.1. A description of the inference structure of the core-task analysis. From "Methodological Considerations in Analysing Anaesthetists' Habits of Action in Clinical Situations," by L. Norros and U. M. Klemola, 1999, *Ergonomics, 42*(11), pp. 1521–1530. Copyright 1999 by Taylor & Francis. Reproduced with permission.

Norros & Klemola, 1999). In addition, documentary data are used for modeling the constraints of the domain. The model depicted in Fig. 28.1 clarifies the different steps of inference in our analysis and explicates the central concepts that are used for representing courses of action in the real world.

In the analysis, we adopt an ecological point of view according to which actions are determined by both the environment and the actor. Thus, action must be contextualized by modeling the domain to determine the situated possibilities and constraints of the action under study (Klemola & Norros, 2000). The activity-system theory of Engeström (1987) and the functionally oriented formative analysis of domains of Rasmussen (1986) and Vicente (1999) provide the conceptual tools for this analysis, captured by the upper part of the model in Fig. 28.1. Specifically, we define the object of activity and decompose it into two components: the tasks, goals, and tools characterizing the situation and the resulting critical functions of the domain. The latter component expresses the general functional significance of the situation-specific tasks. The result is a conception, in formative terms, of a particular performance situation through its functional constraints and possibilities for action (Vicente, 1999).

A second method of contextualization focuses on how the constraints and possibilities are taken into account by the actors. This reveals the actor's beliefs about what is required for an effective action and the logic, which underlies these beliefs (Eskola, 1999). The lower part of the model (Fig. 28.1) depicts the inferences that we make to derive an actor's beliefs from his or her observed actions and its context, based on the pragmatist idea that the situation can be understood as the sensitivity of the person to the situational possibilities and constraints and his or her readiness to act in it (Bourdieu, 1990; Peirce, 1903/1998). The habits reflect the situation as the disposition of the person to act in it, and they become observable in the participant–environment interactions.

The result of this two-pronged contextual analysis is an empirical description of what constitutes the decisive content of a particular work: the core task. In the remainder of this chapter, we illustrate the methodology outlined previously by detailed description of two studies, an interview study summarized in Klemola and Norros (1997) and Norros and Klemola (1999), and a process tracing analysis of anesthesia summarized in Klemola and Norros (2001).

AN INTERVIEW BASED HABIT-CENTERED ANALYSIS OF ANESTHETISTS' ACTIONS

Sixteen specialists in anesthesia with approximately 10 years practice after the specialist education served as participants of this study (Klemola & Norros, 1997, 2001). The experts were queried about their practice in a semistructured extensive interview. The questions were derived from the concept of activity. Questions regarding the object of activity pertained to experts' definition of the patient in anesthesia and their conceptions of their own possibilities to acquire knowledge of the patient's condition during anesthesia. Interviewees were also provided with the opportunity to express their opinions about striking a balance between the contradictory functional demands of inducing, maintaining and controlling the process. Much attention was paid to the experts' conceptions of the monitor-based information and of the role of monitors in their work. Other questions inquired into the use of anesthetic drugs. Questions regarding the participant pertained to the determinants of expertise and methods of coping with the demands of the task. The extensive material was analyzed utilizing a grounded theory approach (Charmaz, 1995). Each participant's protocol was coded and recoded iteratively, beginning with a broad set of data-driven codes that was gradually being reduced to a smaller final set of codes based on codes' frequencies of appearance. The most relevant conceptual categories finally emerged after several iterations of this cycle.

Modeling the Domain: The Anesthesia Activity

The experts shared an extensive formal medical knowledge base and a common institutionalized professional practice. Thus, all actors could be assumed to share the main objective of anesthesia—ensuring a sufficiently deep sleep during all phases of surgery. To achieve this aim, the anesthetist has to strike balance between several critical demands regarding the patient's physiological functioning under the vital constraints set by the anesthetic agents and the surgical stimuli. These constraints also include adequate control of complex, nonlinear drug interactions between the drugs affecting the level of consciousness, pain, and muscle relaxation. Another critical function relates to the nature of knowledge of the patient's physiological reactions to anesthesia. The physiological potentials of the particular patient are only vaguely known before anesthesia, as they have to be inferred on the basis of general theories and of experimental results regarding specific physiological phenomena in controlled settings. Anesthesia, thus, is inherently unpredictable in any particular case. Finally, the process sets demands on the mastery of pharmacokinetic and pharmacodynamic necessities. In particular, the time period that characterizes the effectiveness of the administered drugs requires anticipation of actions.

The three main phases of anesthesia are preoperative evaluation, transformation of the patient's homeostatic state, and maintenance and regulation of the transformed state. These phases were further divided into specific subelements or tasks that constitute the elements of the habits of action.

Analysis of Habits: Orientation and Habits of Action

The central categories that were elaborated from the interview protocols to conceptualize the experts' practices were orientation and the habit of action. Orientation refers to the participant's way of framing the object of activity as situated goal and thus expresses an epistemic attitude of the practitioner toward a particular patient. Two dimensions were identified in our participants' framing of the anesthesia process: recognizing versus not recognizing the uncertainty of the task or not and a communicative versus authoritative attitude toward the patient. These epistemic attitudes can be assumed to influence the way situational information is utilized as a tool in action. Two distinctive approaches to the use of information which expresses habits of action could be identified. The interpretative approach was characterized by the use of several sources of information to form an interpretation of the patient's physiological reaction potentials and ensuring successful anesthesia by using multiple checks of the situation. The reactive approach was characterized by confirming the execution of a preplanned scheme, anticipating the course of the anesthesia by using common classifications based on types of patients, diseases, or surgical interventions. The success of anesthesia was controlled by maintaining predetermined numerical values of parameters, for example, "maintaining a neat curve."

By forming a profile for each interviewee on the basis of these criteria, we obtained two basic orientations: a realistic orientation that recognized each patient's uniqueness and the factor of uncertainty in anesthesia and an objectivistic orientation that treated the patient as a specimen of a certain general category that can be categorized objectively and that did not recognize the uncertainty of anesthesia. These differences in orientations were related to the interpretative and reactive habits of action, respectively.

A hypothesis concerning the core demands of the anesthetist's activity could be formulated on the basis of the findings of the interview study regarding our participants' conceptions of their work. Whereas the patient's physiological potentials in regard to anesthesia are vaguely determined prior to anesthesia, the anesthetist does have an opportunity to elaborate his or her knowledge of the demands of a particular process during two procedures connected with prominent observable changes in the patient's physiology: anesthetic induction and the endotracheal intubation. We hypothesized that the mastery of the anesthesia process was closely connected to the degree to which the anesthetist took advantage of these opportunities for constructing a conception of the particular patient's physiological potentials. We tested this hypothesis in a study (Klemola & Norros, 2001) that focused on revealing the actual work practices that could be identified in real clinical situations.

A PROCESS-TRACING BASED HABIT-CENTERED ANALYSIS OF ANESTHETISTS' ACTIONS

Eight anesthetists randomly selected from the 16 participants of the interview study and equally representing the realistic and objectivistic orientations participated as participants in the Klemola and Norros (2001) study. Participants selected from the weekly operating list patients with physiological reserves that were limited at least to some degree on which a wide range of observational material was collected. In addition to video recordings, a domain expert member of the research team made documented observations of the patients' responses during the anesthesia and the anesthetists' actions, paying special attention to their mutual, temporal relationships. We also conducted three semistructured interviews with each participant, prior to the operation, during the operation, and the most extensive interview immediately after the operation.

Course of Actions

Analysis was comprised of two phases. In the first phase, the anesthetists' tool-using interactions with the patient were identified by combining all available data. Drugs, patient information, and professional concepts were conceived as tools in the interaction of the anesthetist with the patient that aimed at controlling the state of the patient. Graphical representations of the action-perception cycle revealed clearly identifiable performance patterns in the courses of action of different practitioners.

Habits of Action

The most detailed reproductions of observable behavior do not reveal the meaning of the sequence. Further analysis is therefore required in which the meaning of the observed courses of action are inferred. Our method of analysis uses the principle of behavioral inference of reasons (Von Wright, 1998), which states that if a person understands the meaning of a particular sign as a reason for acting, he or she will react to that sign accordingly unless he or she has reasons that overrule the action. This principle provides us with criteria for identifying and testing whether a person shares a particular meaning (Von Wright, 1998; Norros, 2003). Granted that such context-specific meanings can be identified, one can assume that Von Wright's principle holds. We therefore attempted to distinguish meanings in the texture of behavior, employing a procedure that uses the pragmatist notion that meaning is embedded in habit (Peirce, 1958, pp. 369–390; Norros, 2003). In particular, we applied Peirce's triadic structure of meaning that distinguishes between the object (O) of the sign, its material carrier the sign (S), and the interpretant (I) of the sign (i.e., its implications for action) to interpret our data.

Remember that anesthetists can gain information regarding the patient's physiological potentials during the induction phases of anesthesia. For example, heart rate and blood pressure are typical indications of these potentials. Concrete behavioral criteria were sought from our data material to evaluate the extent to which this possibility was used, thereby constituting a habit of action. Figure 28.2 provides an example of the inferential steps that we used to identify meaning structures in behavior that express different habits of action.

The example demonstrates four different habitual relations expressing reasons for action regarding the task of defining the amount of doses and rate of inducing of anesthetic agents manifested by anesthetists during the induction phases of anesthesia. The four S–I–O relations express four different possibilities of attending to a particular patient as the object of activity and of forming a coherent interpretation of the situation. The relation

Sign (S)
S= particular
   indications of
   blood pressure
   and heart rate

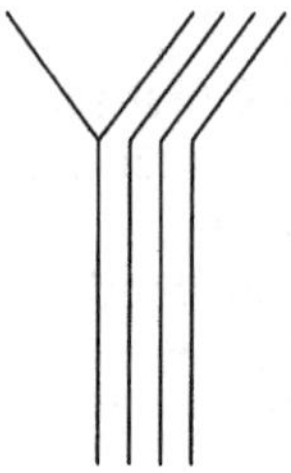

Object (O)
The bases for action:
O1 realisation of a pre-determined plan
O2 control of the level of consciousness
O3 control of reactions as indication of sufficiency
     of sleep
O4 adequacy of dose for this patient

Interpretant (I)
The observable operation:
I1 induction of standard mean doses on a weight basis
I2 induction of a sleeping dose
I3 induction after controlling reaction to laryngoscopy
I4 induction after experimenting

FIG. 28.2. Demonstration of the inference of the meaning of particular operations during anesthetic induction (interpretants I1, I2, I3, I4) and their bases of inference (object O1, O2, O3, and O4) as reactions to the signs of blood pressure and heart rate (S).

S–I1–O1 manifests the weakest and S–I4–O4 the strongest fulfillment of the demand of creating a cumulative interpretation of the patient's physiological potentials, one of the core-task demands of the work.

Through the identification of a considerable number of perception-operation relations in the observational data and the related reasons acquired through the interviews, it was possible to obtain a set of criteria for habits of action. These criteria reflect the extent to which result-critical functions represented by the upper part of Fig. 28.1 provide possible reasons for action in administering anesthesia (Von Wright, 1998). A particular actor may take these constraints and possibilities into account as a relevant reason for action, indicating that they function as efficacious reasons for him or her (Von Wright, 1998). According to Von Wright, comparing the efficacious and possible reasons opens up the possibility to a reason-based evaluation of actions from outside. The defined criteria represent the evaluation categories that are defined as psychological core-task demands in increasing order.

Table 28.1 demonstrates the complete set of criteria that emerged in our study. These criteria are contextually elaborated in Klemola and Norros (2001). The previous example is a slightly extended version of the first item under the task of "Transformation of the homeostatic state of the patient by inducing anesthesia" in Table 28.1. As behavioral expressions of meaning, these criteria provide a possibility for describing anesthetists' actual practices. Using the criteria, we constructed a profile of each anesthetist's habits of action that offered them a mirror for reflecting on their practice. It

TABLE 28.1

Functional Phases of the Anesthesia Process, Elements, and Criteria
for Evaluation of the Anesthetists' Habits of Action

Preoperative evaluation of the patient

Evaluation of physiological condition with regard to anesthesia

　　*Mere enumeration of concurrent diseases

　　*Attempts to evaluate the severity of concurrent diseases

　　*An interpretation of the patient's physiological potential regarding anesthesia

Patient's physiological condition as constraint on the anesthetist's activity

　　*No constraints

　　*Constraints according to common classifications and general rules

　　*Emphasis on situational information as grounds for guiding the administration of anesthetics

Transformation of the homeostatic state of the patient by inducing anesthesia

Interplay between the administration of anesthetic drugs and available information

　　*Anesthetics were given on weight basis or according to a predetermined scheme

　　*The patient's sleeping dose was determined only by following the level of consciousness

　　*Besides consciousness, information from the patient's physiological responses was chosen as
　　grounds for inducing anesthesia

Use of information from cardiovascular intubation response

　　*Information was not used

　　*Information was deliberately reached for

Evaluation of the patient's physiological condition after the transformation phase

　　*Preformed conception was confirmed

　　*Cumulative interpretation of the patient's physiological potential was constructed on the basis
　　of his responses during the transformation phase

Maintenance and regulation of the transformed homeostatic state

Maintenance of balance between the adequate depth of anesthesia and an optimal physiological state

　　*Cardiovascular depression was minimized at the cost of anesthetic depth

　　*Balance was maintained with appropriate means

Maintenance of balance between cardiovascular stability and surgical stimulation

　　*Reactive approach to maintaining balance

　　*Anticipatory approach to maintaining the balance

Regulation of the transformed homeostatic state by using the tools

A. *Use of information*

　　*Controversial reactions to information

　　*Regulation based on information trends and/or on a predetermined scheme

　　*Regulation based on internal tempo of the process in accordance with previous patient
　　responses and situational demands

B. *Adjustment of anesthetic drugs*

　　*Contradiction between the drugs given and the theoretical knowledge referred to by the prac-
　　titioner

　　*Anesthetic drugs adjusted with a mean accuracy or their advantages not exploited

　　*Anesthetic drugs adjusted in accordance with the history of the process and through anticipa-
　　tion of the future situational demands

Note. The process is in bold, the elements are in italics, and criteria are denoted by an asterisk.
From "Process-Based Criteria for Assessment of the Anaesthetists' Habits of Action," by U.-M. Klemola
and L. Norros, 2001, *Medical Education, 35*, pp. 455–464. Copyright 2001 by Blackwell Publishers. Reproduced with permission.

was evident that experts could not report their practices, as expressed by the criteria through an interview.

As mentioned previously, our analysis of different processes of anesthesia revealed differences in actions of different anesthetists. These were not simply due to the differences between patients but rather to the way the anesthetists themselves responded to the peculiarities of each case according to their dispositions. Through the habit-centered analysis of the actions, we could conceptualize these observable differences as indications of two types of practices that were hypothesized on the basis of the interview study: the reactive and interpretative habits of action. When the results of the analysis of habits of action were compared with those of the anesthetists' orientation, the interpretative habit of action was systematically related to the realistic orientation and the reactive habit of action was systematically related to the objectivistic orientation. Thus, the two distinct practices that had already been identified in the interview study could be verified in clinical action. The practice characterized by the objectivistic orientation and reactive habit of action was dominant among the practitioners.

DISCUSSION

Similar to process control operators, anesthetists are required to keep track continuously of an ongoing process, develop awareness of its condition, and be ready to act when necessary. Recent literature conceptualizes these requirements by the construct of situation awareness (SA) and presents empirical measures for evaluating the level of operators' SA during the course of performance (Endsley, 1995a). We conclude our chapter by comparing the construct of SA, which offers empirical tools for naturalistic decision-making research (Endsley, 1997), with our conceptual framework and methodology.

The basic methodological difference between these two approaches stems from their conceptualization of the activity of process control. The SA approach assumes that the process can be observed objectively by the process controller. This is implied by Endsley's (1995b) use of attributes such as "complete" or "adequate" to characterize adequate levels of SA. Thus, SA is a knowledge state with objective reference "out in the world." Such characterizations reflect an epistemic point of view according to which knowledge pertains to an objective external reality of which people (e.g., operators) wish to be certain. Objects of knowledge precede and are independent of actions and perceptions relating to them. Endsley (1995b) emphasized that SA refers to a state of knowledge that should be distinguished from the processes of arriving at this state. Operators are not seen as constructing new knowledge about the process. Rather, they are seen as endeavoring to capture what is already known to exist. The task of the operator is therefore to record as adequately as possible the true state of the process, which is a difficult task owing to the complexities of the connections between the elements of the process and its dynamics.

This conception of process control carries implications for the tools that the analyst uses for its study. Specifically, the tools must be strictly separated from the object so as not to have an impact on it. Indeed, the requirement of impartiality of the participant and his or her actions is consistently taken into account in the empirical methods suggested for the evaluation of SA (Endsley, 1995a). The methodological distinction between the observer-recorder and the interpreter approaches in the study of human action in naturalistic settings, suggested by Lipshitz (chap. 26, this volume) is relevant here. According to the SA approach, operators are observer-recorders of the process whose actions the researcher, in turn, also observes and records.

In our research, we adopt the approach of the observer as an interpreter. Our conception starts by acknowledging that operators have an epistemic uncertainty in regard to the process even when extensive technological defenses are designed to reduce the effects of the natural contingencies. To cope with this uncertainty, the operators might consider the object as a particular, which allows them to construct knowledge for controlling the object through their practical operations. This way of dealing with the world is conceptualized in the operative epistemology proposed by Dewey (1929/1999) in his book *The Quest for Certainty*. Habit is the construct that explains how people gain knowledge about the world and produce continuity of action by inquiring about the object of action and developing generalized ways of attending to particular situations. If the actor finds the uniqueness of the situation interesting and attention is focused on its situational features, the particular phenomena become the objects of inquiry and specific control. This realizes the construct of habit in its full-fledged form as reflective habituality (Kilpinen, 2000), which is qualified both as a repetition of the meaning of an act and its reflection through the interpretation of the object.

Finally, within the SA approach, the adequacy of SA is defined as the goodness of fit between the operator's subjective model of the state of the process and its objectively given true state. Accordingly, it is important for analysts to have a general objective reference for the evaluation of SA. This requirement can be fulfilled, at best, when performance is studied in a full-scope simulator where the values of significant parameters can be precisely measured and compared to the operators' assessments of these values. In our method, the adequacy of the interaction with the object is recorded in real-life working situations and evaluated with respect to the historically and socially produced ways of taking into consideration the functionally relevant constraints and possibilities of the domain. Thus, the two distinct practices that we identified in anesthesia represent both personal evaluations of what is relevant in a situation and what the anesthetists have learned in their communities of practice to be valuable and appropriate. In this perspective, the dominant practice found among the anesthetists, combining the objectivistic orientation with the reactive habit of action, expresses the prevailing epistemology of practice in medicine. This community values scientific experimental knowledge and uses it as standard for trustworthy knowledge. Because clinical practice is perceived as a vague art, it should be standardized for the enhancement of certainty.

An important strategic question in the development of methods of treatment is the consequences of prevalent practices on the welfare of patients. Given the continuing rise in life expectancy, it is safe to assume an increase of future patients requiring surgery with limited physiological reserves to endure it. The resulting narrower margins of error should increase the attractiveness of interpretative habits of action because they are oriented toward the specific demands of each patient.

In conclusion, our habit-centered analysis of situated action revealed differences in the dynamics of dealing with the environment without requiring the identification of discrete events (e.g., errors) in the anesthetists' courses of action. This is a major advantage of this type of analysis, as it opens up the possibility of evaluating the flow of daily practices from the perspective of their developmental potential or their proneness for creating problems. This possibility is particularly valuable in environments in which the likelihood of accidents is low, but their cost, once they occur, is high.

REFERENCES

Bourdieu, P. (1990). *The logic of practice*. Cambridge, England: Polity Press.

Charmaz, K. (1995). Grounded theory. In J. A. Smith, R. Harré, & L. Van Langenhove (Eds.), *Rethinking methods in psychology* (pp. 27–49). London: Sage.

Dewey, J. (1999). *The quest for certainty. A study of the relation of knowledge and action*. (Finnish translation). Helsinki, Finland: Gaudeamus. (Original work published 1929)

Endsley, M. R. (1995a). Measurement of situation awareness in dynamic systems. *Human Factors, 37*, 65–84.

Endsley, M. R. (1995b). Toward a theory of situation awareness in dynamic systems. *Human Factors, 37*, 32–64.

Endsley, M. R. (1997). The role of situation awareness in naturalistic decision making. In C. Zsambok & G. A. Klein (Eds.), *Naturalistic decision making* (pp. 269–284). Mahwah, NJ: Lawrence Erlbaum Associates.

Engeström, Y. (1987). *Learning by expanding*. Jyväskylä, Finland: Orienta.

Eskola, A. (1999). Laws, logics, and human activity. In Y. Engeström, R. Miettinen, & R.-L. Punamäki (Eds.), *Perspectives in activity theory* (pp. 107–114). Cambridge, England: Cambridge University Press.

Gaba, D. M., Maxwell, M., & DeAnda, A. (1987). Anesthetic mishaps: Breaking the chain of accident evolution. *Anesthesiology, 66,* 670–676.

Kilpinen, E. (2000). *The enormous fly-wheel of society. Pragmatism's habitual conception of action and the social theory.* Helsinki, Finland: Department of Sociology, University of Helsinki.

Klemola, U.-M., & Norros, L. (1997). Analysis of the clinical behaviour of anaesthetists: Recognition of uncertainty as a basis for practice. *Medical Education, 31,* 449–456.

Klemola, U.-M., & Norros, L. (2000, May). *Logics of anaesthetic practice—Interdisciplinary methodology for analysing decision making in an open, complex system.* Paper presented at the 5th Conference of Naturalistic Decision-Making in Stockholm, Sweden.

Klemola, U.-M., & Norros, L. (2001). Practice-based criteria for assessment the anaesthetists' habits of action. Outline for a reflexive turn in practice. *Medical Education, 35,* 455–464.

Leontjev, A. N. (1978). *Activity, consciousness, and personality.* Englewood Cliffs, NJ: Prentice Hall.

Luria, A. R. (1976). *Cognitive development: Its cultural and social foundations.* Cambridge, MA: Harvard University Press.

Norros, L. (1995). An orientation-based approach to expertise. In J.-M. Hoc, P. C. Cacciabue, & E. Hollnagel (Eds.), *Cognition and human computer co-operation* (pp. 137–160). Hillsdale, NJ: Lawrence Erlbaum Associates.

Norros, L. (2003). *Acting under uncertainty. The core-task analysis in ecological study of work.* Manuscript submitted for publication.

Norros, L., & Klemola, U.-M. (1999). Methodological considerations in analysing anaesthetists' habits of action in clinical situations. *Ergonomics, 42*(11), 1521–1530.

Peirce, C. S. (Ed.). (1958). *Collected papers of Charles Sanders Peirce.* Cambridge, MA: Harvard University Press.

Peirce, C. S. (1998). *The Harvard lectures on pragmatism. The essential Peirce. Selected philosophical writings, Volume 2.* Bloomington: Indiana University Press. (Original work published 1903)

Rasmussen, J. (1986). *Information processing and the human-machine interaction: An approach to cognitive engineering.* New York: North Holland.

Reason, J. (2000, May). Plenary paper at the 5th Conference of Naturalistic Decision-Making in Stockholm, Sweden.

Severinghaus, J. W., & Kelleher, J. F. (1992). Recent developments in pulse oximetry. *Anesthesiology, 76,* 1018–1038.

Vicente, K. J. (1999). *Cognitive work analysis. Toward safe, productive, and healthy computer-based work.* Mahwah, NJ: Lawrence Erlbaum Associates.

Von Wright, G. H. (1998). *In the shadow of Descartes. Essays in the philosophy of mind.* Dordrecht, The Netherlands: Kluwer Academic.

Vygotsky, L. S. (1978). *Mind in society. The development of higher psychological processes.* Cambridge, MA: Harvard University Press.

Author Index

Subject Index

A

Action/feedback loops, 59, 60, 80
Adaptive strategies
  see Observational methodology
Advances in Naturalistic decision-making
    methodology, 7–10 (chaps. 20–28)
  cognition task analysis (CTA), 8
  generalizability, 7
  observational methodology, 9
  participant-centered studies, 9
  process-tracing, 8
  question-asking/think-aloud methods, 8
  verbal protocols, 8
Anesthetists' clinical practice
  habit-centered analysis, 396, 398–404
  interpreting situated action, 396
  participant-centered approach, 396
  process-tracing course of action, 400–406
  situation as context, 397–399
  situation awareness, 404–406
  uncertainty, 395, 396

B

Behavior modification
  changing innate capacities, 140, 141

  see Expert performance
Bias, 24, 25, 122
  see Economic forecasting

C

Causal interactions
  negative and positive feedback, 8
Causality
  deterministic/probabilistic models
    (representations), 18
Cause-effect relations
  see Anesthetists' clinical practice
Challenger disaster (1986)
  accountability (political/bureaucratic),
    263, 264
  aftermath, 268, 269
  cause and control connections, 256, 257
  causes of accident, 257–259
  culture as connecting link, 256
  danger signals, 260–262
  NASA and its political environment,
    262–264
  normalization of risk assessment, 259, 260
  qualitative methods, 256